Oxidative Stress in Parasites

Oxidative Stress in Parasites

Editor

Serge Ankri

MDPI • Basel • Beijing • Wuhan • Barcelona • Belgrade • Manchester • Tokyo • Cluj • Tianjin

Editor
Serge Ankri
Department of Molecular
Microbiology, Faculty of
Medicine
TECHNION
Haifa
Israel

Editorial Office
MDPI
St. Alban-Anlage 66
4052 Basel, Switzerland

This is a reprint of articles from the Special Issue published online in the open access journal *Antioxidants* (ISSN 2076-3921) (available at: www.mdpi.com/journal/antioxidants/special_issues/Oxidative_Parasites).

For citation purposes, cite each article independently as indicated on the article page online and as indicated below:

LastName, A.A.; LastName, B.B.; LastName, C.C. Article Title. *Journal Name* **Year**, *Volume Number*, Page Range.

ISBN 978-3-0365-7617-6 (Hbk)
ISBN 978-3-0365-7616-9 (PDF)

Contents

About the Editor

Serge Ankri

Prof. Serge Ankri is an associate professor at the B. Rappaport Faculty of Medicine, Technion, in Haifa, Israel. He obtained his Ph.D. in 1996 from the University of Paris XI in France, and completed his postdoctoral work at the Weizmann Institute of Science in Rehovot, Israel. His laboratory at the Technion specializes in the regulation of virulence and adaptation to environmental stresses in the parasite *E. histolytica*, with a focus on the role of epigenetic mechanisms, redox regulation and gut microbiota in infection. He has published more than sixty peer-reviewed papers and received twenty competitive research grants from national and international funding agencies. He serves as a reviewer for high-ranked journals such as *Nature Communications* and *PLoS Pathogens*, as well as for various international funding projects. Prof. Ankri is an Editorial Board Member of *Antioxidants* and *Frontiers in Molecular Biosciences*.

antioxidants

MDPI

Editorial

Insights into the Role of Oxidative Stress and Reactive Oxygen Species in Parasitic Diseases

Serge Ankri

Department of Molecular Microbiology, Ruth and Bruce Rappaport Faculty of Medicine, Technion, Haifa 31096, Israel; sankri@technion.ac.il

Parasitic infections remain a significant public health challenge in many parts of the world, especially in developing countries. Despite significant progress in their treatment, these diseases often cause long-term illness, disability, and mortality. Parasites face various challenges within the host, including oxidative and nitrosative stress generated by the host's immune response. Understanding how parasites respond to these challenges is crucial to our comprehension of parasite–host interactions at both the cellular and molecular levels. This Special Issue includes articles on various protozoan, helminth, and arthropod parasites that highlight recent advances in our understanding of the role of oxidative stress, nitrosative stress, and metabolic pathways in parasitic diseases. These papers cover critical topics, such as the response and adaptation of parasites to oxidative stress induced by drugs [1,2]; their localization in the host during their life cycle [3]; or by nutrition [4], plant-based, and probiotic approaches to modulating parasites' redox responses [5,6], as well as redox talk between parasites and the host's immune defense cells [7–9]. Additionally, this Special Issue addresses nitric oxide resistance by discussing the role of glucose consumption and GSH-mediated redox capability in the resistance of Leishmania to nitrosative stress [10]. These examples underscore the importance of understanding the mechanisms underlying parasitic diseases, and lay the foundation for the development of innovative treatments targeting parasites' defense mechanisms against oxidative stress. I hope that this special edition will serve as a valuable resource for researchers, students, and physicians studying parasitic infections. Finally, I would like to express my gratitude to all the authors who contributed to this Special Issue, and to the *Antioxidants* team for their assistance during the review and editorial process.

Conflicts of Interest: The author declares no conflict of interest.

Citation: Ankri, S. Insights into the Role of Oxidative Stress and Reactive Oxygen Species in Parasitic Diseases. *Antioxidants* **2023**, *12*, 1010. https://doi.org/10.3390/antiox12051010

Received: 25 April 2023
Accepted: 25 April 2023
Published: 27 April 2023

References

1. Shaulov, Y.; Sarid, L.; Trebicz-Geffen, M.; Ankri, S. Entamoeba histolytica Adaption to Auranofin: A Phenotypic and Multi-Omics Characterization. *Antioxidants* **2021**, *10*, 1240. [CrossRef] [PubMed]
2. Egwu, C.O.; Augereau, J.-M.; Reybier, K.; Benoit-Vical, F. Reactive Oxygen Species as the Brainbox in Malaria Treatment. *Antioxidants* **2021**, *10*, 1872. [CrossRef] [PubMed]
3. Martínez-González, J.d.J.; Guevara-Flores, A.; del Arenal Mena, I.P. Evolutionary Adaptations of Parasitic Flatworms to Different Oxygen Tensions. *Antioxidants* **2022**, *11*, 1102. [CrossRef] [PubMed]
4. Hernandez, E.P.; Anisuzzaman; Alim, M.A.; Kawada, H.; Kwofie, K.D.; Ladzekpo, D.; Koike, Y.; Inoue, T.; Sasaki, S.; Mikami, F.; et al. Ambivalent Roles of Oxidative Stress in Triangular Relationships among Arthropod Vectors, Pathogens and Hosts. *Antioxidants* **2022**, *11*, 1254. [CrossRef] [PubMed]
5. Espinel-Mesa, D.X.; González Rugeles, C.I.; Mantilla Hernández, J.C.; Stashenko, E.E.; Villegas-Lanau, C.A.; Quimbaya Ramírez, J.J.; García Sánchez, L.T. Immunomodulation and Antioxidant Activities as Possible Trypanocidal and Cardioprotective Mechanisms of Major Terpenes from Lippia alba Essential Oils in an Experimental Model of Chronic Chagas Disease. *Antioxidants* **2021**, *10*, 1851. [CrossRef] [PubMed]
6. Sarid, L.; Zanditenas, E.; Ye, J.; Trebicz-Geffen, M.; Ankri, S. Insights into the Mechanisms of Lactobacillus acidophilus Activity against Entamoeba histolytica by Using Thiol Redox Proteomics. *Antioxidants* **2022**, *11*, 814. [CrossRef] [PubMed]
7. Calvani, N.E.D.; De Marco Verissimo, C.; Jewhurst, H.L.; Cwiklinski, K.; Flaus, A.; Dalton, J.P. Two Distinct Superoxidase Dismutases (SOD) Secreted by the Helminth Parasite Fasciola hepatica Play Roles in Defence against Metabolic and Host Immune Cell-Derived Reactive Oxygen Species (ROS) during Growth and Development. *Antioxidants* **2022**, *11*, 1968. [CrossRef] [PubMed]
8. Díaz-Godínez, C.; Jorge-Rosas, J.F.; Néquiz, M.; Martínez-Calvillo, S.; Laclette, J.P.; Rosales, C.; Carrero, J.C. New Insights on NETosis Induced by Entamoeba histolytica: Dependence on ROS from Amoebas and Extracellular MPO Activity. *Antioxidants* **2021**, *10*, 974. [CrossRef] [PubMed]
9. Dick, C.F.; Alcantara, C.L.; Carvalho-Kelly, L.F.; Lacerda-Abreu, M.A.; Cunha-e-Silva, N.L.; Meyer-Fernandes, J.R.; Vieyra, A. Iron Uptake Controls Trypanosoma cruzi Metabolic Shift and Cell Proliferation. *Antioxidants* **2023**, *12*, 984. [CrossRef]
10. Pinho, N.; Bombaça, A.C.; Wiśniewski, J.R.; Dias-Lopes, G.; Saboia-Vahia, L.; Cupolillo, E.; de Jesus, J.B.; de Almeida, R.P.; Padrón, G.; Menna-Barreto, R.; et al. Nitric Oxide Resistance in Leishmania (Viannia) braziliensis Involves Regulation of Glucose Consumption, Glutathione Metabolism and Abundance of Pentose Phosphate Pathway Enzymes. *Antioxidants* **2022**, *11*, 277. [CrossRef] [PubMed]

antioxidants

Article

Entamoeba histolytica Adaption to Auranofin: A Phenotypic and Multi-Omics Characterization

Yana Shaulov, Lotem Sarid †, Meirav Trebicz-Geffen † and Serge Ankri *

Department of Molecular Microbiology, Ruth and Bruce Rappaport Faculty of Medicine, Technion, 31096 Haifa, Israel; yana11me@campus.technion.ac.il (Y.S.); Lotemsarid@campus.technion.ac.il (L.S.); meiravg@technion.ac.il (M.T.-G.)
* Correspondence: sankri@technion.ac.il; Tel.: +972-48295453
† These authors contributed equally.

Abstract: Auranofin (AF), an antirheumatic agent, targets mammalian thioredoxin reductase (TrxR), an important enzyme controlling redox homeostasis. AF is also highly effective against a diversity of pathogenic bacteria and protozoan parasites. Here, we report on the resistance of the parasite *Entamoeba histolytica* to 2 μM of AF that was acquired by gradual exposure of the parasite to an increasing amount of the drug. AF-adapted *E. histolytica* trophozoites (AFAT) have impaired growth and cytopathic activity, and are more sensitive to oxidative stress (OS), nitrosative stress (NS), and metronidazole (MNZ) than wild type (WT) trophozoites. Integrated transcriptomics and redoxomics analyses showed that many upregulated genes in AFAT, including genes encoding for dehydrogenase and cytoskeletal proteins, have their product oxidized in wild type trophozoites exposed to AF (acute AF trophozoites) but not in AFAT. We also showed that the level of reactive oxygen species (ROS) and oxidized proteins (OXs) in AFAT is lower than that in acute AF trophozoites. Overexpression of *E. histolytica* TrxR (EhTrxR) did not protect the parasite against AF, which suggests that EhTrxR is not central to the mechanism of adaptation to AF.

Keywords: *Entamoeba histolytica*; auranofin; drug resistance; transcriptomics; redoxomics; thioredoxin reductase

Citation: Shaulov, Y.; Sarid, L.; Trebicz-Geffen, M.; Ankri, S. *Entamoeba histolytica* Adaption to Auranofin: A Phenotypic and Multi-Omics Characterization. *Antioxidants* **2021**, *10*, 1240. https://doi.org/10.3390/antiox10081240

Academic Editor: Simone Carradori

Received: 21 June 2021
Accepted: 30 July 2021
Published: 2 August 2021

1. Introduction

The protozoan parasite *Entamoeba histolytica* is the etiologic agent of amoebiasis, a significant hazard in countries with low socioeconomic status and poor sanitation. This disease accounted for 55,500 deaths and 2.237 million disability-adjusted life years in 2010 [1]. The main symptoms of amoebiasis are inflammation of the large intestine and liver abscesses. Infection occurs following the ingestion of food contaminated with cysts. Trophozoites that emerge from the cysts migrate to the large intestine. Asymptomatic colonization occurs in most cases (90% of all infections). Symptomatic infection is characterized by bloody diarrhea. Metronidazole (MNZ) is the drug currently used for invasive amoebiasis [2]. Inside the parasite, MNZ is reduced through the action of thioredoxin reductase (TrxR) to a nitro radical anion or to a nitroimidazole. This nitro group reduces O_2, leading to the formation of cytotoxic reactive oxygen species (ROS) inside the parasite. The nitroimidazole can also modify cysteine containing proteins such as thioredoxin (Trx), leading to their inactivation [3]. There are numerous common side effects related to MNZ, including dizziness, heartburn, stomach cramps, trouble sleeping, and weight loss [4–6]. Treatment with MNZ is usually highly effective, but resistance to this drug has been reported in various bacteria [7,8] and protozoan parasites [9–11]. To address these drawbacks, new alternatives to MNZ have been initiated and AF has emerged as one of the most potent anti-protozoan parasites drugs. Initially, AF was a gold-containing compound developed in the 1970s for the treatment of rheumatoid arthritis [12]. Its mechanism of action as an antiarthritic gold drug remained controversial but it is assumed that it works by inhibiting the activity of TrxR, a crucial enzyme involved in the maintenance of the redox homeostasis in the cell [13]. AF is also a potent anticancer agent [14] and has been found to be very efficient against a number of pathogens, including *Mycobacterium abscessus* [15], *Clostridium difficile* [16,17], vancomycin-resistant enterococci [18,19], and some additional multidrug resistant bacteria [20]. Auranofin is also very efficient against parasites, including the trematode *Schistosoma mansoni* [21,22], and protozoan parasites, including *Trichomonas vaginalis* [23], *Giardia lamblia* [24], and *E. histolytica* [25]. The mode of action of AF in protozoan parasites is not completely understood although it is assumed that TrxR is the main target of AF in *E. histolytica* [24,25]. In *G. lamblia*, this mechanism of action has been challenged by the significant TrxR activity that occurs in trophozoites exposed to high concentrations of auranofin [26]. Overexpression of TrxR in *G. lamblia* has no effect on the sensitivity of this parasite to AF [26]. AF can also target *E. histolytica* adenosine 5′-phosphosulfate kinase (EhAPSK), an essential enzyme in Entamoeba sulfolipid metabolism [27]. We recently showed that AF induced the formation of more than 500 oxidized proteins (OXs) in *E. histolytica*, including some crucial enzymes for redox homeostasis and cytoskeletal proteins, which are essential for *E. histolytica*'s cytoskeleton dependent virulence [28]. Knowledge about resistance to AF in bacteria and in protozoa is scarce. Recently, toxoplasma trophozoites resistant to AF (2 μM) were successfully generated through chemical mutagenesis. The authors identified point mutations in genes encoding redox-relevant proteins, such as superoxide dismutase and ribonucleotide reductase. However, recapitulation of these mutations in the parasite did not confer resistance to AF, suggesting that the mechanism of resistance is complex [29]. In this work, we used a multi-omics approach to characterize an *E. histolytica* strain that was made resistant to AF (AFAT) by progressively adapting the parasite to 2 μM of AF. At this concentration, the drug is lethal to non-adapted parasites [25,28].

2. Materials and Methods

2.1. *E. histolytica* Culture

E. histolytica trophozoites, the HM-1:IMSS strain (a kind gift of Prof. Samudrala Gourinath, Jawaharlal Nehru University, New Delhi, India), were grown and harvested according to a previously reported protocol [30].

2.2. Adaptation of *E. histolytica* Trophozoites to AF

The concentration of AF in *E. histolytica* trophozoite culture was progressively increased from 0 to 2 μM over a period of one month.

2.3. Growth Rate of WT Trophozoites and AFAT

The growth rate of WT trophozoites or AFAT and their viability were measured according to a previously reported protocol [31].

2.4. Viability of AFAT Exposed to H_2O_2, Paraquat, MNZ or GSNO

The viability of WT trophozoites and AFAT exposed to H_2O_2 (2.5 mM for 30 min), paraquat (2.5 mM for 24 h), MNZ (5 μM for 24 h), or GSNO (350 μM for 2 h) (Sigma-Aldrich, Jerusalem, Israel) was determined by the eosin dye exclusion method [31].

2.5. Measurement of Cytopathic Activity

Cytopathic activity was assayed against HeLa cells (a kind gift from T. Kleinberger, Faculty of Medicine, Technion) (using a previously described protocol [32].

2.6. RNA Extraction

Total RNA was extracted from control trophozoites (WT) and AFAT using a TRI reagent kit, according to the manufacturer instructions (Sigma-Aldrich, Jerusalem, Israel).

2.7. RNA Sequencing (RNAseq): Library Preparation and Data Generation

Six RNAseq libraries were produced according to the manufacturer's protocol (NEBNext UltraII Directional RNA Library Prep Kit, Illumina, NEB, MA, USA) using 800 ng of total RNA. mRNA pull-up was performed using a Magnetic Isolation Module (NEB, MA, USA). All libraries were mixed in a single tube with equal molarity. The RNAseq data was generated on an Illumina NextSeq500, 75 single-end read, high output mode (Illumina). Quality control was assessed using Fastqc (v0.11.5); reads were trimmed for adapters, low quality 3′, and minimum length of 20 using CUTADAPT (v1.12). STAR aligner (v2.6.0a) was used to align 83 bp single-end reads to an *E. histolytica* reference genome (Entamoeba_histolytica.JCVI-ESG2-1.0.dna.toplevel.fa) and annotation file (Entamoeba_histolytica.JCVI-ESG2-1.0.46.gff3), both downloaded from ENSEMBL (strain HM-1:IMSS, imported from the AmoebaDB (https://amoebadb.org/amoeba/app accessed on 28 July 2021)). The number of reads per gene was counted using Htseq-count (v0.9.1) (parameters: -t CDS -i ID -m intersection-nonempty -s reverse).

2.8. Descriptive Analysis

The statistical analysis was preformed using DESeq2 R package (version 1.20.0) [33].

2.9. Differential Expression Analysis

Results of the statistical analysis, i.e., the list of the differentially expressed genes (DEGs) (p-value adjusted (padj) < 0.01) are provided in the DESeq2_results_with_anno.xls file (Table S1). Genes with a fold change >1.5 were taken into account for further bioinformatics analysis. Gene symbol and gene name identification was achieved using Protein ANalysis THrough Evolutionary Relationship (PANTHER) Classification System software (http://www.pantherdb.org/ accessed on 28 July 2021) [34].

2.10. Availability of Data

RNA-Seq data are available at the Gene Expression Omnibus (http://www.ncbi.nlm.nih.gov/geo accessed on 28 July 2021) under the accession number GSE178520.

2.11. Construction of HA-Tagged EhTrxR Trophozoites

For the construction of the pJST4-EhTrxR expression vector that was used to express HA-tagged EhTrxR in the parasite, EhTrxR was amplified from *E. histolytica*'s genomic DNA using the primers 5′EhTrxR_KpnI (ggtaccatgagtaatattcatgatg) and 3′EhTrxR_BamHI (ggatccatgagtttgaagcc). The resulting PCR product was cloned into the pGEM-T Easy vector system (Promega, WI, USA) and then digested with the restriction enzymes, KpnI and BamHI. The digested DNA insert was subcloned into the *E. histolytica* expression vector pJST4, which was previously linearized with KpnI and BamHI. The pJST4 expression vector contains a tandem affinity purification tag for use in protein purification and identification [35]. This CHH tag contains the calmodulin binding protein, hemagglutinin (HA), and histidine (His) residues, and its expression is driven by an actin promoter.

2.12. Immunodetection of (HA)-Tagged EhTrxR

E. histolytica control and HA-tagged EhTrxR trophozoite cytosolic proteins (40 µg) were prepared according to a published method [36] and resolved on a 10% SDS-PAGE in SDS-PAGE running buffer (25 mM Tris, 192 mM glycine, 0.1% SDS). The resultant protein bands were visualized after staining with Ponceau-S (Sigma-Aldrich, USA). Next, proteins were electrotransferred in protein transfer buffer (25 mM Tris, 192 mM glycine, 20% methanol, pH 8.3) to nitrocellulose membranes (Whatman, Protran BA83). The blots were first blocked using 3% skim milk and then probed with 1:500 mouse monoclonal HA antibody clone 12CA5 (a kind gift from Prof. Ami Aronheim) for 16 h at 4 °C. After incubation with the primary antibody, the blots were incubated with 1:5000 secondary antibody for one hour at RT (Jackson ImmunoResearch, PA, USA), and then developed using enhanced chemiluminescence (Bio RAD, Rishon Le Zion, Israel).

2.13. Viability Assay

E. histolytica trophozoite controls and EhTrxR overexpressing trophozoites (2.5×10^4) were cultivated or not in the presence of 2 μM AF for 24 h. The cells were harvested at 400× *g* for 5 min, stained with Propidium iodide (1 μg/mL), and analyzed by flow cytometry. Flow cytometry was performed using Cyan ADP (Agilent Dako, CA, USA) and data from 10,000 cells were collected for each condition.

2.14. Detection of ROS

WT trophozoites, AFAT, and WT trophozoites that were cultivated with AF (2 μM) for 24 h (WT + AF acute) were incubated with 0.4 mM H2DCFDA for 15 min at 37 °C. The trophozoites were washed twice with PBS, and the level of oxidation was analyzed by flow cytometry. Flow cytometry was performed using Cyan ADP flow cytometer (Agilent Dako, CA, USA) and data from 10,000 cells were collected for each condition.

2.15. Detection of OXs by RAC (OX-RAC)

The detection of OXs by OX-RAC was performed using a previously described protocol [31]. A protein was considered to be oxidized when its relative amount in the DTT-treated lysates was at least two times greater than that in the untreated lysates ($p < 0.05$ according to the results of an unpaired *t*-test).

2.16. In-Gel Proteolysis and MS Analysis

In-gel proteolysis, MS, and data analysis were performed according to a previously reported protocol [31,37].

2.17. Classification of OXs According to Their Protein Class

The OXs were classified according to their protein class using PANTHER Classification System software (http://www.pantherdb.org/ accessed on 28 July 2021) [34].

2.18. Immunofluorescence Microscopy Analysis

The formation of F-actin in WT trophozoites, acute AF trophozoites, and AFAT was determined as described previously [28].

3. Results

3.1. Generation of AFAT

Resistance to 2 μM of AF in the parasite *T. gondi* can be generated by exposure of the parasite to the mutagenic compound *N*-Ethyl-*N*-nitrosourea [29]. To the best of our knowledge, the development by natural selection of parasites resistant to AF has never been attempted. To address this knowledge gap, we adapted *E. histolytica* trophozoites to AF by progressively increasing the drug concentration over a period of one month to 2 μM. To check whether adaptation to AF has an effect on the growth of the parasite, we compared the doubling time of AF-adapted trophozoites (AFAT) with that of wild type (WT) trophozoites. We found that the doubling time of AFAT (13 ± 0.32 h) was significantly higher than the doubling time of WT trophozoites (10.6 ± 0.24 h).

3.2. Reponse of AFAT to OS, NS, and Cytopathic Activity

The response of AFAT to OS was tested by exposing them to H_2O_2 (2.5 mM for 30 min) or to paraquat (2.5 mM for 12 h). We observed that AFAT are significantly more sensitive to H_2O_2 or to paraquat than WT trophozoites (Figure 1A). We also examined the resistance of AFAT to MNZ (5 μM for 24 h) and found that AFAT are significantly more sensitive to MNZ than WT trophozoites (Figure 1A). The sensitivity of AFAT to nitrosative stress (NS) was tested by exposing them to the NO donor S-nitrosoglutathion (GSNO) (350 μM for 2 h). We observed that AFAT are significantly more sensitive to NS than WT trophozoites (Figure 1A). The ability of AFAT to destroy a monolayer of mammalian cells (cytopathic activity) was also determined (Figure 1B). We observed that the cytopathic activity of AFAT is impaired compared to that of WT trophozoites. Overall, these results indicate that, for *E. histolytica* trophozoites, adaptation to AF results in a loss of fitness.

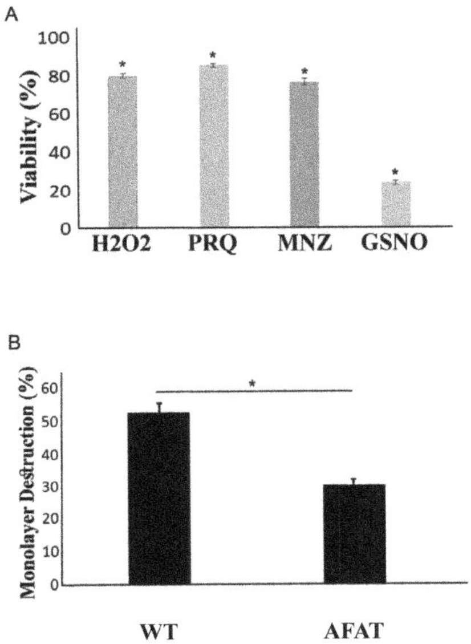

Figure 1. (**A**) Viability of AFAT exposed to H_2O_2, paraquat, MNZ, or GSNO. WT and AFAT were exposed to 2.5 mM H_2O_2 for 30 min, 2.5 mM paraquat (PRQ) and 5 µM metronidazole (MNZ) for 24 h, or 350 µM GSNO for 2 h. All experiments were undertaken at 37 °C. Data are expressed as the mean ± standard deviation of three independent experiments that were performed in triplicate. The graph represents the ratio percentage of viable amoebas compared to WT. The viability of AFAT exposed to H_2O_2, PRQ, MNZ, or GSNO was significantly different (* $p < 0.05$) from that of the WT according to the results of an unpaired Student's *t* test. (**B**) Cytopathic activity of AFAT. Data are displayed as the mean ± standard deviation of four independent experiments that were performed in triplicate. The cytopathic activity of AFAT was significantly different (* $p < 0.05$) from that of the WT according to the results of an unpaired Student's *t* test.

3.3. *Transcriptomics of AFAT*

We used RNA sequencing (RNA-seq) to examine the mechanism of adaptation to AF. Transcriptomics of WT trophozoites vs. AFAT was compared. Our comparisons revealed that adaptation to AF has a strong effect on the *E. histolytica* transcriptome, with more than 500 upregulated and downregulated genes (Table S1).

3.4. *Gene Categories Modulated in AFAT*

The differentially regulated genes in AFAT vs. WT trophozoites were classified, according to the protein class they encode, using PANTHER. The categories for functional classification of genes upregulated in AFAT are shown in Figure 2A. The most abundant classes are the gene-encoded protein-binding activity modulator (PC00095), such as AIG1-type G domain-containing protein (EHI_176590); metabolite interconversion enzyme (PC00262), such as Lecithin:cholesterol acyltransferase (EHI_065250); protein modifying enzyme (PC00260), such as Leucine rich repeat protein phosphatase 2C domain containing protein (EHI_178020); and cytoskeletal protein (PC00085), such as F-actin-capping protein subunit beta (EHI_134490). Of the upregulated genes in AFAT, genes that encode for actin or for actin-binding cytoskeletal proteins, such as actin (EHI_107290) or EHI_172960 (Actin-related protein 2/3 complex subunit 3); dehydrogenase (PC00092), such as NAD (FAD)-dependent dehydrogenase (EHI_099700) or Aldehyde-alcohol dehydrogenase 2

(EHI_024240); and guanyl-nucleotide exchange factor, such as Ras guanine nucleotide exchange factor (EHI_023270) or Rho guanine nucleotide exchange factor (EHI_005910) are significantly enriched according to the PANTHER statistical overrepresentation test (Figure 2B).

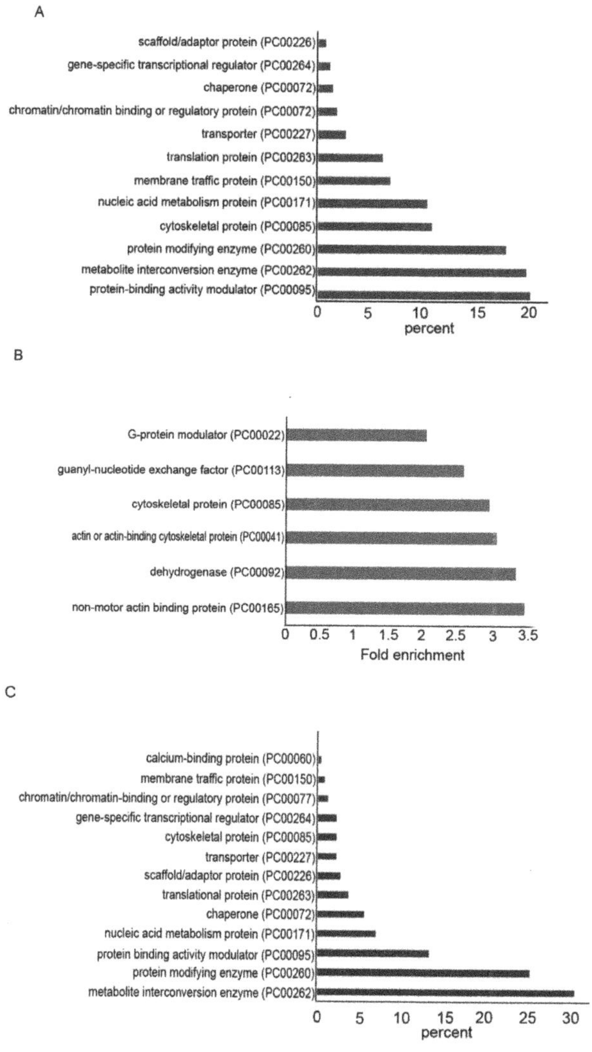

Figure 2. (**A**) PANTHER sequence classification of genes upregulated in AFAT; (**B**) PANTHER statistical overrepresentation test of upregulated genes in AFAT; (**C**) PANTHER sequence classification of genes downregulated in AFAT.

The categories for functional classification of genes downregulated in AFAT are shown in Figure 2C. The most abundant class of gene encoded proteins are metabolite interconversion enzyme (PC00262), such as alpha-amylase (EHI_152880); protein modifying enzyme (PC00260), such as Gal/GalNAc lectin Igl2 (EHI_183000); and protein-binding activity modulator (PC00095); such as guanylate binding protein (EHI_175080). Of the downregulated genes in AFAT, no enrichment of a specific biological process was detected according to the PANTHER statistical overrepresentation test.

3.5. Redoxomics of AFAT

Using OX-RAC, we previously detected 583 OXs in acute AF trophozoites [28]. Here, we also used OX-RAC to detect OXs in the lysate of AFAT (Figure 3A). We identified 96 OXs in AFAT (Table S2), which were classified using PANTHER. The most abundant OX families belong to metabolite interconversion enzyme (PC00262), such as Purine nucleoside phosphorylase (EHI_200080); protein modifying enzyme (PC00260), such as NEDD8-activating enzyme E1 (EHI_098550); chaperone (PC00072), such as Peptidylprolyl isomerase (EHI_044850); and Protein-binding activity modulator (PC00095), such as glucosidase 2 subunit beta (EHI_135420) (Figure 3B).

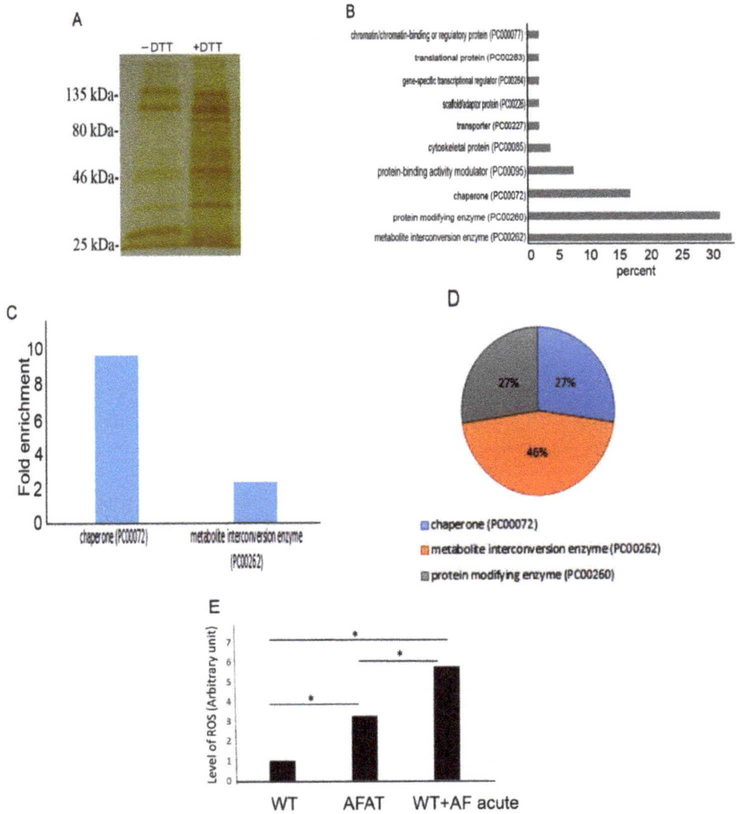

Figure 3. Detection of oxidized proteins by resin-assisted capture (OX-RAC) analysis of AFAT. (**A**) Silver staining of OXs. OXs in the AFAT lysates were subjected to RAC in the presence of 10 mM DTT (+DTT) or the absence of DTT (−DTT). (**B**) Protein ANalysis THrough Evolutionary Relationships (PANTHER) sequence classification of the OXs identified in AFAT. (**C**) PANTHER statistical overrepresentation test of the OXs identified in AFAT. (**D**) PANTHER sequence classification of the 17 OXs common between trophozoites exposed to an acute AF treatment [28] and AFAT. (**E**) Level of ROS in AFAT and acute AF trophozoites. WT trophozoites, AFAT, and WT trophozoites that were cultivated with AF (2 μM) for 24 h (WT + AF acute) were incubated with 0.4 mM H2DCFDA for 15 min at 37 °C. The trophozoites were washed twice with PBS, and the level of oxidation was analyzed by flow cytometry. Flow cytometry was performed using Cyan ADP (Agilent Dako, CA, USA) and data from 10,000 cells were collected for each condition. Data are expressed as the mean ± standard deviation of three independent experiments. The level of ROS in AFAT was significantly different from that of the WT + AF acute trophozoites according to the results of an unpaired Student's *t* test (* p value < 0.05).

Of the OXs in AFAT (Table S2), chaperone (PC00072), such as HSP16 (EHI_125830) or Trx (EHI_110350), and metabolite interconversion enzyme (PC00262), such as Aminotran_5 domain-containing protein EhnifS (EHI_136380) or alpha-amylase EHI_152880, are significantly enriched according to the PANTHER statistical overrepresentation test (Figure 3C).

Seventeen OXs are shared between acute AF trophozoites [28] and AFAT (Table S3). These common OXs belong to chaperone (PC00072), metabolite interconversion enzyme (PC00262), and protein modifying enzyme (PC00260) (Figure 3D).

3.6. Level of ROS in AFAT

The lower quantity of OXs in AFAT compared to the quantity of OXs in acute AF trophozoites [28] suggests that AFAT are less exposed to ROS. Consequently, we measured the level of ROS with dichloro-fluorescein (H_2DCDFC) in acute AF trophozoites and in AFAT. We observed that the ROS level in AFAT is significantly lower than that in acute AF trophozoites (Figure 3E).

3.7. Comparison between Transcriptomics and Redoxomics of AFAT

We found that only two genes upregulated in AFAT (Gal/GalNAc lectin Igl1 EHI_006980 and SNF7 family protein EHI_077530) have their product oxidized (Table S3). None of the genes downregulated in AFAT have their product oxidized (Table S3).

3.8. Comparison between Transcriptomics of AFAT and Redoxomics of Acute AF Trophozoites

We found that 77 genes upregulated in AFAT have their product oxidized in acute AF trophozoites [28] (Table S3). The most abundant OXs belong to metabolite interconversion enzyme (PC00262), protein-binding activity modulator (PC00095), protein modifying enzyme (PC00260), and cytoskeletal protein (PC00085) (Figure 4A). Of the upregulated genes in AFAT that have their product oxidized in acute AF trophozoites, genes that encode for dehydrogenase (PC00092), such as NAD(FAD)-dependent dehydrogenase (EHI_099700); oxydoreductase (PC00176), such as Pyruvate:ferredoxin oxidoreductase (EHI_051060); and metabolite interconversion enzymes (PC00262), such as isopentenyl phosphate kinase (EHI_178490), are significantly enriched according to the PANTHER statistical overrepresentation test (Figure 4B).

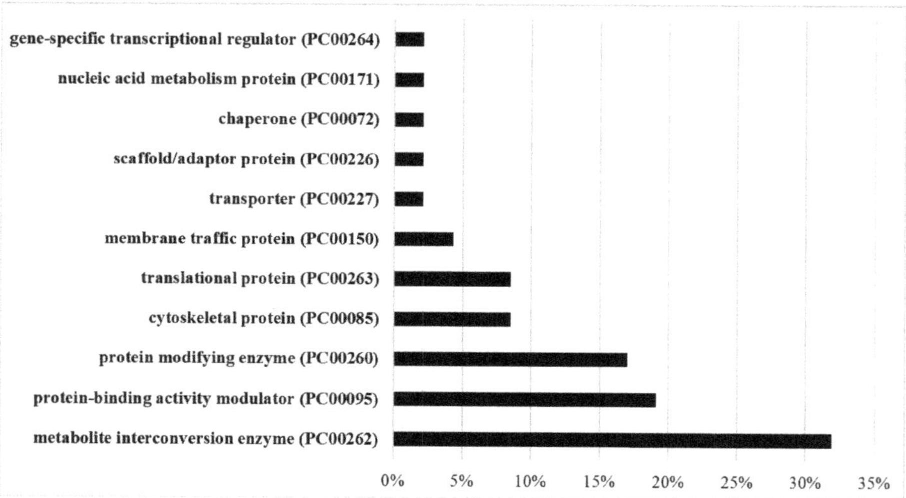

(A)

Figure 4. *Cont.*

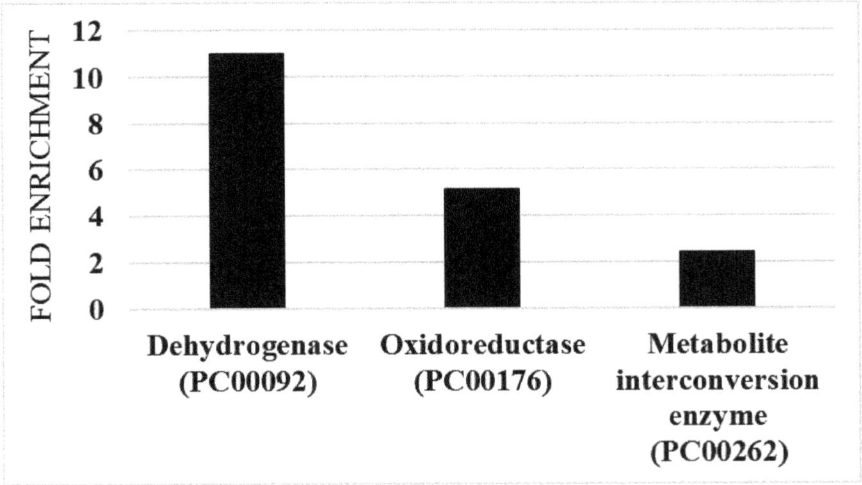

(B)

Figure 4. Comparison between transcriptomics of AFAT and redoxomics of acute AF trophozoites. (**A**) PANTHER sequence classification of genes upregulated in AFAT that have their product oxidized in acute AF trophozoites. (**B**) PANTHER statistical overrepresentation test of upregulated genes in AFAT that have their product oxidized in acute AF trophozoites.

Eight genes that are downregulated in AFAT have their product oxidized in acute AF trophozoites [28] (Table S3). These OXs are the uncharacterized proteins (EHI_008120, EHI_065710, and EHI_110780), Asparagine–tRNA ligase (EHI_126920), Cytosolic Fe-S cluster assembly factor NUBP1 (EHI_047750), ribonuclease (EHI_156310), and Flavodoxin-like domain-containing protein (EHI_096710).

3.9. Formation of F-Actin in AFAT

We have previously shown that AF leads to the oxidation of cytoskeletal proteins and inhibits the formation of actin filaments (F-actin) [28]. In contrast, cytoskeletal proteins in AFAT are not significantly enriched among OXs according to the PANTHER statistical overrepresentation test (Table S2). In order to confirm this observation, we looked at the level of F-actin in WT trophozoites, acute AF trophozoites, and AFAT. As described previously [28], the F-actin signal in acute AF trophozoites was significantly less intense than that in WT trophozoites. In contrast, the F-actin signal was identical in WT trophozoites and AFAT (Figure 5A,B). These results confirm that the formation of F-actin is impaired in acute AF trophozoites [28], but is not impaired in AFAT.

3.10. Overexpression of EhTrxR Does Not Protect E. histolytica Trophozoites against AF

Overexpression of TrxR in the parasite *Giardia lamblia* has no effect on its resistance to AF [26]. In *E. histolytica*, Debnath et al. found that AF inhibits the amebic TrxR and its reduction, leading to a higher sensitivity of trophozoites to ROS-mediated killing [39]. Our observations regarding the level of TrxR expression, which was the same in WT trophozoites and in AFAT (Table S1), and the fact that Trxs are enriched OXs in AFAT, strongly suggest that *E. histolytica* TrxR is not central to the mechanism of adaptation of the parasite to AF. To test this hypothesis, we overexpressed EhTrxR in *E. histolytica* trophozoites. Overexpression of EhTrxR was confirmed by Western blotting and its level of expression in *E. histolytica* was proportional to the amount of G418 used for selection (Figure 6A–C) [40]. Next, we determined the level of resistance to AF of HA-tagged EhTrxR trophozoites. We observed that the level of resistance to AF of HA-tagged EhTrxR trophozoites did not differ significantly from the level of resistance of the control trophozoites (trophozoites transfected with pEhExGFP (a kind gift from Dr. Tomoyoshi Nozaki [41])) (Figure 6D). pEhExGFP allows the constitutive expression of the green fluorescent protein (GFP).

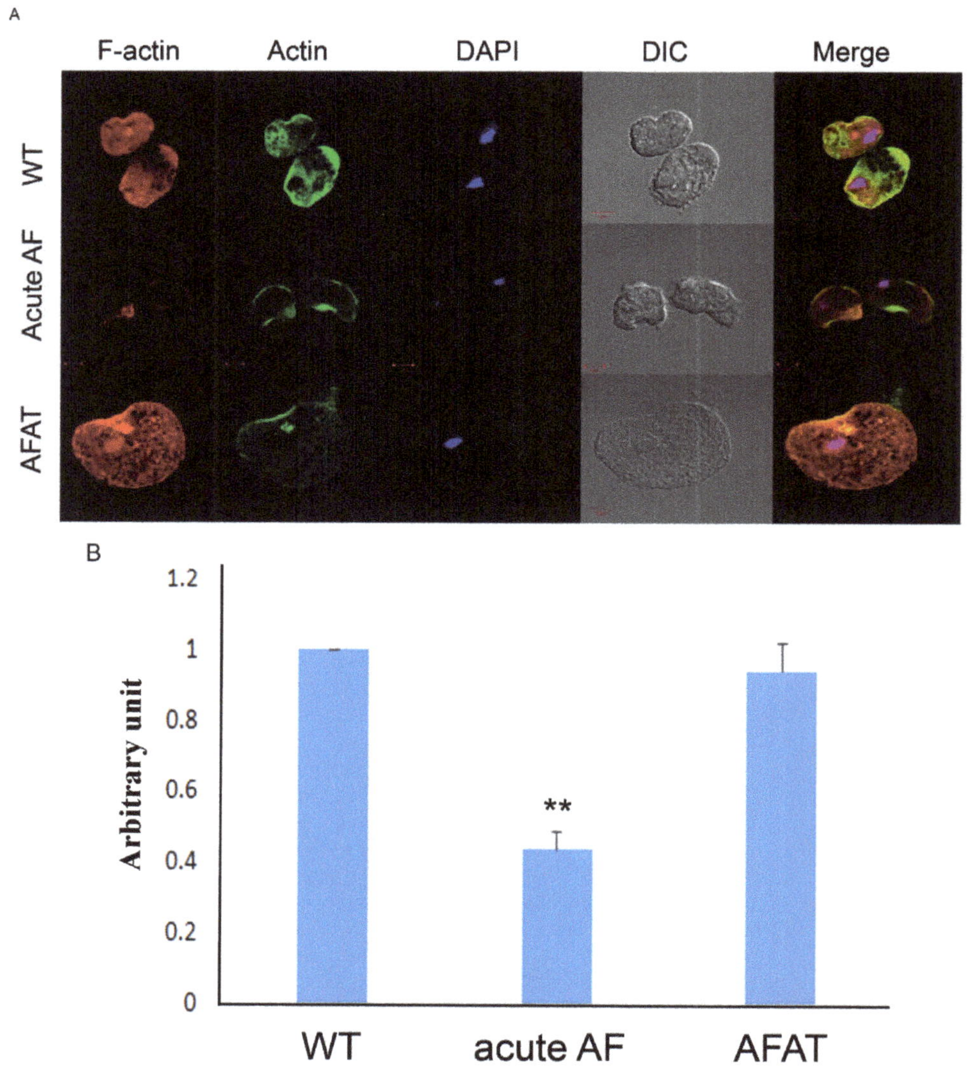

Figure 5. Formation of F-actin in WT trophozoites, acute AF trophozoites, and AFAT. (**A**) Confocal laser scanning microscopy of F-actin and total actin in WT trophozoites, acute AF trophozoites, and AFAT showed that F-actin was detected using rhodamine-conjugated phalloidin. Total actin was detected using a primary actin antibody and a secondary Cy2-conjugated immunoglobulin G (IgG) antibody. The nuclei (blue) were stained by 4′,6-diamidino-2-phenylindole (DAPI). (**B**) A computer-assisted image was overlaid on the signal emitted by the actin antibody, phalloidin, and DAPI. Fluorescence quantification was performed using Fiji software [38] on 10 trophozoites and the F-actin signal was normalized to the total actin signal. The level of F-actin in WT was arbitrary defined as 1. Data are expressed as the mean ± standard deviation of two independent experiments. The level of F-actin in acute AF trophozoites was significantly different from that in WT and AFAT according to the results of an unpaired Student's *t* test (** *p* value < 0.01). No difference of F-actin level between WT and AFAT was observed according to the results of an unpaired Student's *t* test (*p* value > 0.05).

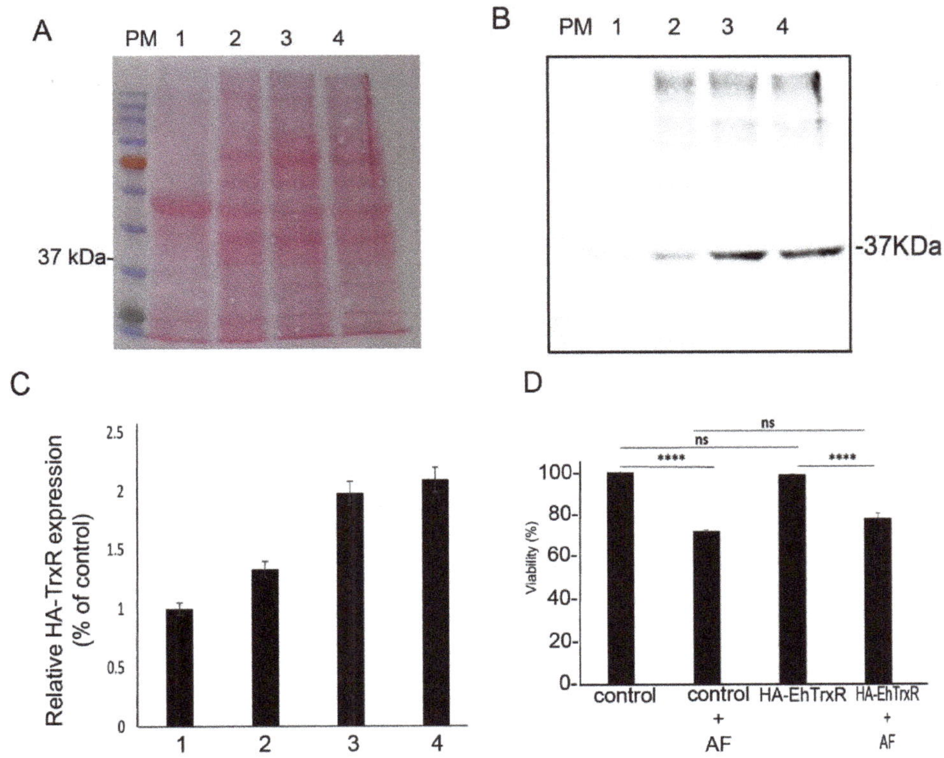

Figure 6. Western blot analysis of *E. histolytica* trophozoites that overexpress a hemagglutinin (HA)-tagged EhTrxR and viability assay. Legend: Protein molecular weight marker (PM). Control trophozoites (lane 1). HA-tagged EhTrxR trophozoites cultivated in the presence of an increasing concentration of G418 (lane 2: 6 μg/mL, lane 3: 30 μg/mL, lane 4: 48 μg/mL). (**A**) Ponceau staining of a nitrocellulose membrane containing cytosolic proteins (40 μg) separated by SDS PAGE of control trophozoites and of HA-tagged EhTrxR trophozoites cultivated in the presence of an increasing concentration of G418. (**B**) Immunodetection of (HA)-tagged EHTrxR with an HA monoclonal antibody (1:500) in HA-tagged EhTrxR trophozoites cultivated in the presence of an increasing concentration of G418. (**C**) Relative quantification of the HA EhTrxR signal following its normalization with the level of total protein in each well with ImageJ software. Normalized values for control trophozoites were taken as 100%. These results are representative of two independent experiments. (**D**) Viability assay. *E. histolytica* trophozoite controls and HA-tagged EhTrxR trophozoites were cultivated in the presence of 2 μM AF for 24 h. The cells were harvested at 400× *g* for 5 min, stained with Propidium iodide, and analyzed by flow cytometry. Flow cytometry was performed using Cyan ADP (Agilent Dako, CA, USA) and data from 10,000 cells were collected for each condition. Data are expressed as the mean ± standard deviation of three independent experiments that were performed in triplicate. The viability of the control trophozoites was defined as 100%. The viability of control trophozoites was not significantly different (ns) from that of the HA-tagged EhTrxR trophozoites according to the results of an unpaired Student's *t* test (p value < 0.05). The viability of control trophozoites was significantly different from that of the control trophozoites exposed to AF according to the results of an unpaired Student's *t* test (**** p value < 0.001). The viability of HA-tagged EhTrxR trophozoites was significantly different from that of the HA-tagged EhTrxR trophozoites exposed to AF according to the results of an unpaired Student's *t* test (**** p value < 0.001).

4. Discussion

In our previous work, we demonstrated that AF triggers OS inside *E. histolytica* trophozoites, resulting in the oxidation of more than 500 proteins, including many redox enzymes that are essential for controlling the intracellular levels of ROS in the parasite [28,42,43]. Here, we characterized *E. histolytica* trophozoites that were adapted to 2 μM AF. Adaptation of *E. histolytica* to AF leads to the upregulation and downregulation of hundreds of genes, which suggests that the mechanism of adaptation is complex. Drug resistance is often mediated by a drug's molecular target gene overexpression [44,45]. Consequently, we expected that *E. histolytica* TrxR (EhTrxR), the assumed main target of AF [25], would be one of the upregulated genes in AFAT. However, transcriptomics of AFAT indicates that this was not the case. Indeed, the overexpression of EhTrxR did not confer to *E. histolytica* resistance to AF. This information raises a question about why EhTrxR expression is not upregulated as a simple mechanism to resist AF. One possible answer is that, as for *Giardia lamblia*, TrxR is not the primary target of AF in *E. histolytica* [26]. This is supported by the absence of the detection of EhTrxR among OXs in AFAT (this work) and acute AF trophozoites [28].

It is also possible that the fitness cost for *E. histolytica* to overexpress TrxR during adaptation to AF resistance is too high. EhTrxR can generate H_2O_2 from molecular oxygen, leading to the formation of reactive species [46]. Therefore, it is possible that the production of H_2O_2 resulting from EhTrxR overexpression combined with OS triggered by AF [28] during the adaptation process cannot be tolerated by the parasite.

In this work, we found that only two genes upregulated in AFAT have their products oxidized in AFAT. In contrast, 77 genes upregulated in AFAT have their product oxidized in acute AF trophozoites [28]. The upregulation of these 77 genes in AFAT may be essential for the adaption of the parasite to AF by replacing their oxidized-inactivated products by reduced-activated proteins. The relevance of this mechanism for some of these 77 genes is discussed in the following.

Pyruvate:ferredoxin oxidoreductase (EHI_051060), NADP-dependent alcohol dehydrogenase (EHI_107210), and Fe-ADH domain-containing protein (EHI_198760), which encode for proteins involved in redox regulation: These redox enzymes depend on cysteine residues for their activity [47–49]. The oxidation of these cysteine residues impairs their activity [47,50].

Genes that encode the protein-binding activity modulator, such as Ras guanine nucleotide exchange factor (EHI_035800), Rho guanine nucleotide exchange factor (EHI_005910), or Ras GTPase-activating protein (EHI_105250): These proteins have their product oxidized in acute AF trophozoites [28]. G proteins are involved in vesicular trafficking and cytoskeleton regulation [51]. Redox regulation of G-proteins have been well documented [52] and their oxidation impairs *E. histolytica*'s motility [28].

Genes that encode protein-modifying enzymes such as protein kinase domain-containing proteins (EHI_186820) (EHI_101280) and Protein kinase (EHI_188110), which are also oxidized in acute AF trophozoites [28]: Protein kinases have been associated with the virulence and phagocytic activity of *E. histolytica* [53]. The redox regulation of protein kinases is well established [54], and it has been demonstrated that AF can directly inhibit protein kinase C by interacting with thiol groups present in the catalytic site [55].

Genes that encode actin or actin-binding cytoskeletal proteins are upregulated in AFAT and oxidized in acute AF trophozoites [28]: In our previous work, we showed that AF induces the oxidation of *E. histolytica* cytoskeletal proteins and consequently inhibits the formation of F-actin [28]. Consequently, it appears that the parasite upregulated the expression of actin-binding cytoskeletal proteins as a mechanism to adapt to AF by replacing oxidized cytoskeletal proteins that were formed during the process of adaptation to AF. The low level of F-actin in acute AF trophozoites and the normal level of F-actin in AFAT (this work) support this hypothesis.

The fact that *E. histolytica* can adapt to AF illustrates the remarkable ability of *E. histolytica* to adapt to drugs [56,57] and environmental stresses [32,58]. The fitness cost paid by the parasite to adapt to AF resembles collateral sensitivity, which occurs when the acquisition of resistance to one antibiotic produces increased susceptibility to a second antibiotic [59]. AFAT are more sensitive to OS, paraquat, MNZ, and GSNO than WT trophozoites. Resistance to OS in *E. histolytica* involves the upregulation of 29 kDa peroxiredoxin [60] and iron-containing peroxide dismutase expression, which is also involved in the resistance to MNZ [10,61]. The level of expression of 29 kDa peroxiredoxin and iron-containing peroxide dismutase is globally the same in WT and in AFAT, which suggests that the sensitivity of AFAT to OS and MNZ is not caused by a reduced level of these redox enzymes' expressions. As discussed above, many oxidized proteins in AFAT have their level of expression upregulated. The fitness cost observed in AFAT may be due to numerous factors, including the rerouting of protein synthesis toward oxidized proteins, or substrate wasting that results from target overexpression [62]. In hydroxamic acid analog pan-histone deacetylase inhibitor-resistant leukemia cells, overexpression of the target protein heat shock protein 90 (HSP90) revealed collateral sensitivity to the HSP90 inhibitor 17-*N*-allylamino-17-demethoxygeldanamycin [63].

5. Conclusions

We showed that *E. histolytica* trophozoites can be easily selected to resist toxic concentrations of AF in vitro. Adaptation to AF reduces the fitness of *E. histolytica*, as seen in a decreased growth rate and virulence, and a sensitivity to OS, NS, and MNZ. Overexpression of genes whose products are sensitive to AF-mediated oxidation may represent an important step in the adaptation process to AF, and EhTrxR does not appear to be central to this process.

AF is FDA approved for the treatment of rheumatoid arthritis but has not been yet used as an antimicrobial drug in the field. The ability of *E. histolytica* to adapt to amebicidal concentrations of AF raises concerns about the future use of this drug as an antiamebic compound. Our omics data provide the basis for the development of strategies to limit the emergence of resistance against AF. One possible strategy suggested by our data is to promote dual antibiotic therapy (AF + MNZ) vs. single AF therapy, because adaptation to AF leads to more MNZ sensitivity in *E. histolytica*.

Supplementary Materials: The following are available online at https://www.mdpi.com/article/10.3390/antiox10081240/s1. Table S1: transcriptomics of WT trophozoites vs. AFAT. Table S2: legend of Table S1. Table S3: comparative analysis of OXs in AFAT vs. OXs in acute AF; OXs in AFAT vs. gene products upregulated in AFAT and OXs in acute AF vs. gene products upregulated or downregulated in AFAT.

Author Contributions: Conceptualization: Y.S., S.A.; methodology, Y.S., S.A.; software, Y.S., S.A.; validation, Y.S., L.S., M.T.-G., S.A.; formal analysis, Y.S., L.S., M.T.-G., S.A.; investigation, Y.S., L.S., M.T.-G., S.A.; resources, S.A.; data curation, Y.S., S.A.; writing—original draft preparation, Y.S., S.A.; writing—review and editing, Y.S., L.S., M.T.-G., S.A.; visualization, S.A.; supervision, S.A.; project administration, S.A.; funding acquisition, S.A. All authors have read and agreed to the published version of the manuscript.

Funding: This research was supported by the Israel Science Foundation (260/16), the ISF-NRF program (3208/19), the US–Israel Binational Science Foundation (2015211), and the Niedersachsen program.

Institutional Review Board Statement: Not applicable.

Informed Consent Statement: Not applicable.

Data Availability Statement: RNA-Seq data have been deposited at the Gene Expression Omnibus (http://www.ncbi.nlm.nih.gov/geo accessed on 28 July 2021) under the accession number GSE178520.

Acknowledgments: We thank the staff of the Microscopy Imaging facility, the genomics Core Facility Laboratory, and the Smoler Proteomics Center at the Technion for their technical help.

Conflicts of Interest: The authors declare no conflict of interest.

References

1. Turkeltaub, J.A.; McCarty, T.R., 3rd; Hotez, P.J. The intestinal protozoa: Emerging impact on global health and development. *Curr. Opin. Gastroenterol.* **2015**, *31*, 38–44. [CrossRef] [PubMed]
2. Powell, S.J.; MacLeod, I.; Wilmot, A.J.; Elsdon-Dew, R. Metronidazole in amoebic dysentery and amoebic liver abscess. *Lancet* **1966**, *2*, 1329–1331. [CrossRef] [PubMed]
3. Leitsch, D.; Kolarich, D.; Binder, M.; Wilson, I.B.H.; Altmann, F.; Duchene, M. Nitroimidazole action in *Entamoeba histolytica*: A central role for thioredoxin reductase. *PLoS Biol.* **2007**, *5*, e211. [CrossRef] [PubMed]

4. Cowdrey, S.C. Letter: Hazards of metronidazole. *N. Engl. J. Med.* **1975**, *293*, 455.
5. Andersson, K.E. Pharmacokinetics of Nitroimidazoles-Spectrum of Adverse Reactions. *Scand. J. Infect. Dis. Suppl.* **1981**, *26*, 60–67.
6. Roe, F.J. Metronidazole: Review of uses and toxicity. *J. Antimicrob. Chemother.* **1977**, *3*, 205–212. [CrossRef] [PubMed]
7. Camacho, N.; Espinoza, C.; Rodriguez, C.; Rodriguez, E. Isolates of Clostridium perfringens recovered from Costa Rican patients with antibiotic-associated diarrhoea are mostly enterotoxin-negative and susceptible to first-choice antimicrobials. *J. Med. Microbiol.* **2008**, *57*, 343–347. [CrossRef] [PubMed]
8. Hashemi, S.J.; Sheikh, A.F.; Goodarzi, H.; Yadyad, M.J.; Seyedian, S.S.; Aslani, S. Genetic basis for metronidazole and clarithromycin resistance in Helicobacter pylori strains isolated from patients with gastroduodenal disorders. *Infect. Drug Resist.* **2019**, *12*, 535–543. [CrossRef]
9. Wassmann, C.; Bruchhaus, I. Superoxide dismutase reduces susceptibility to metronidazole of the pathogenic protozoan *Entamoeba histolytica* under microaerophilic but not under anaerobic conditions. *Arch. Biochem. Biophys.* **2000**, *376*, 236–238. [CrossRef]
10. Wassmann, C.; Hellberg, A.; Tannich, E.; Bruchhaus, I. Metronidazole resistance in the protozoan parasite Entamoeba *histolytica* is associated with increased expression of iron-containing superoxide dismutase and peroxiredoxin and decreased expression of ferredoxin 1 and flavin reductase. *J. Biol. Chem.* **1999**, *274*, 26051–26056. [CrossRef]
11. Upcroft, J.A.; Upcroft, P. Drug susceptibility testing of anaerobic protozoa. *Antimicrob. Agents Chemother.* **2001**, *45*, 1810–1814. [CrossRef] [PubMed]
12. Finkelstein, A.E.; Walz, D.T.; Batista, V.; Mizraji, M.; Roisman, F.; Misher, A. Auranofin. New oral gold compound for treatment of rheumatoid arthritis. *Ann. Rheum. Dis.* **1976**, *35*, 251–257. [CrossRef]
13. Gromer, S.; Arscott, L.D.; Williams, C.H., Jr.; Schirmer, R.H.; Becker, K. Human placenta thioredoxin reductase. Isolation of the selenoenzyme, steady state kinetics, and inhibition by therapeutic gold compounds. *J. Biol. Chem.* **1998**, *273*, 20096–20101. [CrossRef] [PubMed]
14. Onodera, T.; Momose, I.; Kawada, M. Potential Anticancer Activity of Auranofin. *Chem. Pharm. Bull.* **2019**, *67*, 186–191. [CrossRef]
15. Ruth, M.M.; van Rossum, M.; Koeken, V.; Pennings, L.J.; Svensson, E.M.; Ruesen, C. Auranofin Activity Exposes Thioredoxin Reductase as a Viable Drug Target in Mycobacterium abscessus. *Antimicrob. Agents Chemother.* **2019**, *63*, e00449-19. [CrossRef]
16. AbdelKhalek, A.; Abutaleb, N.S.; Mohammad, H.; Seleem, M.N. Antibacterial and antivirulence activities of auranofin against Clostridium difficile. *Int. J. Antimicrob. Agents* **2019**, *53*, 54–62. [CrossRef] [PubMed]
17. Jackson-Rosario, S.; Cowart, D.; Myers, A.; Tarrien, R.; Levine, R.L.; Scott, R.A. Auranofin disrupts selenium metabolism in Clostridium difficile by forming a stable Au-Se adduct. *J. Biol. Inorg. Chem.* **2009**, *14*, 507–519. [CrossRef]
18. Abutaleb, N.S.; Seleem, M.N. Antivirulence activity of auranofin against vancomycin-resistant enterococci: In vitro and in vivo studies. *Int. J. Antimicrob. Agents* **2020**, *55*, 105828. [CrossRef]
19. AbdelKhalek, A.; Abutaleb, N.S.; Elmagarmid, K.A.; Seleem, M.N. Repurposing auranofin as an intestinal decolonizing agent for vancomycin-resistant enterococci. *Sci. Rep.* **2018**, *8*, 8353. [CrossRef]
20. Thangamani, S.; Mohammad, H.; Abushahba, M.F.; Sobreira, T.J.; Hedrick, V.E.; Paul, L.N. Antibacterial activity and mechanism of action of auranofin against multi-drug resistant bacterial pathogens. *Sci. Rep.* **2016**, *6*, 22571. [CrossRef]
21. Angelucci, F.; Sayed, A.A.; Williams, D.L.; Boumis, G.; Brunori, M.; Dimastrogiovanni, D. Inhibition of Schistosoma mansoni thioredoxin-glutathione reductase by auranofin: Structural and kinetic aspects. *J. Biol. Chem.* **2009**, *284*, 28977–28985. [CrossRef] [PubMed]
22. Kuntz, A.N.; Davioud-Charvet, E.; Sayed, A.A.; Califf, L.L.; Dessolin, J.; Arner, E.S. Thioredoxin glutathione reductase from Schistosoma mansoni: An essential parasite enzyme and a key drug target. *PLoS Med.* **2007**, *4*, e206. [CrossRef]
23. Hopper, M.; Yun, J.F.; Zhou, B.; Le, C.; Kehoe, K.; Le, R. Auranofin inactivates Trichomonas vaginalis thioredoxin reductase and is effective against trichomonads in vitro and in vivo. *Int. J. Antimicrob. Agents* **2016**, *48*, 690–694. [CrossRef] [PubMed]
24. Tejman-Yarden, N.; Miyamoto, Y.; Leitsch, D.; Santini, J.; Debnath, A.; Gut, J. A reprofiled drug, auranofin, is effective against metronidazole-resistant Giardia lamblia. *Antimicrob. Agents Chemother.* **2013**, *57*, 2029–2035. [CrossRef] [PubMed]
25. Debnath, A.; Parsonage, D.; Andrade, R.M.; He, C.; Cobo, E.R.; Hirata, K. A high-throughput drug screen for *Entamoeba histolytica* identifies a new lead and target. *Nat. Med.* **2012**, *18*, 956–960. [CrossRef] [PubMed]
26. Leitsch, D.; Muller, J.; Muller, N. Evaluation of Giardia lamblia thioredoxin reductase as drug activating enzyme and as drug target. *Int. J. Parasitol. Drugs Drug Resist.* **2016**, *6*, 148–153. [CrossRef] [PubMed]
27. Mi-Ichi, F.; Ishikawa, T.; Tam, V.K.; Deloer, S.; Hamano, S.; Hamada, T.; Yoshida, H. Characterization of *Entamoeba histolytica* adenosine 5′-phosphosulfate (APS) kinase; validation as a target and provision of leads for the development of new drugs against amoebiasis. *PLoS Negl. Trop. Dis.* **2019**, *13*, e0007633. [CrossRef] [PubMed]
28. Shaulov, Y.; Nagaraja, S.; Sarid, L.; Trebicz-Geffen, M.; Ankri, S. Formation of oxidised (OX) proteins in *Entamoeba histolytica* exposed to auranofin and consequences on the parasite virulence. *Cell. Microbiol.* **2020**, *22*, e13174. [CrossRef] [PubMed]
29. Ma, C.I.; Tirtorahardjo, J.A.; Jan, S.; Schweizer, S.S.; Rosario, S.A.C.; Du, Y. Auranofin Resistance in Toxoplasma gondii Decreases the Accumulation of Reactive Oxygen Species but Does Not Target Parasite Thioredoxin Reductase. *Front. Cell. Infect. Microbiol.* **2021**, *11*, 618994. [CrossRef]
30. Diamond, L.S.; Harlow, D.R.; Cunnick, C.C. A new medium for the axenic cultivation of *Entamoeba histolytica* and other Entamoeba. *Trans. R. Soc. Trop. Med. Hyg.* **1978**, *72*, 431–432. [CrossRef]

31. Shahi, P.; Trebicz-Geffen, M.; Nagaraja, S.; Alterzon-Baumel, S.; Hertz, R.; Methling, K. Proteomic Identification of Oxidized Proteins in *Entamoeba histolytica* by Resin-Assisted Capture: Insights into the Role of Arginase in Resistance to Oxidative Stress. *PLoS Negl. Trop. Dis.* **2016**, *10*, e0004340. [CrossRef]
32. Trebicz-Geffen, M.; Shahi, P.; Nagaraja, S.; Vanunu, S.; Manor, S.; Avrahami, A.; Ankri, S. Identification of S-Nitrosylated (SNO) Proteins in *Entamoeba histolytica* Adapted to Nitrosative Stress: Insights into the Role of SNO Actin and In vitro Virulence. *Front. Cell. Infect. Microbiol.* **2017**, *7*, 192. [CrossRef]
33. Love, M.I.; Huber, W.; Anders, S. Moderated estimation of fold change and dispersion for RNA-seq data with DESeq2. *Genome Biol.* **2014**, *15*, 550. [CrossRef]
34. Mi, H.; Ebert, D.; Muruganujan, A.; Mills, C.; Albou, L.P.; Mushayamaha, T. PANTHER version 16: A revised family classification, tree-based classification tool, enhancer regions and extensive API. *Nucleic Acids Res.* **2021**, *49*, D394–D403. [CrossRef]
35. Dastidar, P.G.; Majumder, S.; Lohia, A. Eh Klp5 is a divergent member of the kinesin 5 family that regulates genome content and microtubular assembly in *Entamoeba histolytica*. *Cell. Microbiol.* **2007**, *9*, 316–328. [CrossRef] [PubMed]
36. Lavi, T.; Isakov, E.; Harony, H.; Fisher, O.; Siman-Tov, R.; Ankri, S. Sensing DNA methylation in the protozoan parasite *Entamoeba histolytica*. *Mol. Microbiol.* **2006**, *62*, 1373–1386. [CrossRef] [PubMed]
37. Cox, J.; Mann, M. MaxQuant enables high peptide identification rates, individualized p.p.b.-range mass accuracies and proteome-wide protein quantification. *Nat. Biotechnol.* **2008**, *26*, 1367–1372. [CrossRef]
38. Schindelin, J.; Arganda-Carreras, I.; Frise, E.; Kaynig, V.; Longair, M.; Pietzsch, T.; Preibisch, S.; Rueden, C.; Saalfeld, S.; Schmid, B.; et al. Fiji: An open-source platform for biological-image analysis. *Nat. Methods* **2012**, *9*, 676–682. [CrossRef]
39. Parsonage, D.; Sheng, F.; Hirata, K.; Debnath, A.; McKerrow, J.H.; Reed, S.L.; Abagyan, R.; Poole, L.B.; Podust, L.M. X-ray structures of thioredoxin and thioredoxin reductase from *Entamoeba histolytica* and prevailing hypothesis of the mechanism of Auranofin action. *J. Struct. Biol.* **2016**, *194*, 180–190. [CrossRef] [PubMed]
40. Hamann, L.; Nickel, R.; Tannich, E. Transfection and continuous expression of heterologous genes in the protozoan parasite *Entamoeba histolytica*. *Proc. Natl. Acad. Sci. USA* **1995**, *92*, 8975–8979. [CrossRef] [PubMed]
41. Yousuf, M.A.; Mi-ichi, F.; Nakada-Tsukui, K.; Nozaki, T. Localization and targeting of an unusual pyridine nucleotide transhydrogenase in *Entamoeba histolytica*. *Eukaryot. Cell* **2010**, *9*, 926–933. [CrossRef] [PubMed]
42. Ankri, S. *Entamoeba histolytica*-Gut Microbiota Interaction: More Than Meets the Eye. *Microorganisms* **2021**, *9*, 581. [CrossRef]
43. Pineda, E.; Perdomo, D. *Entamoeba histolytica* under Oxidative Stress: What Countermeasure Mechanisms Are in Place? *Cells* **2017**, *6*, 44. [CrossRef]
44. Chopra, I. Over-expression of target genes as a mechanism of antibiotic resistance in bacteria. *J. Antimicrob. Chemother.* **1998**, *41*, 584–588. [CrossRef]
45. Capela, R.; Moreira, R.; Lopes, F. An Overview of Drug Resistance in Protozoal Diseases. *Int. J. Mol. Sci.* **2019**, *20*, 5748. [CrossRef]
46. Arias, D.G.; Regner, E.L.; Iglesias, A.A.; Guerrero, S.A. *Entamoeba histolytica* thioredoxin reductase: Molecular and functional characterization of its atypical properties. *Biochim. Et Biophys. Acta* **2012**, *1820*, 1859–1866. [CrossRef]
47. Pineda, E.; Encalada, R.; Rodriguez-Zavala, J.S.; Olivos-Garcia, A.; Moreno-Sanchez, R.; Saavedra, E. Pyruvate:ferredoxin oxidoreductase and bifunctional aldehyde-alcohol dehydrogenase are essential for energy metabolism under oxidative stress in *Entamoeba histolytica*. *FEBS J.* **2010**, *277*, 3382–3395. [CrossRef]
48. Kumar, A.; Shen, P.S.; Descoteaux, S.; Pohl, J.; Bailey, G.; Samuelson, J. Cloning and expression of an NADP(+)-dependent alcohol dehydrogenase gene of *Entamoeba histolytica*. *Proc. Natl. Acad. Sci. USA* **1992**, *89*, 10188–10192. [CrossRef]
49. König, C.; Meyer, M.; Lender, C.; Nehls, S.; Wallaschkowski, T.; Holm, T.; Matthies, T.; Lercher, D.; Matthiesen, J.; Fehling, H.; et al. An Alcohol Dehydrogenase 3 (ADH3) from *Entamoeba histolytica* Is Involved in the Detoxification of Toxic Aldehydes. *Microorganisms* **2020**, *8*, 1608. [CrossRef]
50. Klomsiri, C.; Karplus, P.A.; Poole, L.B. Cysteine-based redox switches in enzymes. *Antioxid. Redox Signal.* **2011**, *14*, 1065–1077. [CrossRef]
51. Bosch, D.E.; Siderovski, D.P. G protein signaling in the parasite *Entamoeba histolytica*. *Exp. Mol. Med.* **2013**, *45*, e15. [CrossRef] [PubMed]
52. Accorsi, K.; Giglione, C.; Vanoni, M.; Parmeggiani, A. The Ras GDP/GTP cycle is regulated by oxidizing agents at the level of Ras regulators and effectors. *FEBS Lett.* **2001**, *492*, 139–145. [CrossRef]
53. Ahmad, A.; Mishra, S.; Som, L.; Gourinath, S. Role of kinases in virulence and pathogenesis of protozoan parasite *E. Histolytica*. *Front. Biosci.* **2020**, *25*, 1617–1635.
54. Corcoran, A.; Cotter, T.G. Redox regulation of protein kinases. *FEBS J.* **2013**, *280*, 1944–1965. [CrossRef]
55. Froscio, M.; Murray, A.W.; Hurst, N.P. Inhibition of protein kinase C activity by the antirheumatic drug auranofin. *Biochem. Pharmacol.* **1989**, *38*, 2087–2089. [CrossRef]
56. Penuliar, G.M.; Nakada-Tsukui, K.; Nozaki, T. Phenotypic and transcriptional profiling in *Entamoeba histolytica* reveal costs to fitness and adaptive responses associated with metronidazole resistance. *Front. Microbiol.* **2015**, *6*, 354. [CrossRef]
57. Ehrenkaufer, G.M.; Suresh, S.; Solow-Cordero, D.; Singh, U. High-Throughput Screening of Entamoeba Identifies Compounds Which Target Both Life Cycle Stages and Which Are Effective Against Metronidazole Resistant Parasites. *Front. Cell. Infect. Microbiol.* **2018**, *8*, 276. [CrossRef]
58. Baumel-Alterzon, S.; Weber, C.; Guillen, N.; Ankri, S. Identification of dihydropyrimidine dehydrogenase as a virulence factor essential for the survival of *Entamoeba histolytica* in glucose-poor environments. *Cell. Microbiol.* **2013**, *15*, 130–144. [CrossRef]

59. Herencias, C.; Rodriguez-Beltran, J.; Leon-Sampedro, R.; Alonso-Del Valle, A.; Palkovicova, J.; Canton, R. Collateral sensitivity associated with antibiotic resistance plasmids. *eLife* **2021**, *10*, e65130. [CrossRef]
60. Tekwani, B.L.; Mehlotra, R.K. Molecular basis of defence against oxidative stress in *Entamoeba histolytica* and Giardia lamblia. *Microbes Infect* **1999**, *1*, 385–394. [CrossRef]
61. Bruchhaus, I.; Tannich, E. Induction of the iron-containing superoxide dismutase in *Entamoeba histolytica* by a superoxide anion-generating system or by iron chelation. *Mol. Biochem. Parasitol.* **1994**, *67*, 281–288. [CrossRef]
62. Palmer, A.C.; Kishony, R. Opposing effects of target overexpression reveal drug mechanisms. *Nat. Commun.* **2014**, *5*, 4296. [CrossRef]
63. Fiskus, W.; Rao, R.; Fernandez, P.; Herger, B.; Yang, Y.; Chen, J. Molecular and biologic characterization and drug sensitivity of pan-histone deacetylase inhibitor-resistant acute myeloid leukemia cells. *Blood* **2008**, *112*, 2896–2905. [CrossRef]

antioxidants

MDPI

Review

Reactive Oxygen Species as the Brainbox in Malaria Treatment

Chinedu Ogbonnia Egwu [1,2,3,4,5], Jean-Michel Augereau [3,4,5], Karine Reybier [1,*,†] and Françoise Benoit-Vical [3,4,5,*,†]

1 PharmaDev, UMR 152, Université de Toulouse, IRD, UPS, 31400 Toulouse, France; chinedu.egwu@funai.edu.ng
2 Medical Biochemistry, College of Medicine, Alex-Ekwueme Federal University, Ndufu-Alike Ikwo, P.M.B. 1010, Abakaliki, Nigeria
3 CNRS, LCC, Laboratoire de Chimie de Coordination, Université de Toulouse, 31077 Toulouse, France; jean-michel.augereau@lcc-toulouse.fr
4 Institut de Pharmacologie et de Biologie Structurale, IPBS, Université de Toulouse, CNRS, UPS, 31077 Toulouse, France
5 MAAP, New Antimalarial Molecules and Pharmacological Approaches, ERL 1289 Inserm, 31077 Toulouse, France
* Correspondence: karine.reybier-vuattoux@univ-tlse3.fr (K.R.); Francoise.Vical@inserm.fr (F.B.-V.)
† F.B.-V. and K.R. share senior authorship.

Abstract: Several measures are in place to combat the worldwide spread of malaria, especially in regions of high endemicity. In part, most common antimalarials, such as quinolines and artemisinin and its derivatives, deploy an ROS-mediated approach to kill malaria parasites. Although some antimalarials may share similar targets and mechanisms of action, varying levels of reactive oxygen species (ROS) generation may account for their varying pharmacological activities. Regardless of the numerous approaches employed currently and in development to treat malaria, concerningly, there has been increasing development of resistance by *Plasmodium falciparum*, which can be connected to the ability of the parasites to manage the oxidative stress from ROS produced under steady or treatment states. ROS generation has remained the mainstay in enforcing the antiparasitic activity of most conventional antimalarials. However, a combination of conventional drugs with ROS-generating ability and newer drugs that exploit vital metabolic pathways, such antioxidant machinery, could be the way forward in effective malaria control.

Keywords: ROS; antimalarials; *Plasmodium falciparum*; malaria and oxidative stress

Citation: Egwu, C.O.; Augereau, J.-M.; Reybier, K.; Benoit-Vical, F. Reactive Oxygen Species as the Brainbox in Malaria Treatment. *Antioxidants* **2021**, *10*, 1872. https://doi.org/10.3390/antiox10121872

Academic Editor: Serge Ankri

Received: 15 October 2021
Accepted: 16 November 2021
Published: 24 November 2021

1. Introduction

Malaria is a vector-transmitted parasite disease that continues to plague mankind. It is caused in humans by five main species of *Plasmodium*. The World Health Organization estimated a global prevalence of 229 million cases in 2019, with Sub-Saharan Africa taking the top spot, with more than 90% of the global burden and most of the deaths being due to the parasite *Plasmodium falciparum*. *Plasmodium vivax* is also a notable species and, although it is less deadly, it still has a very significant economic impact [1]. The physiopathology of *P. falciparum* and *P. vivax* malaria relies to a large extent on the oxidative stress generated by the parasites during their erythrocytic cycle [2–4].

It is worrisome to note that the control of malaria has stalled since 2014 [1,5], which calls for a doubling of efforts by all stakeholders. Several recommendations have been made with regard to malaria control. These recommendations range from the use of insecticide spray and sleeping under insecticide-treated nets to the use of chemotherapeutic agents.

Regardless of these numerous interventions, the seeming lack of progress may be attributed to many factors, which may include, but are not limited to, poverty, poor sanitation, and inadequate or nonoperational health policies [6,7]. Moreover, overuse, inadequate or incomplete treatment regimens, and counterfeit drugs, which lead to drug failures and the development of resistance by both mosquitoes and malaria parasites to insecticides and antimalarials, respectively, are also responsible for this scenario [1,8,9]. Chloroquine and most antiplasmodial drugs have lost their usefulness in malaria control due to parasite resistance development [10,11], and it is worrisome to note that the continued relevance of the current gold-standard drug artemisinin and its derivatives is threatened by resistance development, which was first reported in 2008 [12,13]. Resistance of the parasite to antimalarials could be a result of its ability to develop diverse mechanisms to evade death, which can be preventive, reductive, or reparative.

As for any cell, the maintenance of redox homeostasis is vital for parasites. The overall level of ROS in a biological system is determined by its level of production and/or elimination by the antioxidant machinery of the cells [14]. The ROS generated are very reactive and oxidize any biomolecules around their site of production [15], leading to intense cellular damage.

Since the majority of conventional antimalarials kill parasites via direct or indirect overproduction of reactive oxygen species (ROS) [16], the present article seeks to review the generation and modulation of ROS in steady state and during treatment with different antimalarials, especially artemisinin and its derivatives in relation to *P. falciparum* resistance to these drugs. An in silico comparison of the enzymes of the *P. falciparum* redox system with the *P. vivax* proteins has shown a certain level of homology between these two parasite species [17]. However, as the in vitro culture of *P. vivax* has not yet been mastered, the redox system of *Plasmodium* has almost exclusively been studied in *P. falciparum*.

2. Defining ROS in Living Cells

ROS are defined as oxygen-containing reactive species. This term includes superoxide radicals ($O_2^{\bullet-}$), hydrogen peroxide (H_2O_2), hydroxyl radicals ($^{\bullet}OH$), singlet oxygen ($^1O_2$), peroxyl radicals ($LOO^{\bullet}$), alkoxyl radicals ($LO^{\bullet}$), lipid hydroperoxide (LOOH), and peroxynitrite ($ONOO^-$), among others [15]. Their generation can first occur as a result of the partial reduction of oxygen, as described in Figure 1. Among the ROS, some species are radicals, i.e., have unpaired electrons in their outer orbit—for example, superoxide and hydroxyl radicals. The different forms of ROS have varying levels of reactivity depending on their oxidation potential, with H_2O_2 being the least reactive and $^{\bullet}OH$ being the most reactive.

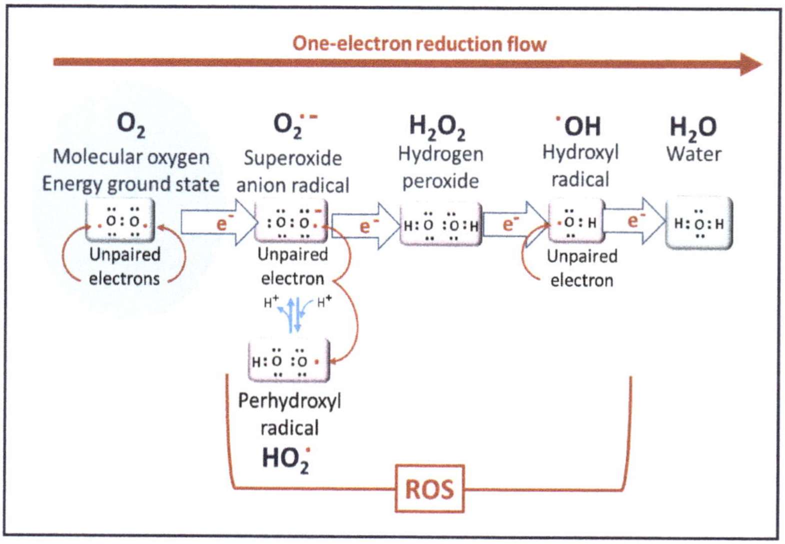

Figure 1. Molecular oxygen to ROS. Superoxide radicals are the primary ROS but have poor reactivity. Hydrogen peroxide is the least reactive ROS. At the end of the reduction flow, hydroxyl radicals are the most reactive [14].

3. The Biochemical Impacts of ROS in Living Cells: Is There Any Peculiarity in Malaria?

ROS generated at low concentrations under physiological conditions are often beneficial, playing important roles as regulatory mediators in signaling processes [15,18]. Nevertheless, at high concentrations, they become deleterious for the cells. The toxicity of free radicals stems from their unstable nature and predisposition to donate or abstract electrons from nearby molecules, triggering a chain reaction that can be terminated by another molecule with unpaired electrons. Lipids, proteins, and nucleic acids are attacked and damaged by ROS, which affects the survival of living organisms [14,15] (Figure 2).

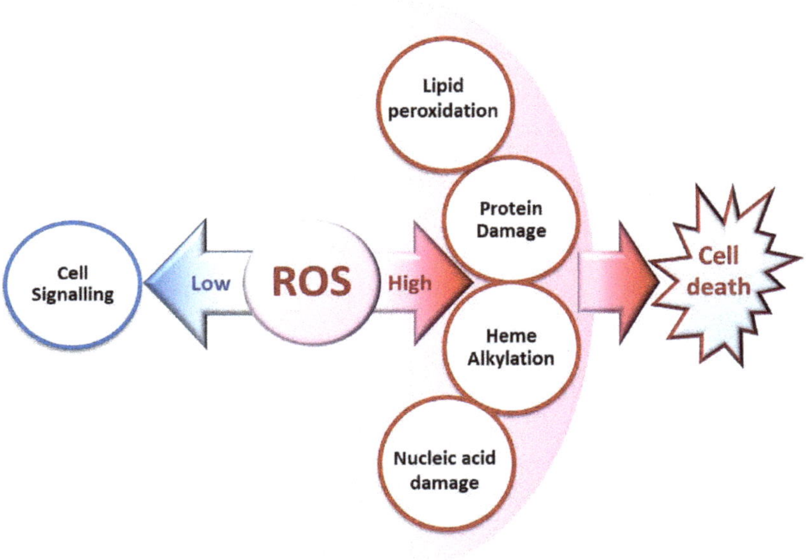

Figure 2. Biochemical impacts of ROS depending on their concentrations.

This overproduction of ROS can be endogenous due to the dysregulation associated with pathological processes such as aging, inflammation [19,20], or cardiovascular diseases [21] and can also be exogenous with the administration of xenobiotics such as antimalarials [22].

3.1. Cell Signaling

ROS play important physiological roles, such as in the induction of apoptosis and suppression of some genes' expression, at low concentrations [18]. Studies have reported the production of ROS by specialized plasma membrane oxidases and nicotinamide adenine dinucleotide phosphate (NADPH) oxidases in normal physiological signaling by growth factors and cytokines [23]. For example, NADPH oxidase activity plays defensive roles in phagocytic cells [18]. The small size and ability of ROS (such as H_2O_2) to traverse membranes make them suitable for cell signaling; however, this role is not yet fully understood in *Plasmodium* [24].

3.2. Lipid Peroxidation

Lipids are prone to attacks from reactive species, resulting in their oxidation and the formation of lipid peroxides [25] (Figure 3). This leads to cellular dysfunction, especially as lipids are major components of cell membranes. In *Plasmodium*, artemisinin and its derivatives accumulate in neutral lipid bodies, especially in the digestive vacuole, where they can trigger oxidative damage after their heme–iron activation [26], via a lipid peroxidation process that, once initiated, is propagated by autocatalysis to free fatty acids [26]. This oxidative damage leads to a loss of Plasmodium membrane integrity and, consequently, parasite death. Moreover, tetraoxanes oxidatively damage phospholipids more than artemisinin [27], which may account for the differential antimalarial effect of these endoperoxides.

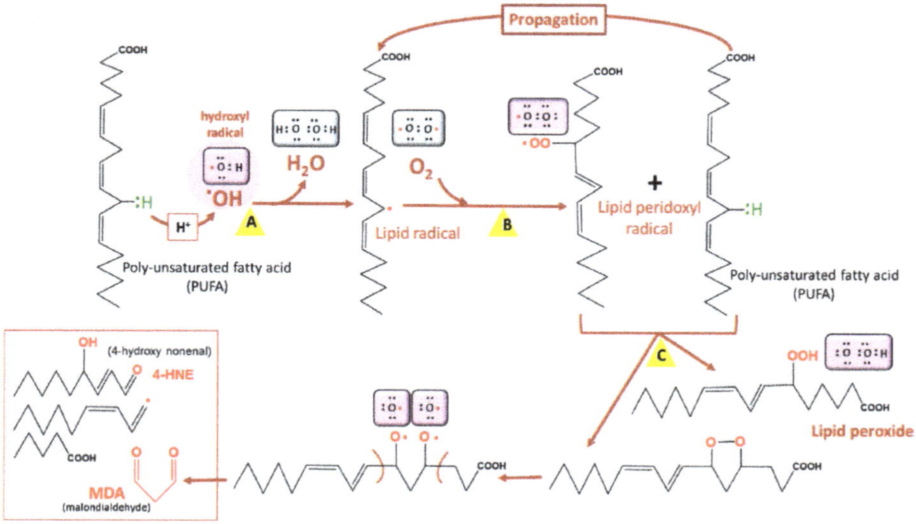

Figure 3. Lipid peroxidation. Lipid peroxidation occurs in three phases: initiation (A), propagation (B), and termination (C). Malondialdehyde (MDA) is a biomarker of lipid peroxidation in living cells. Among lipids, polyunsaturated fatty acids (PUFAs) are the most vulnerable to lipid peroxidation. COOH = carboxyl group, OOH = hydroperoxyl.

3.3. Protein Damage

Proteins are vulnerable to oxidation by ROS, a phenomenon generally referred to as protein carbonylation. Protein carbonyls are reactive aldehydes and ketone adducts formed through the α-amidation pathway, the formation of protein–protein cross-linked derivatives, the oxidative cleavage of glutamyl residues, and cell membrane damage by lipid oxidation products [28]. Protein carbonyls have been used as biomarkers of oxidative stress due to their relative stability and early formation [29]. ROS generate cytotoxic protein carbonyls in two main and irreversible ways:

(i) by a metal-catalyzed oxidative (MCO) attack targeting the amino acid moiety of arginine, lysine, threonine, and proline (Figure 4);
(ii) by secondary reactions on lysine, cysteine, and histidine with reactive carbonyl derivatives, resulting, among other things, from lipid peroxidation.

These post-translational modifications due to the oxidization of amino acid side chains, which results in aldehyde, ketone, and lactam formation, can also lead to damage to proteins [30,31].

The endoperoxides alkylate numerous vital proteins, including enzymes in the parasite, particularly the cysteine residue of cysteine proteases, which play a vast role in *P. falciparum*, ranging from hemoglobin (Hb) uptake and digestion to aiding red blood cell rupture [8,32,33]. Therefore, their alteration, leading to the inhibition of their enzymatic properties, can cause severe setbacks for parasite survival, considering the huge dependence of the parasite on Hb digestion products. Endoperoxides also disrupt the activities of sarcoplasmic–endoplasmic reticulum Ca^{2+}-ATPase- (SERCA-) type protein, encoded by the *pfatpase6* gene, presumably from the ROS that they generate [34]. More recently, approximately 124 proteins were identified as covalent binding targets of artemisinins [35].

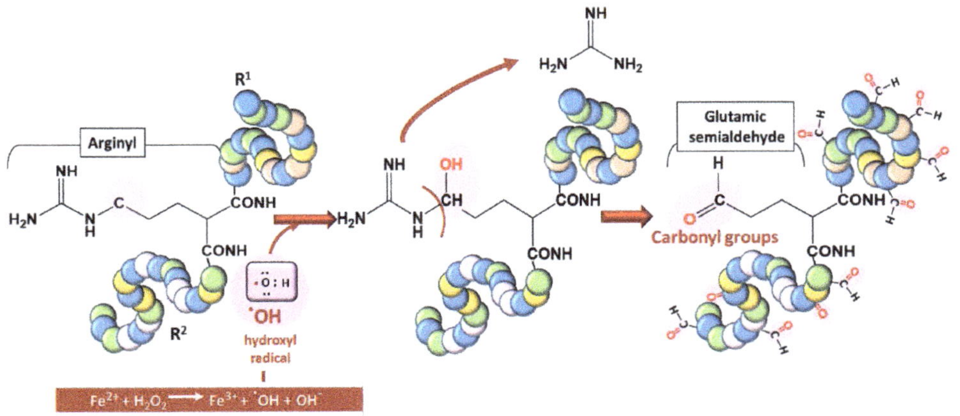

Figure 4. Typical carbonylation of proteins based on a metal-catalyzed oxidation (MCO) attack involving a transition metal (Fe^{2+}) for hydroxyl radical generation.

3.4. Nucleic Acid Damage

The molecular integrity of DNA and other genetic material is necessary for the continued existence and survival of all living organisms, including *Plasmodium*. ROS cause structural damage to DNA by attacking mainly one of its bases, guanine, due to its lower oxidation potential [36,37] (Figure 5). Reports have revealed that artesunate can cause double-stranded breaks of plasmodial DNA in less than one hour and that it is linked to an increase in ROS generation [38].

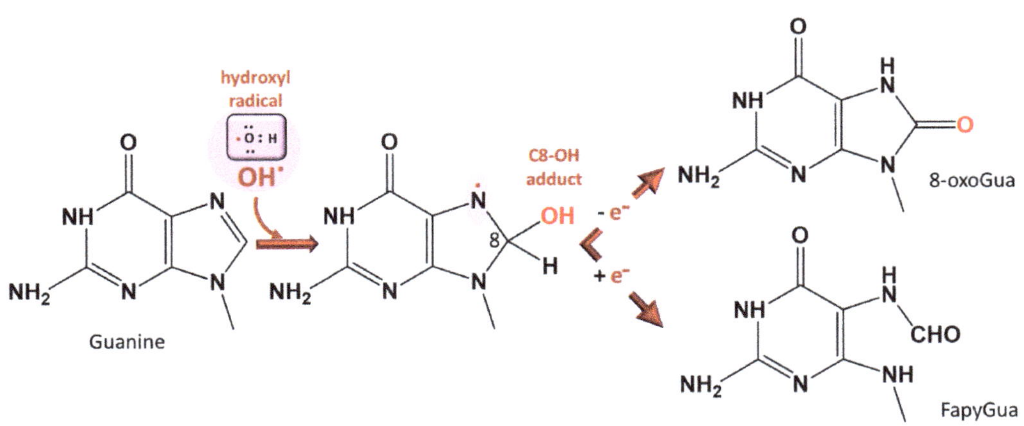

Figure 5. DNA base damage by ROS at the purine base guanine. The C8-OH adduct 7-hydro8-oxo-2′-deoxyguanosine is used as a marker of oxidative stress. A similar reaction occurs with adenine (8-oxoAde and Fapy-Ade). FapyGua: formamidopyrimidine, 8-oxoGua: 8-oxoguanine [36,39].

Disruption of the conformation of biomolecules by artemisinins (endoperoxides) or other ROS-generating antimalarials kills parasites. Although they have a similar activation process generating carbon- and oxygen-centered radicals, the efficacy of endoperoxides varies, which can be explained, in part, by their different pharmacological properties, notably the stage of the parasite erythrocyte cycle, the nature of their target, and their location in the parasite, i.e., trioxolanes vs. artemisinin and its derivatives [26,40].

4. Sources and Management of ROS in *Plasmodium*-Infected Erythrocytes under Steady State

ROS arise from both the metabolism of the parasite and the host defense system [41]. In erythrocytic parasites, there are two important sources of ROS: the mitochondrial electron transport chain and the degradation of hemoglobin [24,42]. The involvement of the antioxidant machinery and the elimination of heme produced during the digestion of hemoglobin are the two strategies by which the parasite maintains its redox homeostasis (Table 1).

Table 1. Plasmodial heme elimination and antioxidant machinery.

Agent	Site of Production	Role in Oxidative Homeostasis
HRP	FV	Binding with heme for polymerization
HDP	FV	Heme polymerization to hemozoin
H_2O_2	Cytosol and FV	Degrades heme
SOD	Cytosol, Mitochondria	Dismutation of $O_2^{\bullet-}$
Prx	Cytosol, Mitochondria, Apicoplast	Reduction of H_2O_2 to H_2O
Trx	Cytosol, Mitochondria	Reduction of Prx
GSH	Cytosol	Degradation of heme, reduction of proteins and ROS
Vit B6	Cytosol	Role unclear

H_2O_2: hydrogen peroxide, HRP: histidine-rich protein, HDP: heme detoxification protein, SOD: superoxide dismutase, Trx: thioredoxin, Prx: peroxiredoxin, GSH: reduced glutathione, Vit B6: vitamin B6, FV: food vacuole [43–50].

4.1. ROS Production from Mitochondrial Electron Transport Chain

The mitochondrion constitutes a very important source of free radicals in aerobic organisms. Approximately 1–3% of electrons escape from the electron transport chain directly to react with molecular oxygen, leading to the formation of superoxide radicals ($O_2^{\bullet-}$) [15]. This is all the more important as the mitochondrion of the parasite responsible for malaria has cytochromes with heme as a cofactor in the electron transport chain [51]. Although there is a paucity of data on the involvement of the parasitic mitochondrion in ROS generation, the complexity and diversity of its antioxidant system is a very good pointer, especially the report of the expression of cytosolic PfSOD-1 and mitochondrial PfSOD-2 throughout the blood stages of the parasite [52]. Superoxide dismutases (SODs) quickly dismutate the formed superoxide radical ($O_2^{\bullet-}$) to hydrogen peroxide (H_2O_2), which can be reduced to water by the peroxiredoxin 2-Cys Prx (TPx-2) [53,54].

4.2. ROS Production from Hemoglobin Digestion

During the blood stage, parasites take up and break down approximately 75% of the hemoglobin (Hb) of red blood cells to obtain essential amino acids for their development and replication, releasing free heme as residual toxic waste [44,55]. This free heme contains reactive ferrous iron (Fe^{2+}) that can readily reoxidize by transferring electrons to oxygen to form superoxide radicals ($O_2^{\bullet-}$) [16] and then hydrogen peroxide, finally leading to the production of the highly deleterious hydroxyl radical ($^{\bullet}OH$) through the Fenton reaction involving a new molecule of heme (Fe^{2+}) (Figure 6) [3].

Because of the important role that heme plays in ROS generation, immediately after its production, a detoxification process takes place where approximately 95% of the produced heme (Fe^{2+}) is polymerized into hemozoin, a nontoxic crystal [46]. This reaction involves the heme detoxification protein (HDP), whose action is aided by histidine-rich protein-2 (HRP2) (Figure 7) [48,56]. HRP2 is said to have a very high affinity for heme [57,58], and elevated HRP2 reduces the vulnerability of the parasite. Kapishnikov et al. showed that mature parasites have approximately 70% of the total iron from red blood cells in the hemozoin crystals and therefore suggested a coupling of the rate of Hb digestion to that of heme polymerization [59].

Although *Plasmodium* lacks the heme oxygenase enzyme, which is deployed by most organisms to degrade heme [60], it has a system in place that mimics the heme oxygenase [44]. During the digestion of hemoglobin, H_2O_2 is also generated in the food vacuole as a result of the immediate conversion of oxyhemoglobin to methemoglobin due to heme reduction at the prevailing pH of 5.2 [61,62]. The heme is peroxidatively degraded by the reaction with H_2O_2, leading to the formation of a ferryl intermediate (Fe(IV)=O) [44].

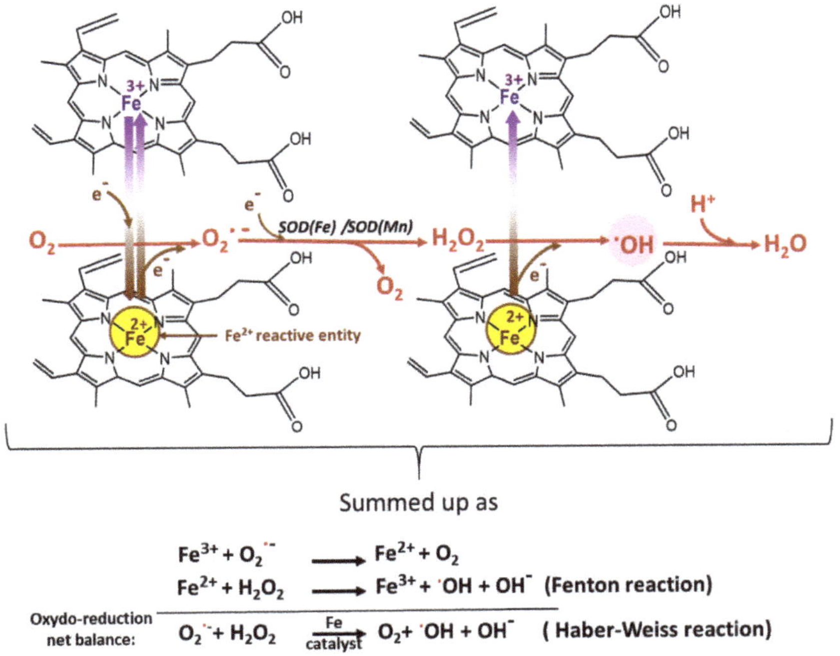

Figure 6. Heme Fe^{2+} reaction with $O_2^{\bullet-}$ leading to ROS production, especially the most deleterious one, the hydroxyl radical $^{\bullet}OH$ [3,63].

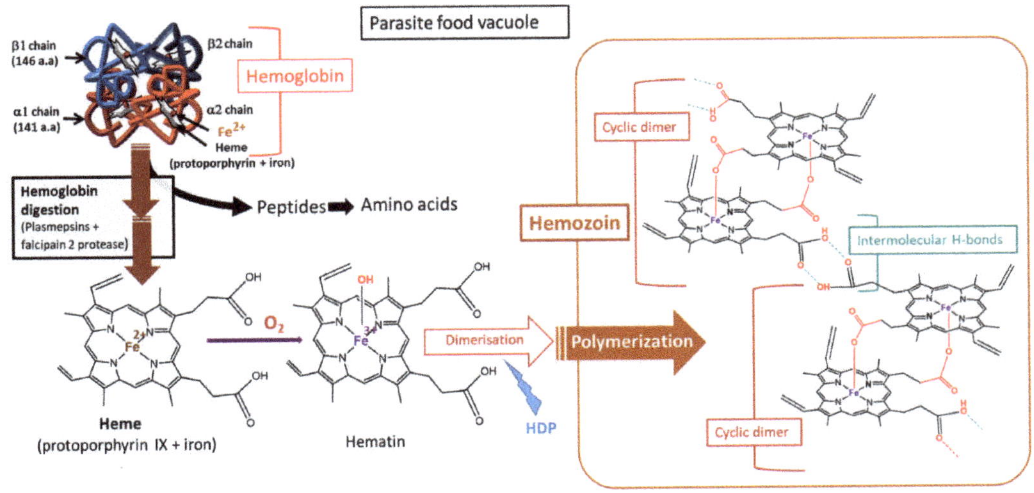

Figure 7. Hemoglobin digestion and hemozoin production in *Plasmodium*-infected red blood cells. HDP, heme detoxification protein [59,64].

Plasmodium uses a family of hemoglobinases to degrade hemoglobin into amino acids. *Plasmodium* also requires heme from the host to synthesize proteins such as the heme-dependent cytochromes in the mitochondrial electron transport chain [51], although genetic studies have revealed that the parasites encode and express all the enzymes for heme production [60]. Moreover, the digestion of hemoglobin ensures that the infected red blood cells are osmotically stable throughout the intraerythrocytic stages of the parasite [65]. The digestion of hemoglobin is therefore essential for *Plasmodium* during its intraerythrocytic cycle. In addition to the formation of hemozoin, there are other antioxidant defense mechanisms (discussed below) that can address the toxic effect of resting unpolymerized heme. The significant role played by hemoglobin in oxygen transport makes red blood cells and their vicinity highly vulnerable to ROS formation [66].

Box 1. The role of nitric oxide in potentiation of reactive species [15,67,68].

Potentiation of reactive species: Nitric oxide in view
Nitric oxide ($^{\bullet}NO$), which is also considered a ROS, is produced by macrophages against *Plasmodium*. The free radical nitric oxide reacts with the primary radical, $O_2^{\bullet-}$, to give peroxynitrite, a more lethal reactive species, which can also damage biomolecules such as proteins and DNA. It remains to be seen if $^{\bullet}NO$, produced by human phagocytes, diffuses into the parasitic cytosol. However, $^{\bullet}NO$ production has also been reported in the parasite by an isoform of nitric oxide synthase (NOS), which is calcium-independent and whose inhibition affects the growth of *P. falciparum*. The $^{\bullet}NO$ production is higher in the trophozoite than in the ring stage. Although $^{\bullet}NO$ may not be very deleterious, its reaction with H_2O_2 to form peroxynitrite will potentiate their radicality

4.3. Management of ROS by Plasmodium under Steady State: The Antioxidant Machinery

To match the multiplicity of the sources of ROS, including nitric oxide (Box 1), from both the host and the parasites themselves, malaria parasites have several antioxidant machineries in addition to heme elimination processes, which may be additive or synergistic in ensuring their survival in red blood cells. *Plasmodium* species express antioxidant proteins/enzymes such as superoxide dismutases (SODs), glutathione/glutathione-dependent proteins, thioredoxin/thioredoxin-dependent proteins and thioredoxin reductase, and peroxiredoxins [53,69–71].

- The SODs act fundamentally to enable the spontaneous dismutation of $O_2^{\bullet-}$ to H_2O_2, which can then be reduced by other enzymes to water, preventing its reaction with iron (Fe^{2+}) to form highly toxic $^{\bullet}OH$. *Plasmodium* expresses two forms of SOD: cytosolic PfSOD-1 (Fe-SOD) and mitochondrial PfSOD-2 (Mn-SOD). The uptake of SOD from the host may also be complementary [52,72].
- Glutathione, a tripeptide composed of cysteine, glycine, and glutamate, plays a vital role in the defense system of malaria parasites. Although the level of heme that escapes polymerization is only 5%, it can still mount oxidative stress, which can be eliminated by GSH-dependent mechanisms [45,46,73]. GSH is generated by the reduction of GSSG by glutathione reductase, which can be amplified by other proteins with thiol groups, such as thioredoxin (Trx) [74] (Figure 8). This could be further boosted by the de novo synthesis of GSH, a predominant pathway under oxidative stress [75,76]. Suggestions of possible uptake from the host were rebuffed by the findings of Patzewitz et al., who demonstrated that GSH uptake from the host is not statistically significant [77]. In addition to the nonenzymatic degradation of heme by GSH to generate nonheme iron, *Plasmodium* also possesses a thiol enzyme, glutathione-S-transferase (GST), which binds and sequesters heme (Fe^{2+}) [78]. To affect its detoxification roles, GSH also serves as a cofactor for GSH-dependent enzymes such as GST, which helps in the reduction of H_2O_2 and glutathione-peroxidase-like proteins. Overall, GSH acts as a redox buffer to provide redox homeostasis [79]. This is essential in preventing the escalation of ROS generation, which is often aided by the redox cycling of transition metals such as iron.

- Thioredoxins (Trxs) are disulfide oxidoreductases, expressed in all organisms, that interact with a myriad of proteins through their cysteine moiety to donate electrons [80,81]. Trxs keep biomolecules in their reduced and active conformations, consequently promoting parasite survival under oxidative stress. The regeneration of reduced thioredoxin is achieved through the action of thioredoxin reductase (a FAD-dependent enzyme) using NADPH [82]. Three isoforms of Trx have been characterized: a cytosolic isoform, Trx1, and two isoforms in the apicoplast, Trx2 and Trx3 [50].
- *Plasmodium* lacks catalase and glutathione peroxidase and therefore will likely depend greatly on peroxiredoxins (Prx) to reduce H_2O_2 generated by SOD [52,53]. *P. falciparum* expresses five different isoforms of Prx that are strategically localized in the cytosol, mitochondria, and perhaps the apicoplast. These are 1-Cys Prx, two typical 2-Cys Prxs (Prx1 and Prx2), a 1-Cys antioxidant protein (AOP), and a GSH peroxidase-like thioredoxin peroxidase (TPx_{Gl}) [50]. It is known that 1-CysPrx and Prx1 are localized in the cytosol, while Prx2 is localized in the mitochondria. The localizations of AOP and TPx_{Gl} are not yet fully understood. Prx1, one of the most abundant peroxidases, is constitutively expressed across all blood parasitic stages and has a very high affinity for H_2O_2, hence being judged to serve a housekeeping role [53,83,84]. Prxs are kept in reduced form through reduction by thioredoxin-dependent systems, comprising three Trxs and Trx-like proteins (TLPs), Tlp1 and Tlp2, which have been identified genomically in the parasite [50]. The parasite, in addition to the expressed Trxs, has approximately 50% of its thioredoxin peroxidase activity depending on the imported human Prx-2 (hPrx-2) [85]. Moreover, it has been demonstrated that H_2O_2 can diffuse into the host and be handled by the host's catalase enzyme [86]. The import of human antioxidant machinery may be applicable to other antioxidants, although it has not been fully verified.
- NADPH is involved in maintaining vital antioxidant enzymes, such as glutathione and thioredoxin reductases, in their active conformations. However, the accumulation of heme shifts the equilibrium toward the oxidized form NADP+, increasing the pressure on the reduced form NADPH necessary to reduce the oxidized glutathione (GSSG) to glutathione (GSH). This can lead to the suppression of the main antioxidant mechanisms, such as glutathione reductase, thioredoxin reductase (TrxR), glyceraldehyde-3-phosphate dehydrogenase, and other proteins, leading to the elevation of oxidative stress [87].
- Vitamin B6, in addition to playing several biochemical roles, has also been seen to act as a potent antioxidant [88,89]. The enzymes governing vitamin B6 synthesis are elevated in parasites exposed to oxidative stress, e.g., generated by methylene blue and hence seen as a potential pharmacological target [49]. Although its antioxidant mechanism is still subject to debate, it is believed to scavenge ROS and prevent lipid peroxidation [90–92].

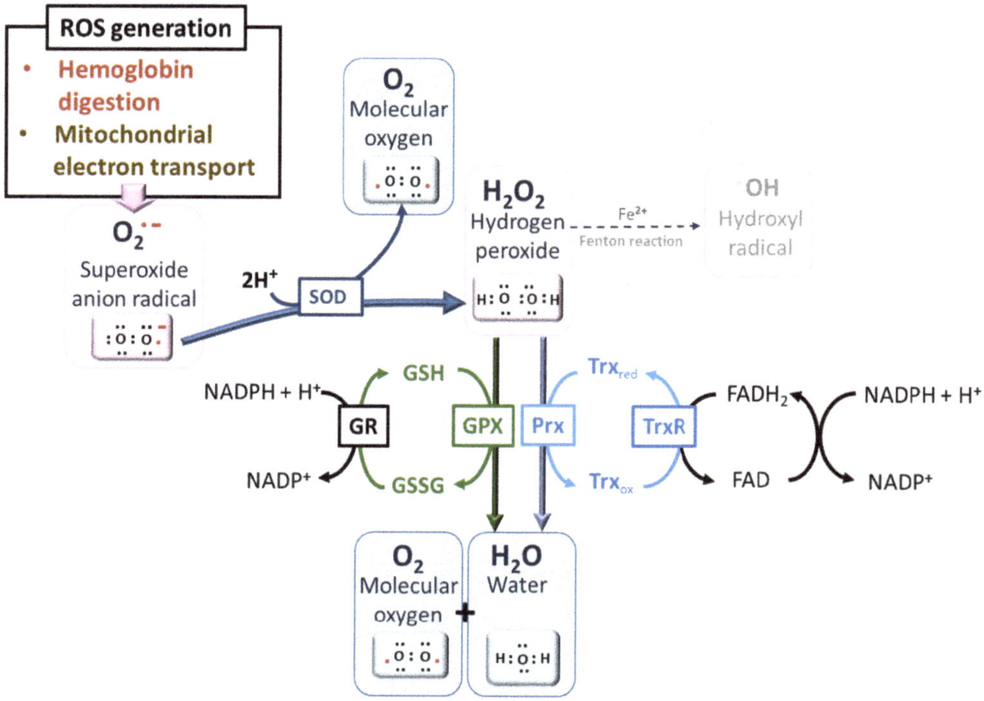

Figure 8. Major pathways of ROS detoxification. A network of redox cycling cofactors, such as flavin adenine dinucleotide (FADH2) and nicotinamide adenine dinucleotide phosphate (NADPH), aids in the reduction of ROS, avoiding the ultimate and deleterious production of hydroxyl radical •OH from H_2O_2 by the Fenton reaction in the presence of reduced iron [52]. Abbreviations—SOD: superoxide dismutase, GR: glutathione reductase, GPX: glutathione peroxidases, Prx: peroxiredoxins, TrxR: thioredoxin reductase, GSH: reduced glutathione, GSSG: oxidized glutathione dipeptide, Trx_{red}: reduced thioredoxin, Trx_{ox}: oxidized thioredoxin.

5. ROS Production in *Plasmodium*-Infected Erythrocytes under Antimalarial Treatment

Oxidative stress is not only an important clinical and pathobiochemical factor but also an effective therapeutic principle in malaria. Indeed, the pharmacological activity of some of the most important antimalarials, such as quinolines, atovaquones, and artemisinin derivatives, is mediated by the high production of ROS, exceeding the oxidative stress management capacities of the parasite [16,93].

5.1. Mode of Action of Chloroquine and Other Quinolines

Quinolines have been reported to inhibit the polymerization of heme to hemozoin in food vacuoles, leading to the accumulation of heme [94]. Moreover, chloroquine and other quinolines differently inhibit the peroxidative degradation of heme by H_2O_2, hence increasing heme accumulation and consequently enhancing heme-catalyzed reactions, leading to the production of ROS and parasite death [44,95]. The formation of the heme–Fe^{III} adduct with chloroquine in the food vacuole leads to its diffusion into the cytosol, where the complex is dissociated, releasing heme in the parasite cytosol and hence promoting the redox cycling of Fe^{III} to Fe^{II}, which enhances ROS generation [96]. This has also been reported for quinine [96] and may apply to other quinolines. Other purported modes of action of quinolines include the inhibition of the peroxidase-like activities of heme, leading to the accumulation of H_2O_2, which interacts with heme to form radicals [97]. Reviewed reports about the mode of action of quinolines have revealed their varying levels of activity that trigger ROS generation [44,94,96].

5.2. Mode of Action of Atovaquone and Hydroxynaphtoquinones

The mode of action of atovaquone and other hydroxynaphtoquinones is based on the inhibition of mitochondrial cytochrome *bc1* complex via the competitive inhibition of ubiquinol binding [98,99]. Consequently, atovaquone induces the collapse of the mitochondrial membrane potential, blocking the energy supply of the parasites, leading to parasite death [100,101]. Concomitantly, the ubiquinol accumulation due to the inhibition of its binding site generates also a significant amount of superoxide radicals (ROS), as was shown in cancer cells and *Plasmodium* [16,102].

5.3. Mode of Action of Artemisinin: The Role of Endoperoxide

Artemisinins are at the forefront of the war against malaria and are used in combination with other antimalarial agents in artemisinin-based combination therapies (ACTs) [103,104]. This class of compounds, which includes arteether, artemether, artesunate, and dihydroartemisinin [105], is characterized by the presence of an endoperoxide group that is responsible for the antimalarial activity.

5.3.1. Activation of the Endoperoxide to Generate ROS

The endoperoxides are converted into carbon-centered radicals, which mediate most of their effects. Activation is performed by the transfer of electrons from transition metals, mainly iron (Box 2), leading to the homolytic cleavage of the endoperoxide bridge [106–109]. The carbon-centered radical formed can alkylate heme, preventing its polymerization into hemozoin and thus inducing the formation of toxic ROS. They also alkylate several essential biomolecules, rendering them dysfunctional and consequently leading to plasmodial death. The iron could be heme (nonchelatable iron) (Figure 9) or freely circulating iron (chelatable iron).

Available evidence shows that artemisinin reacts more efficiently with heme than other forms of iron [110], thus pointing to the huge role played by heme in the pharmacological activity of endoperoxides. The dependence on iron is affected by the stage of the parasite. However, artemisinin activation at the early ring stages depends more on chelatable iron than on the trophozoite stage (35% vs. 15%), for which artemisinin activation depends more on heme generation from hemoglobin digestion [111]. This could be explained by a hemoglobinase system that is not fully developed at the ring stage but has a low level of falcipain 2 and 3 activity sufficient to sustain hemoglobin degradation that permits normal growth [111]. The molecular structure of an endoperoxide also determines its level of dependence on heme for activation [111]. Although the most studied source of heme for artemisinin activation is hemoglobin, the debate continues on whether other heme-containing proteins may not do the same [107,112]. Some of these proteins are cytochrome p450, catalases, peroxidases, and other heme-containing proteins [113–116]. The use of chelators of free iron and inhibitors of hemoglobin digestion revealed that artemisinin activity at the early ring and trophozoite stages depends more than 80% on heme activation, while, at the mid-ring stage, it is 40% dependent. This means that artemisinin and, by extension, other endoperoxides may have other mechanisms of action that may not be iron-dependent [111].

5.3.2. Depolarization of the Mitochondrial Membrane Potential

Mitochondria can play an essential role in the indirect killing of the parasite from ROS generation [107,112]. ROS production occurs as a result of membrane depolarization, dissipating its membrane potential, which plays vital roles in maintaining parasite cellular integrity, hence causing death [117,118]. This depolarization can be caused by the direct inhibition of PfATPase6 by artemisinins and other endoperoxides [119]. Artemisinins can also generate ROS at the mitochondrial level by altering the transfer of electrons from complex III to molecular oxygen, forming superoxide radicals [16,38,69,112].

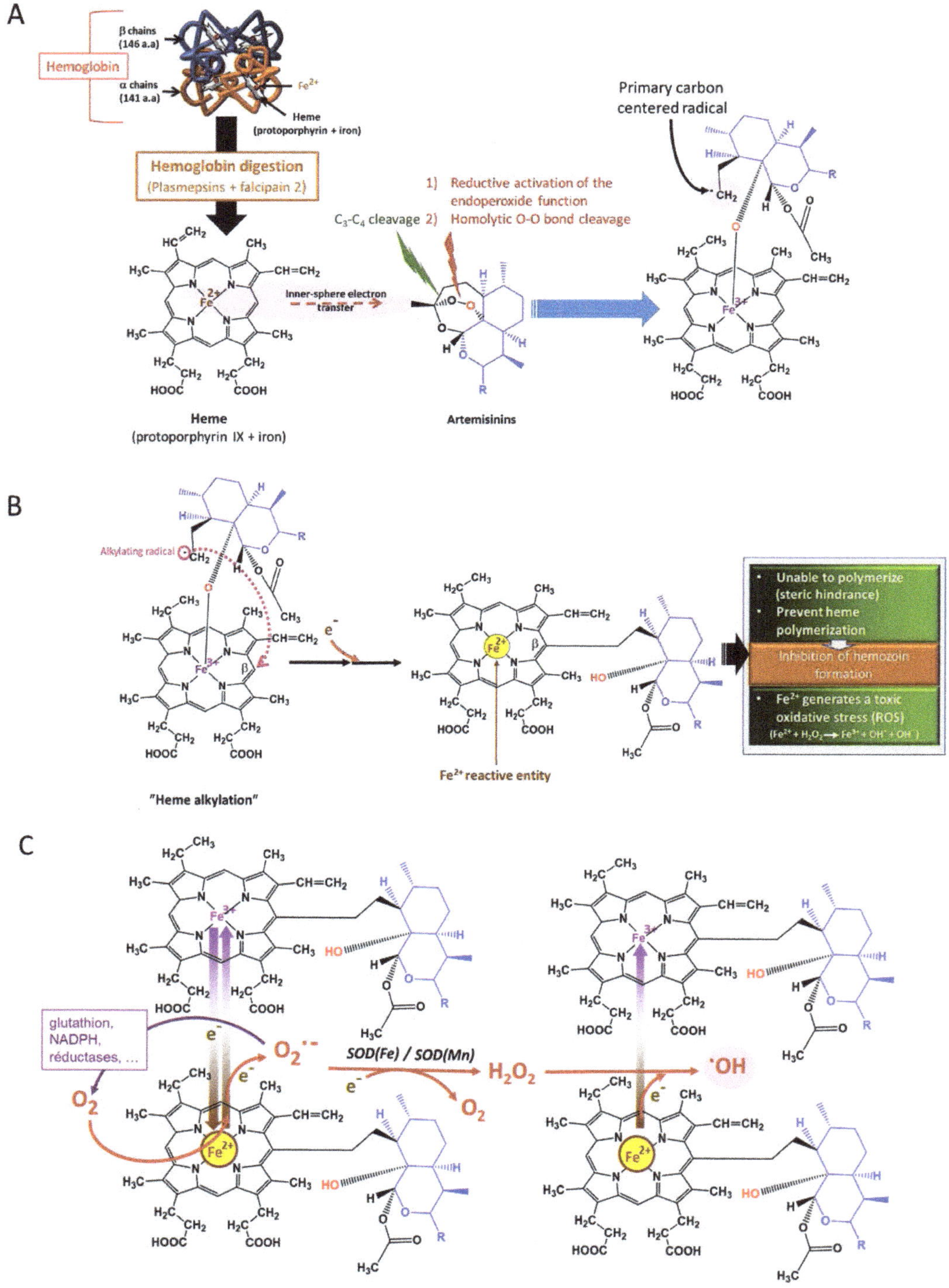

Figure 9. Mode of action of artemisinins [109]. (**A**) Primary carbon-centered radical formation; (**B**) heme alkylation and adduct formation of heme–artemisinin; (**C**) ROS generation, especially the most deleterious hydroxy radical, •OH.

Box 2. Probable roles of other transition metals as cofactors in endoperoxide activities [15,96].

Transition metals potentiate the artemisinin activity
Apart from iron, other metals, especially transition metals such as copper, which are essential components of some proteins, also play a crucial role in ROS generation in the malaria parasite by catalyzing redox cycling in their unbound forms. The ROS generation by antimalarials is potentiated through the redox recycling of Fe^{3+} to Fe^{2+} with the aid of reduced flavin cofactors and probably NADPH-Fe (flavin reductase). Among the transition metals are copper and zinc.

5.4. Artemisinin-Based Combination Therapies

As discussed earlier, quinolines and artemisinins act essentially via the production of ROS. The WHO has recommended six artemisinin-based combination therapies (ACTs): artemether–lumefantrine (AL), dihydroartemisinin–piperaquine (DHA-PPQ), artesunate–amodiaquine (AS-AQ), artesunate–mefloquine (AS-MQ), artesunate–sulfadoxine-primetha mine (AS-SP), and artesunate–pyronaridine (AS-PY) [120]. ACTs drastically clear the parasite and resolve malaria symptoms (such as fever) [121], hence the need for their continued use. Synthetic hybrids of endoperoxides and quinolines have been developed to enhance their pharmacological activities and cost-effectiveness in malaria treatment [105,122]. Resistance to partner drugs used in ACTs in Southeast Asia raises concern over the long-term usefulness of these antiplasmodial combinations [120,123,124].

5.5. ROS Evasive Mechanisms under Treatment

To handle the exacerbated ROS generation by the presence of antimalarials, the parasites upregulate their survival mechanisms, in addition to the already discussed mechanisms. The parasite's evasive mechanisms could simply be classified as preventive, reductive, and reparative.

5.5.1. Preventive Mechanisms

Reduction of self-generated ROS: The parasite is said to reduce its own production of ROS and develop new evolutionary mechanisms to curb the effect of ROS [125]. This evolutionary mechanism undergoes apicoplast metabolism with lipoic acid production, which has antioxidant properties. Lipoic acids are disulfide-containing derivatives of octanoic acid that can exist in oxidized and reduced forms. This ability enables them to play antioxidant roles. Lipoic acids serve as cofactors to several enzymes involved in energy and amino acid metabolism [126]. In the case of artemisinin resistance, reduced hemoglobin import leads to a reduction in the activation of artemisinin and reduced ROS generation [127].

Reduced or mutated expression of hemoglobinases and reduced Hb endocytosis: Endoperoxide activation largely depends on heme production from Hb [110,111]. Downregulation of hemoglobinases (falcipain 2 and 3) in the early ring stage could participate in artemisinin resistance by enhancing the effect of the K13 mutation [111]. Moreover, it has been established that a reduction in the import of hemoglobin, especially at the ring stage, in K13 mutants is responsible for artemisinin resistance [127,128]. Quiescence state was also described as a parasite response to face the oxidative stress generated by artemisinin [129–131]. The parasites recuperate rapidly from dormancy artemisinin-induced once the treatment is removed [130–132].

5.5.2. Reductive Mechanisms

Import of biomolecules: The host boasts more developed antioxidant machinery against ROS than *Plasmodium*. Regardless of this, *Plasmodium* can develop evolutionary measures to eliminate the marauding ROS in its cytosol. Among these measures is the importation of some human antioxidant machinery, such as human peroxiredoxin-2, which uses PfTrx as a reducing agent. Importation is reported to be elevated under treatment with ROS-generating antimalarials [85]. Other forms of human proteins (SOD, catalase, aminolevulinic acid dehydratase, and ferrochelatase) that may augment parasitic defense mechanisms have also been reportedly imported by *Plasmodium* [72,133–135].

Increased expression of antioxidants: Because of the elevated ROS generation due to antimalarial treatment, parasites increase their expression of vital antioxidant enzymes. Such enzymes include iron-superoxide synthetase (Fe-SOD), glutathione-S-transferase (GST), glutathione synthetase (GS), γ-glutamylcysteine synthetase (γ-GCS), thioredoxin reductase (TrxR), and peroxiredoxins (nPrx) [69,70,136]. This overexpression, which is higher in the artemisinins-resistant strains, especially SOD and GST, is associated with parasite resistance characterized by a lower level of ROS and less oxidized proteins [69].

5.5.3. Reparative Mechanisms

ROS generated by antimalarials damage many biomolecules, such as proteins, lipids, and nucleic acids [15]. Increased repair is therefore necessary for continued survival under elevated ROS production. The unfolded protein response (UPR), often referred to as a stress gene, is essential for parasite survival and is upregulated in K13 mutants. This confers upon the parasite an increased ability to repair or degrade proteins damaged by alkylation and oxidation generated by artemisinin [137]. Moreover, K13 mutations have been associated with the elevation of *P. falciparum* phosphatidylinositol-3-kinase (PfPI3K) due to its reduced association with PfKelch13 and polyubiquitination [138]. This consequently leads to the elevation of phosphatidylinositol-3-phosphate (PI3P), which is said to be essential in the trafficking of proteins and lipids toward the apicoplast, where they are needed [138,139]. K13 gene mutations have demonstrated a global spread and attracted increased attention, especially in endemic regions, as molecular markers for ART resistance [129], although not all have been linked to resistance to artemisinin [140,141]. Single-nucleotide polymorphisms (SNPs) in *P. falciparum mlh1*, *pms1*, and *exo1* lead to increased expression of thioredoxin (PfTrx) and signal peptide peptidase, which increases adaptation to oxidative stress and protein damage, leading to the upregulation of the DNA repair mechanisms of the parasite with a consequent decrease in the antimalarial effect of artemisinin [136,142]. These mechanisms keep the vital biomolecules of parasites functional and, as a result, improve their chances of survival under treatment, i.e., resistance development.

6. Conclusions

It is now apparent that conventional antimalarials deploy ROS, ultimately responsible for their parasiticidal activity. However, the evolutionary dynamism of *Plasmodium* seeks to systematically counteract the antimalarial activity mediated by ROS. Given this observation, it is unsurprising that the time required for *P. falciparum* to acquire resistance to antimalarial drugs, whose mode of action is based on the generation of ROS, is becoming increasingly shorter. In this way, the development of artemisinin resistance to new endoperoxide-based hybrid molecules indicates a shared pathway of resistance to the K13 propeller gene mutation [143]. To ensure the continued relevance of ROS-producing antimalarials, the ROS-managing machinery of the parasite could be thwarted to preserve and enhance the activities of the antimalarials. ACTs with newer molecules that disrupt the resistance mechanism of the parasite should be sought. Such molecules, companion drugs of artemisinin, could target key antioxidant enzymes linked to glutathione-dependent and thioredoxin-dependent systems, phosphatidyl inositol 4-kinase enzyme (PI4K), and mRNA translation pathways in *Plasmodium* [69,144–147]. ROS generation thus remains a key element in the strategy in the fight against *Plasmodium* and sustainable malaria control.

Author Contributions: Conceptualization and review process, C.O.E., J.-M.A., K.R. and F.B.-V.; figure and table preparation, C.O.E., J.-M.A., K.R. and F.B.-V.; review and editing of final manuscript, C.O.E., J.-M.A., K.R. and F.B.-V. All authors have read and agreed to the published version of the manuscript.

Funding: This work was supported through a PhD fellowship program to C.O. Egwu from Alex-Ekwueme Federal University, Ndufu-Alike, Ikwo (AE-FUNAI), Nigeria and Campus France, France. CNRS (Centre National de la Recherche Scientifique, France), Université Paul Sabatier, Toulouse III (France) and the IRD (Institut de Recherche pour le Développement, France) participated to the funding of the work.

Acknowledgments: None.

Conflicts of Interest: The authors have no conflict of interest to declare.

References

1. WHO World Malaria Report 2020: 20 Years of Global Progress and Challenges. Available online: https://www.who.int/teams/global-malaria-programme/reports/world-malaria-report-2020 (accessed on 30 November 2020).
2. Bilgin, R.; Yalcin, M.S.; Yucebilgic, G.; Koltas, I.S.; Yazar, S. Oxidative stress in vivax malaria. *Korean J. Parasitol.* **2012**, *50*, 375–377. [CrossRef] [PubMed]
3. Becker, K.; Tilley, L.; Vennerstrom, J.L.; Roberts, D.; Rogerson, S.; Ginsburg, H. Oxidative stress in malaria parasite-infected erythrocytes: Host-parasite interactions. *Int. J. Parasitol.* **2004**, *34*, 163–189. [CrossRef] [PubMed]
4. Erel, O.; Vural, H.; Aksoy, N.; Aslan, G.; Ulukanligil, M. Oxidative stress of platelets and thrombocytopenia in patients with vivax malaria. *Clin. Biochem.* **2001**, *34*, 341–344. [CrossRef]
5. WHO. *World Malaria Report 2019*; WHO: Geneva, Switzerland, 2019; ISBN 9789241565721.
6. Sahu, M.; Tediosi, F.; Noor, A.M.; Aponte, J.J.; Fink, G. Health systems and global progress towards malaria elimination, 2000–2016. *Malar. J.* **2020**, *19*, 141. [CrossRef]
7. Dako-Gyeke, M.; Kofie, H.M. Factors influencing prevention and control of malaria among pregnant women resident in Urban slums, Southern Ghana. *Afr. J. Reprod. Health* **2015**, *19*, 44–53.
8. Thomas, J.A.; Tan, M.S.Y.; Bisson, C.; Borg, A.; Umrekar, T.R.; Hackett, F.; Hale, V.L.; Vizcay-Barrena, G.; Fleck, R.A.; Snijders, A.P.; et al. A protease cascade regulates release of the human malaria parasite Plasmodium falciparum from host red blood cells. *Nat. Microbiol.* **2018**, *3*, 447–455. [CrossRef]
9. Yewhalaw, D.; Wassie, F.; Steurbaut, W.; Spanoghe, P.; Van Bortel, W.; Denis, L.; Tessema, D.A.; Getachew, Y.; Coosemans, M.; Duchateau, L.; et al. Multiple Insecticide Resistance: An Impediment to Insecticide-Based Malaria Vector Control Program. *PLoS ONE* **2011**, *6*, e16066. [CrossRef]
10. Shibeshi, M.A.; Kifle, Z.D.; Atnafie, S.A. Antimalarial Drug Resistance and Novel Targets for Antimalarial Drug Discovery. *Infect. Drug Resist.* **2020**, *13*, 4047–4060. [CrossRef]
11. Payne, D. Spread of chloroquine resistance in Plasmodium falciparum. *Parasitol. Today* **1987**, *3*, 241–246. [CrossRef]
12. Noedl, H.; Se, Y.; Schaecher, K.; Smith, B.L.; Socheat, D.; Fukuda, M.M. Evidence of artemisinin-resistant malaria in Western Cambodia. *N. Engl. J. Med.* **2008**, *359*, 2619–2620. [CrossRef]
13. Dondorp, A.M.; Nosten, F.; Yi, P.; Das, D.; Phyo, A.P.; Tarning, J.; Lwin, K.M.; Ariey, F.; Hanpithakpong, W.; Lee, S.J.; et al. Artemisinin Resistance in Plasmodium falciparum Malaria. *N. Engl. J. Med.* **2009**, *361*, 455–467. [CrossRef]
14. Li, Y.R.; Trush, M. Defining ROS in Biology and Medicine. *React. Oxyg. Species* **2016**, *1*, 9–21. [CrossRef]
15. Santo, A.; Zhu, H.; Li, Y.R. Free Radicals: From Health to Disease. *React. Oxyg. Species* **2016**, *2*, 245–263. [CrossRef]
16. Egwu, C.O.; Tsamesidis, I.; Pério, P.; Augereau, J.-M.; Benoit-Vical, F.; Reybier, K. Superoxide: A major role in the mechanism of action of essential antimalarial drugs. *Free Radic. Biol. Med.* **2021**, *167*, 271–275. [CrossRef]
17. Mohring, F.; Pretzel, J.; Jortzik, E.; Becker, K. The redox systems of Plasmodium falciparum and Plasmodium vivax: Comparison, in silico analyses and inhibitor studies. *Curr. Med. Chem.* **2014**, *21*, 1728–1756. [CrossRef]
18. Hancock, J.T.; Desikan, R.; Neill, S.J. Role of reactive oxygen species in cell signalling pathways. *Biochem. Soc. Trans.* **2001**, *2*, 345–350. [CrossRef]
19. Mittal, M.; Siddiqui, M.R.; Tran, K.; Reddy, S.P.; Malik, A.B. Reactive Oxygen Species in Inflammation and Tissue Injury. *Antioxid. Redox Signal.* **2014**, *20*, 1126. [CrossRef]
20. Stefanatos, R.; Sanz, A. The role of mitochondrial ROS in the aging brain. *FEBS Lett.* **2018**, *592*, 743–758. [CrossRef]
21. Sugamura, K.; Keaney, J.F. Reactive oxygen species in cardiovascular disease. *Free Radic. Biol. Med.* **2011**, *51*, 978–992. [CrossRef]
22. Deavall, D.G.; Martin, E.A.; Horner, J.M.; Roberts, R. Drug-induced oxidative stress and toxicity. *J. Toxicol.* **2012**, *2012*, 645460. [CrossRef]
23. Thannickal, V.J.; Fanburg, B.L. Reactive oxygen species in cell signaling. *Am. J. Physiol. Lung Cell. Mol. Physiol.* **2000**, *279*, L1005–L1028. [CrossRef] [PubMed]
24. Rahbari, M.; Rahlfs, S.; Jortzik, E.; Bogeski, I.; Becker, K. H2O2 dynamics in the malaria parasite Plasmodium falciparum. *PLoS ONE* **2017**, *12*, e0174837. [CrossRef]
25. Sampson, C.; Keens, R.H.; Kattnig, D.R. On the magnetosensitivity of lipid peroxidation: Two- versus three-radical dynamics. *Phys. Chem. Chem. Phys.* **2019**, *21*, 13526–13538. [CrossRef] [PubMed]
26. Hartwig, C.L.; Rosenthal, A.S.; D'Angelo, J.; Griffin, C.E.; Posner, G.H.; Cooper, R.A. Accumulation of artemisinin trioxane derivatives within neutral lipids of Plasmodium falciparum malaria parasites is endoperoxide-dependent. *Biochem. Pharmacol.* **2009**, *77*, 322–336. [CrossRef]

27. Kumura, N.; Furukawa, H.; Onyango, A.N.; Izumi, M.; Nakajima, S.; Ito, H.; Hatano, T.; Kim, H.S.; Wataya, Y.; Baba, N. Different behavior of artemisinin and tetraoxane in the oxidative degradation of phospholipid. *Chem. Phys. Lipids* **2009**, *160*, 114–120. [CrossRef]
28. Fernando, N.; Wickremesinghe, S.; Niloofa, R.; Rodrigo, C.; Karunanayake, L.; De Silva, H.J.; Wickremesinghe, A.R.; Premawansa, S.; Rajapakse, S.; Handunnetti, S.M. Protein carbonyl as a biomarker of oxidative stress in severe leptospirosis, and its usefulness in differentiating leptospirosis from dengue infections. *PLoS ONE* **2016**, *11*, e0156085. [CrossRef]
29. Dalle-Donne, I.; Rossi, R.; Giustarini, D.; Milzani, A.; Colombo, R. Protein carbonyl groups as biomarkers of oxidative stress. *Clin. Chim. Acta* **2003**, *329*, 23–38. [CrossRef]
30. Nyström, T. Role of oxidative carbonylation in protein quality control and senescence. *EMBO J.* **2005**, *24*, 1311–1317. [CrossRef]
31. Fedorova, M.; Bollineni, R.C.; Hoffmann, R. Protein carbonylation as a major hallmark of oxidative damage. *Mass Spectrom Rev* **2014**, *33*, 79–97. [CrossRef]
32. Rosenthal, P.J. Cysteine proteases of malaria parasites. *Int. J. Parasitol.* **2004**, *34*, 1489–1499. [CrossRef]
33. Wu, W.M.; Chen, Y.L.; Zhai, Z.; Xiao, S.H.; Wu, Y.L. Study on the mechanism of action of artemether against schistosomes: The identification of cysteine adducts of both carbon-centred free radicals derived from artemether. *Bioorg. Med. Chem. Lett.* **2003**, *13*, 1645–1647. [CrossRef]
34. Eckstein-Ludwig, U.; Webb, R.J.; Van Goethem, I.D.A.; East, J.M.; Lee, A.G.; Kimura, M.; O'Neill, P.M.; Bray, P.G.; Ward, S.A.; Krishna, S. Artemisinins target the SERCA of Plasmodium falciparum. *Nature* **2003**, *424*, 957–961. [CrossRef]
35. Wang, J.; Xu, C.; Lun, Z.R.; Meshnick, S.R. Unpacking 'Artemisinin Resistance'. *Trends Pharmacol. Sci.* **2017**, *38*, 506–511. [CrossRef]
36. Jena, N.R. DNA damage by reactive species: Mechanisms, mutation and repair. *J. Biosci.* **2012**, *37*, 503–517. [CrossRef]
37. Cadet, J.; Wagner, J.R. DNA base damage by reactive oxygen species, oxidizing agents, and UV radiation. *Cold Spring Harb. Perspect. Biol.* **2013**, *5*, a012559. [CrossRef]
38. Gopalakrishnan, A.M.; Kumar, N. Antimalarial action of artesunate involves DNA damage mediated by reactive oxygen species. *Antimicrob. Agents Chemother.* **2015**, *59*, 317–325. [CrossRef]
39. Freese, R. Markers of oxidative DNA damage in human interventions with fruit and berries. *Nutr. Cancer* **2006**, *54*, 143–147. [CrossRef]
40. Uhlemann, A.C.; Wittlin, S.; Matile, H.; Bustamante, L.Y.; Krishna, S. Mechanism of antimalarial action of the synthetic trioxolane RBX11160 (OZ277). *Antimicrob. Agents Chemother.* **2007**, *51*, 667–672. [CrossRef]
41. Postma, N.S.; Mommers, E.C.; Eling, W.M.C.; Zuidema, J. Oxidative stress in malaria; implications for prevention and therapy. *Pharm. World Sci.* **1996**, *18*, 121–129. [CrossRef]
42. Percário, S.; Moreira, D.R.; Gomes, B.A.Q.; Ferreira, M.E.S.; Gonçalves, A.C.M.; Laurindo, P.S.O.C.; Vilhena, T.C.; Dolabela, M.F.; Green, M.D. Oxidative stress in Malaria. *Int. J. Mol. Sci.* **2012**, *13*, 16346–16372. [CrossRef]
43. Tiwari, S.; Sharma, N.; Sharma, G.P.; Mishra, N. Redox interactome in malaria parasite Plasmodium falciparum. *Parasitol. Res.* **2021**, *120*, 423–434. [CrossRef] [PubMed]
44. Loria, P.; Miller, S.; Foley, M.; Tilley, L. Inhibition of the peroxidative degradation of haem as the basis of action of chloroquine and other quinoline antimalarials. *Biochem. J.* **1999**, *339*, 363–370. [CrossRef] [PubMed]
45. Zhang, J.; Krugliak, M.; Ginsburg, H. The fate of ferriprotorphyrin IX in malaria infected erythrocytes in conjunction with the mode of action of antimalarial drugs. *Mol. Biochem. Parasitol.* **1999**, *99*, 129–141. [CrossRef]
46. Egan, T.J.; Combrinck, J.M.; Egan, J.; Hearne, G.R.; Marques, H.M.; Ntenteni, S.; Sewell, B.T.; Smith, P.J.; Taylor, D.; Van Schalkwyk, D.A.; et al. Fate of haem iron in the malaria parasite Plasmodium falciparum. *Biochem. J.* **2002**, *365*, 343–347. [CrossRef]
47. Rahlfs, S.; Schirmer, R.H.; Becker, K. The thioredoxin system of Plasmodium falciparum and other parasites. *Cell. Mol. Life Sci.* **2002**, *59*, 1024–1041. [CrossRef]
48. Desakorn, V.; Dondorp, A.M.; Silamut, K.; Pongtavornpinyo, W.; Sahassananda, D.; Chotivanich, K.; Pitisuttithum, P.; Smithyman, A.M.; Day, N.P.J.; White, N.J. Stage-dependent production and release of histidine-rich protein 2 by Plasmodium falciparum. *Trans. R. Soc. Trop. Med. Hyg.* **2005**, *99*, 517–524. [CrossRef]
49. Wrenger, C.; Eschbach, M.L.; Müller, I.B.; Warnecke, D.; Walter, R.D. Analysis of the vitamin B6 biosynthesis pathway in the human malaria parasite Plasmodium falciparum. *J. Biol. Chem.* **2005**, *280*, 5242–5248. [CrossRef]
50. Nickel, C.; Rahlfs, S.; Deponte, M.; Koncarevic, S.; Becker, K. Thioredoxin networks in the malarial parasite Plasmodium falciparum. *Antioxid. Redox Signal.* **2006**, *8*, 1227–1239. [CrossRef]
51. Goldberg, D.E.; Sigala, P.A. Plasmodium heme biosynthesis: To be or not to be essential? *PLOS Pathog.* **2017**, *13*, e1006511. [CrossRef]
52. Müller, S. Redox and antioxidant systems of the malaria parasite Plasmodium falciparum. *Mol. Microbiol.* **2004**, *53*, 1291–1305. [CrossRef]
53. Kawazu, S.; Komaki-Yasuda, K.; Oku, H.; Kano, S. Peroxiredoxins in malaria parasites: Parasitologic aspects. *Parasitol. Int.* **2008**, *57*, 1–7. [CrossRef]
54. Mustacich, D.; Powis, G. Thioredoxin reductase. *Biochem. J.* **2000**, *346*, 1–8. [CrossRef]
55. Slater, A.F.G.; Cerami, A. Inhibition by chloroquine of a novel haem polymerase enzyme activity in malaria trophozoites. *Nature* **1992**, *355*, 167–169. [CrossRef]

56. Gupta, P.; Mehrotra, S.; Sharma, A.; Chugh, M.; Pandey, R.; Kaushik, A.; Khurana, S.; Srivastava, N.; Srivastava, T.; Deshmukh, A.; et al. Exploring Heme and Hemoglobin Binding Regions of Plasmodium Heme Detoxification Protein for New Antimalarial Discovery. *J. Med. Chem.* **2017**, *60*, 8298–8308. [CrossRef]
57. Choi, C.Y.H.; Cerda, J.F.; Hsiu-An, C.; Babcock, G.T.; Marletta, M.A. Spectroscopic characterization of the heme-binding sites in Plasmodium falciparum histidine-rich protein 2. *Biochemistry* **1999**, *38*, 16916–16924. [CrossRef]
58. Sullivan, D.J.; Gluzman, I.Y.; Goldberg, D.E. Plasmodium Hemozoin Formation Mediated by Histidine-Rich Proteins. *Science* **1996**, *271*, 219–222. [CrossRef]
59. Kapishnikov, S.; Grolimund, D.; Schneider, G.; Pereiro, E.; McNally, J.G.; Als-Nielsen, J.; Leiserowitz, L. Unraveling heme detoxification in the malaria parasite by in situ correlative X-ray fluorescence microscopy and soft X-ray tomography. *Sci. Rep.* **2017**, *7*, 1–12. [CrossRef]
60. Sigala, P.A.; Goldberg, D.E. The peculiarities and paradoxes of Plasmodium heme metabolism. *Annu. Rev. Microbiol.* **2014**, *68*, 259–278. [CrossRef]
61. Wallace, W.J.; Houtchens, R.A.; Maxwell, J.C.; Caughey, W.S. Mechanism of autooxidation for hemoglobins and myoglobins. Promotion of superoxide production by protons and anions. *J. Biol. Chem.* **1982**, *257*, 4966–4977. [CrossRef]
62. Carrell, R.W.; Winterbourn, C.C.; Rachmilewitz, E.A. Annotation: Activated oxygen and haemolysis. *Br. J. Haematol.* **1975**, *30*, 259–264. [CrossRef]
63. Thomas, C.; Mackey, M.; Diaz, A.; Cox, D. Hydroxyl radical is produced via the Fenton reaction in submitochondrial particles under oxidative stress: Implications for diseases associated with iron accumulation. *Redox Rep.* **2009**, *14*, 102–108. [CrossRef] [PubMed]
64. Jani, D.; Nagarkatti, R.; Beatty, W.; Angel, R.; Slebodnick, C.; Andersen, J.; Kumar, S.; Rathore, D. HDP—A novel heme detoxification protein from the malaria parasite. *PLoS Pathog.* **2008**, *4*, e1000053. [CrossRef] [PubMed]
65. Lew, V.L.; Tiffert, T.; Ginsburg, H. Excess hemoglobin digestion and the osmotic stability of Plasmodium falciparum-Infected red blood cells. *Blood* **2003**, *101*, 4189–4194. [CrossRef] [PubMed]
66. Mairbäurl, H.; Weber, R.E. Oxygen transport by hemoglobin. *Compr. Physiol.* **2012**, *2*, 1463–1489. [CrossRef] [PubMed]
67. Clark, I.A.; Rockett, K.A. Nitric oxide and parasitic disease. *Adv. Parasitol.* **1996**, *37*, 1–56. [CrossRef]
68. Ghigo, D.; Todde, R.; Ginsburg, H.; Costamagna, C.; Gautret, P.; Bussolino, F.; Ulliers, D.; Giribaldi, G.; Deharo, E.; Gabrielli, G.; et al. Erythrocyte stages of Plasmodium falciparum exhibit a high nitric oxide synthase (NOS) activity and release an NOS-inducing soluble factor. *J. Exp. Med.* **1995**, *182*, 677–688. [CrossRef]
69. Egwu, C.O.; Pério, P.; Augereau, J.-M.; Tsamesidis, I.; Benoit-Vical, F.; Reybier, K. Resistance to artemisinin in falciparum malaria parasites: A redox-mediated phenomenon. *Free Radic. Biol. Med.* **2021**, *S0891-5849*, 00476–00477. [CrossRef]
70. Nogueira, F.; Diez, A.; Radfar, A.; Pérez-Benavente, S.; Rosario, V.E.; Puyet, A.; Bautista, J.M. Early transcriptional response to chloroquine of the Plasmodium falciparum antioxidant defence in sensitive and resistant clones. *Acta Trop.* **2010**, *114*, 109–115. [CrossRef]
71. Wang, W.; Huang, P.; Jiang, N.; Lu, H.; Zhang, D.; Wang, D.; Zhang, K.; Wahlgren, M.; Chen, Q. A thioredoxin homologous protein of Plasmodium falciparum participates in erythrocyte invasion. *Infect. Immun.* **2018**, *86*, e00289-18. [CrossRef]
72. Fairfield, A.S.; Meshnick, S.R.; Eaton, J.W. Malaria parasites adopt host cell superoxide dismutase. *Science* **1983**, *221*, 764–766. [CrossRef]
73. Ginsburg, H.; Famin, O.; Zhang, J.; Krugliak, M. Inhibition of glutathione-dependent degradation of heme by chloroquine and amodiaquine as a possible basis for their antimalarial mode of action. *Biochem. Pharmacol.* **1998**, *56*, 1305–1313. [CrossRef]
74. Porras, P.; Pedrajas, J.R.; Martínez-Galisteo, E.; Alicia Padilla, C.; Johansson, C.; Holmgren, A.; Antonio Bárcena, J. Glutaredoxins catalyze the reduction of glutathione by dihydrolipoamide with high efficiency. *Biochem. Biophys. Res. Commun.* **2002**, *295*, 1046–1051. [CrossRef]
75. Patzewitz, E.-M.; Müller, S. Glutathione biosynthesis and metabolism in Plasmodium falciparum. *Malar. J.* **2010**, *9*, 37. [CrossRef]
76. Lüersen, K.; Walter, R.D.; Müller, S. Plasmodium falciparum-infected red blood cells depend on a functional glutathione de novo synthesis attributable to an enhanced loss of glutathione. *Biochem. J.* **2000**, *346*, 545–552. [CrossRef]
77. Patzewitz, E.M.; Wong, E.H.; Müller, S. Dissecting the role of glutathione biosynthesis in Plasmodium falciparum. *Mol. Microbiol.* **2012**, *83*, 304–318. [CrossRef]
78. Harvey, J.W.; Beutler, E. Binding of heme by glutathione S-transferase: A possible role of the erythrocyte enzyme. *Blood* **1982**, *60*, 1227–1230. [CrossRef]
79. Müller, S. Role and regulation of glutathione metabolism in Plasmodium falciparum. *Molecules* **2015**, *20*, 10511–10534. [CrossRef]
80. Hirt, R.P.; Müller, S.; Martin Embley, T.; Coombs, G.H. The diversity and evolution of thioredoxin reductase: New perspectives. *Trends Parasitol.* **2002**, *18*, 302–308. [CrossRef]
81. Bozdech, Z.; Ginsburg, H. Antioxidant defense in Plasmodium falciparum - Data mining of the transcriptome. *Malar. J.* **2004**, *3*, 1–10. [CrossRef]
82. Lillig, C.H.; Holmgren, A. Thioredoxin and related molecules - From biology to health and disease. *Antioxid. Redox Signal.* **2007**, *9*, 25–47. [CrossRef]
83. Yano, K.; Komaki-Yasuda, K.; Kobayashi, T.; Takemae, H.; Kita, K.; Kano, S.; Kawazu, S.I. Expression of mRNAs and proteins for peroxiredoxins in Plasmodium falciparum erythrocytic stage. *Parasitol. Int.* **2005**, *54*, 35–41. [CrossRef] [PubMed]

84. Akerman, S.E.; Müller, S. 2-Cys peroxiredoxin PfTrx-Px1 is involved in the antioxidant defence of Plasmodium falciparum. *Mol. Biochem. Parasitol.* **2003**, *130*, 75–81. [CrossRef]
85. Koncarevic, S.; Rohrbach, P.; Deponte, M.; Krohne, G.; Prieto, J.H.; Yates, J.; Rahlfs, S.; Becker, K. The malarial parasite Plasmodium falciparum imports the human protein peroxiredoxin 2 for peroxide detoxification. *Proc. Natl. Acad. Sci. USA* **2009**, *106*, 13323–13328. [CrossRef] [PubMed]
86. Atamna, H.; Pascarmona, G.; Ginsburg, H. Hexose-monophosphate shunt activity in intact Plasmodium falciparum-infected erythrocytes and in free parasites. *Mol. Biochem. Parasitol.* **1994**, *67*, 79–89. [CrossRef]
87. Campanale, N.; Nickel, C.; Daubenberger, C.A.; Wehlan, D.A.; Gorman, J.J.; Klonis, N.; Becker, K.; Tilley, L. Identification and characterization of heme-interacting proteins in the malaria parasite, Plasmodium falciparum. *J. Biol. Chem.* **2003**, *278*, 27354–27361. [CrossRef]
88. Ehrenshaft, M.; Bilski, P.; Li, M.; Chignell, C.F.; Daub, M.E. A highly conserved sequence is a novel gene involved in de novo vitamin B6 biosynthesis. *Proc. Natl. Acad. Sci. USA* **1999**, *96*, 9374–9378. [CrossRef]
89. Bilski, P.; Li, M.Y.; Ehrenshaft, M.; Daub, M.E.; Chignell, C.F. Vitamin B6 (Pyridoxine) and Its Derivatives Are Efficient Singlet Oxygen Quenchers and Potential Fungal Antioxidants. *Photochem. Photobiol.* **2000**, *71*, 129–134. [CrossRef]
90. Matxain, J.M.; Padro, D.; Ristilä, M.; Strid, Å.; Eriksson, L.A. Evidence of high ·OH radical quenching efficiency by vitamin B 6. *J. Phys. Chem. B* **2009**, *113*, 9629–9632. [CrossRef]
91. Natera, J.; Massad, W.; García, N.A. The role of vitamin B6 as an antioxidant in the presence of vitamin B2-photogenerated reactive oxygen species. A kinetic and mechanistic study. *Photochem. Photobiol. Sci.* **2012**, *11*, 938. [CrossRef]
92. Kannan, K.; Jain, S.K. Effect of vitamin B 6 on oxygen radicals, mitochondrial membrane potential, and lipid peroxidation in H 2 O 2 -treated U937 monocytes. *Free Radic. Biol. Med.* **2004**, *36*, 423–428. [CrossRef]
93. Tsamesidis, I.; Egwu, C.O.; Pério, P.; Augereau, J.M.; Benoit-Vical, F.; Reybier, K. An LC–MS assay to measure superoxide radicals and hydrogen peroxide in the blood system. *Metabolites* **2020**, *10*, 175. [CrossRef]
94. Sullivan, D.J.; Matile, H.; Ridley, R.G.; Goldberg, D.E. A common mechanism for blockade of heme polymerization by antimalarial quinolines. *J. Biol. Chem.* **1998**, *273*, 31103–31107. [CrossRef]
95. Sugioka, Y.; Suzuki, M.; Sugioka, K.; Nakano, M. A ferriprotoporphyrin IX-chloroquine complex promotes membrane phospholipid peroxidation A possible mechanism for antimalarial action. *FEBS Lett.* **1987**, *223*, 251–254. [CrossRef]
96. Haynes, R.K.; Cheu, K.W.; Chan, H.W.; Wong, H.N.; Li, K.Y.; Tang, M.M.K.; Chen, M.J.; Guo, Z.F.; Guo, Z.H.; Sinniah, K.; et al. Interactions between Artemisinins and other Antimalarial Drugs in Relation to the Cofactor Model-A Unifying Proposal for Drug Action. *ChemMedChem* **2012**, *7*, 2204–2226. [CrossRef]
97. De Almeida Ribeiro, M.C.; Augusto, O.; Da Costa Ferreira, A.M. Inhibitory effect of chloroquine on the peroxidase activity of ferriprotoporphyrin IX. *J. Chem. Soc. Dalt. Trans.* **1995**, 3759–3766. [CrossRef]
98. Birth, D.; Kao, W.C.; Hunte, C. Structural analysis of atovaquone-inhibited cytochrome bc 1 complex reveals the molecular basis of antimalarial drug action. *Nat. Commun.* **2014**, *5*, 1–11. [CrossRef]
99. Fry, M.; Pudney, M. Site of action of the antimalarial hydroxynaphthoquinone, 2-[trans-4-(4′-chlorophenyl) cyclohexyl]-3-hydroxy-1,4-naphthoquinone (566C80). *Biochem. Pharmacol.* **1992**, *43*, 1545–1553. [CrossRef]
100. Srivastava, I.K.; Rottenberg, H.; Vaidya, A.B. Atovaquone, a broad spectrum antiparasitic drug, collapses mitochondrial membrane potential in a malarial parasite. *J. Biol. Chem.* **1997**, *272*, 3961–3966. [CrossRef]
101. Barton, V.; Fisher, N.; Biagini, G.A.; Ward, S.A.; O'Neill, P.M. Inhibiting Plasmodium cytochrome bc1: A complex issue. *Curr. Opin. Chem. Biol.* **2010**, *14*, 440–446. [CrossRef]
102. Fiorillo, M.; Lamb, R.; Tanowitz, H.B.; Mutti, L.; Krstic-Demonacos, M.; Cappello, A.R.; Martinez-Outschoorn, U.E.; Sotgia, F.; Lisanti, M.P. Repurposing atovaquone: Targeting mitochondrial complex III and OXPHOS to eradicate cancer stem cells. *Oncotarget* **2016**, *7*, 34084–34099. [CrossRef]
103. Kevin Park, B.; O'Neill, P.M.; Maggs, J.L.; Pirmohamed, M. Safety assessment of peroxide antimalarials: Clinical and chemical perspectives. *Br. J. Clin. Pharmacol.* **1998**, *46*, 521–529. [CrossRef] [PubMed]
104. Nosten, F.; White, N.J. Artemisinin-based combination treatment of falciparum malaria. *Am. J. Trop. Med. Hyg.* **2007**, *77*, 181–192. [CrossRef] [PubMed]
105. Rudrapal, M.; Chetia, D. Endoperoxide antimalarials: Development, structural diversity and pharmacodynamic aspects with reference to 1,2,4-trioxane-based structural scaffold. *Drug Des. Devel. Ther.* **2016**, *10*, 3575–3590. [CrossRef] [PubMed]
106. Posner, G.H.; Oh, C.H. A Regiospecifically Oxygen-18 Labeled 1,2,4-Trioxane: A Simple Chemical Model System To Probe the Mechanism(s) for the Antimalarial Activity of Artemisinin (Qinghaosu). *J. Am. Chem. Soc.* **1992**, *114*, 8328–8329. [CrossRef]
107. Mercer, A.E.; Copple, I.M.; Maggs, J.L.; O'Neill, P.M.; Park, B.K. The role of heme and the mitochondrion in the chemical and molecular mechanisms of mammalian cell death induced by the artemisinin antimalarials. *J. Biol. Chem.* **2011**, *286*, 987–996. [CrossRef]
108. Posner, G.H.; Oh, C.H.; Wang, D.; Gerena, L.; Milhous, W.K.; Meshnick, S.R.; Asawamahasadka, W. Mechanism-Based Design, Synthesis, and in Vitro Antimalarial Testing of New 4-Methylated Trioxanes Structurally Related to Artemisinin: The Importance of a Carbon-Centered Radical for Antimalarial Activity. *J. Med. Chem.* **1994**, *37*, 1256–1258. [CrossRef]
109. Robert, A.; Benoit-Vical, F.; Claparols, C.; Meunier, B. The antimalarial drug artemisinin alkylates heme in infected mice. *Proc. Natl. Acad. Sci. USA* **2005**, *102*, 13676–13680. [CrossRef]

110. Zhang, S.; Gerhard, G.S. Heme activates artemisinin more efficiently than hemin, inorganic iron, or hemoglobin. *Bioorg. Med. Chem.* **2008**, *16*, 7853–7861. [CrossRef]
111. Xie, S.C.; Dogovski, C.; Hanssen, E.; Chiu, F.; Yang, T.; Crespo, M.P.; Stafford, C.; Batinovic, S.; Teguh, S.; Charman, S.; et al. Haemoglobin degradation underpins the sensitivity of early ring stage Plasmodium falciparum to artemisinins. *J. Cell Sci.* **2016**, *129*, 406–416. [CrossRef]
112. Wang, J.; Huang, L.; Li, J.; Fan, Q.; Long, Y.; Li, Y.; Zhou, B. Artemisinin directly targets malarial mitochondria through its specific mitochondrial activation. *PLoS ONE* **2010**, *5*, e9582. [CrossRef]
113. Meunier, B.; de Visser, S.P.; Shaik, S. Mechanism of oxidation reactions catalyzed by cytochrome P450 enzymes. *Chem. Rev.* **2004**, *104*, 3947–3980. [CrossRef]
114. Dawson, J.H. Probing structure-function relations in heme-containing oxygenases and peroxidases. *Science* **1988**, *240*, 433–439. [CrossRef]
115. Fita, I.; Rossmann, M.G. The active center of catalase. *J. Mol. Biol.* **1985**, *185*, 21–37. [CrossRef]
116. Edwards, S.L.; Kraut, J.; Poulos, T.L. Crystal Structure of Nitric Oxide Inhibited Cytochrome c Peroxidase. *Biochemistry* **1988**, *27*, 8074–8081. [CrossRef]
117. Allen, R.J.W.; Kirk, K. The Membrane Potential of the Intraerythrocytic Malaria Parasite Plasmodium falciparum. *J. Biol. Chem.* **2004**, *279*, 11264–11272. [CrossRef]
118. Biagini, G.A.; Viriyavejakul, P.; O'Neill, P.M.; Bray, P.G.; Ward, S.A. Functional characterization and target validation of alternative complex I of Plasmodium falciparum mitochondria. *Antimicrob. Agents Chemother.* **2006**, *50*, 1841–1851. [CrossRef]
119. Antoine, T.; Fisher, N.; Amewu, R.; O'Neill, P.M.; Ward, S.A.; Biagini, G.A. Rapid kill of malaria parasites by artemisinin and semi-synthetic endoperoxides involves ROS-dependent depolarization of the membrane potential. *J. Antimicrob. Chemother.* **2014**, *69*, 1005–1016. [CrossRef]
120. WHO. Artemisinin Resistance and Artemisinin-Based Combination Therapy Efficacy (Status Report—August 2018). Available online: https://apps.who.int/iris/bitstream/handle/10665/274362/WHO-CDS-GMP-2018.18-eng.pdf?ua=1 (accessed on 26 December 2020).
121. Van Vugt, M.; Brockman, A.; Gemperli, B.; Luxemburger, C.; Gathmann, I.; Royce, C.; Slight, T.; Looareesuwan, S.; White, N.J.; Nosten, F. Randomized comparison of artemether-benflumetol and artesunate-mefloquine in treatment of multidrug-resistant falciparum malaria. *Antimicrob. Agents Chemother.* **1998**, *42*, 135–139. [CrossRef]
122. Benoit-Vical, F.; Lelièvre, J.; Berry, A.; Deymier, C.; Dechy-Cabaret, O.; Cazelles, J.; Loup, C.; Robert, A.; Magnaval, J.F.; Meunier, B. Trioxaquines are new antimalarial agents active on all erythrocytic forms, including gametocytes. *Antimicrob. Agents Chemother.* **2007**, *51*, 1463–1472. [CrossRef]
123. Leang, R.; Barrette, A.; Bouth, D.M.; Menard, D.; Abdur, R.; Duong, S.; Ringwald, P. Efficacy of dihydroartemisinin-piperaquine for treatment of uncomplicated Plasmodium falciparum and Plasmodium vivax in Cambodia, 2008 to 2010. *Antimicrob. Agents Chemother.* **2013**, *57*, 818–826. [CrossRef]
124. Bopp, S.; Magistrado, P.; Wong, W.; Schaffner, S.F.; Mukherjee, A.; Lim, P.; Dhorda, M.; Amaratunga, C.; Woodrow, C.J.; Ashley, E.A.; et al. Plasmepsin II-III copy number accounts for bimodal piperaquine resistance among Cambodian Plasmodium falciparum. *Nat. Commun.* **2018**, *9*, 1769. [CrossRef] [PubMed]
125. Toler, S. The plasmodial apicoplast was retained under evolutionary selective pressure to assuage blood stage oxidative stress. *Med. Hypotheses* **2005**, *65*, 683–690. [CrossRef] [PubMed]
126. Storm, J. Lipoic Acid Metabolism of Plasmodium—A Suitable Drug Target. *Curr. Pharm. Des.* **2012**, *18*, 3480. [CrossRef] [PubMed]
127. Birnbaum, J.; Scharf, S.; Schmidt, S.; Jonscher, E.; Maria Hoeijmakers, W.A.; Flemming, S.; Toenhake, C.G.; Schmitt, M.; Sabitzki, R.; Bergmann, B.; et al. A Kelch13-defined endocytosis pathway mediates artemisinin resistance in malaria parasites. *Science* **2020**, *367*, 51–59. [CrossRef]
128. Klonis, N.; Crespo-Ortiz, M.P.; Bottova, I.; Abu-Bakar, N.; Kenny, S.; Rosenthal, P.J.; Tilley, L. Artemisinin activity against Plasmodium falciparum requires hemoglobin uptake and digestion. *Proc. Natl. Acad. Sci. USA* **2011**, *108*, 11405–11410. [CrossRef]
129. Ariey, F.; Witkowski, B.; Amaratunga, C.; Beghain, J.; Langlois, A.C.; Khim, N.; Kim, S.; Duru, V.; Bouchier, C.; Ma, L.; et al. A molecular marker of artemisinin-resistant Plasmodium falciparum malaria. *Nature* **2014**, *505*, 50–55. [CrossRef]
130. Witkowski, B.; Lelièvre, J.; Barragán, M.J.L.; Laurent, V.; Su, X.Z.; Berry, A.; Benoit-Vical, F. Increased tolerance to artemisinin in Plasmodium falciparum is mediated by a quiescence mechanism. *Antimicrob. Agents Chemother.* **2010**, *54*, 1872–1877. [CrossRef]
131. Witkowski, B.; Khim, N.; Chim, P.; Kim, S.; Ke, S.; Kloeung, N.; Chy, S.; Duong, S.; Leang, R.; Ringwald, P.; et al. Reduced artemisinin susceptibility of Plasmodium falciparum ring stages in western cambodia. *Antimicrob. Agents Chemother.* **2013**, *57*, 914–923. [CrossRef]
132. Peatey, C.; Chen, N.; Gresty, K.; Anderson, K.; Pickering, P.; Watts, R.; Gatton, M.L.; McCarthy, J.; Cheng, Q. Dormant Plasmodium falciparum Parasites in Human Infections Following Artesunate Therapy. *J. Infect. Dis.* **2021**, *223*, 1631–1638. [CrossRef]
133. Clarebout, G.; Slomianny, C.; Delcourt, P.; Leu, B.; Masset, A.; Camus, D.; Dive, D. Status of Plasmodium falciparum towards catalase. *Br. J. Haematol.* **1998**, *103*, 52–59. [CrossRef]
134. Varadharajan, S.; Sagar, B.K.C.; Rangarajan, P.N.; Padmanaban, G. Localization of ferrochelatase in Plasmodium falciparum. *Biochem. J.* **2004**, *384*, 429–436. [CrossRef]
135. Bonday, Z.Q.; Dhanasekaran, S.; Rangarajan, P.N.; Padmanaban, G. Import of host δ-aminolevulinate dehydratase into the malarial parasite: Identification of a new drug target. *Nat. Med.* **2000**, *6*, 898–903. [CrossRef]

136. Rocamora, F.; Zhu, L.; Liong, K.Y.; Dondorp, A.; Miotto, O.; Mok, S.; Bozdech, Z. Oxidative stress and protein damage responses mediate artemisinin resistance in malaria parasites. *PLoS Pathog.* **2018**, *14*, e1006930. [CrossRef]
137. Mok, S.; Ashley, E.A.; Ferreira, P.E.; Zhu, L.; Lin, Z.; Yeo, T.; Chotivanich, K.; Imwong, M.; Pukrittayakamee, S.; Dhorda, M.; et al. Population transcriptomics of human malaria parasites reveals the mechanism of artemisinin resistance. *Science* **2015**, *347*, 431–435. [CrossRef]
138. Mbengue, A.; Bhattacharjee, S.; Pandharkar, T.; Liu, H.; Estiu, G.; Stahelin, R.V.; Rizk, S.S.; Njimoh, D.L.; Ryan, Y.; Chotivanich, K.; et al. A molecular mechanism of artemisinin resistance in Plasmodium falciparum malaria. *Nature* **2015**, *520*, 683–687. [CrossRef]
139. Tawk, L.; Chicanne, G.; Dubremetz, J.F.; Richard, V.; Payrastre, B.; Vial, H.J.; Roy, C.; Wengelnik, K. Phosphatidylinositol 3-phosphate, an essential lipid in Plasmodium, localizes to the food vacuole membrane and the apicoplast. *Eukaryot. Cell* **2010**, *9*, 1519–1530. [CrossRef]
140. Isozumi, R.; Uemura, H.; Kimata, I.; Ichinose, Y.; Logedi, J.; Omar, A.H.; Kaneko, A. Novel mutations in k13 propeller gene of artemisinin-resistant Plasmodium falciparum. *Emerg. Infect. Dis.* **2015**, *21*, 490–492. [CrossRef]
141. Balikagala, B.; Mita, T.; Ikeda, M.; Sakurai, M.; Yatsushiro, S.; Takahashi, N.; Tachibana, S.I.; Auma, M.; Ntege, E.H.; Ito, D.; et al. Absence of in vivo selection for K13 mutations after artemether-lumefantrine treatment in Uganda. *Malar. J.* **2017**, *16*, 23. [CrossRef]
142. Lee, A.H.; Fidock, D.A. Evidence of a mild mutator phenotype in cambodian Plasmodium falciparum malaria parasites. *PLoS ONE* **2016**, *11*, e0154166. [CrossRef]
143. Paloque, L.; Witkowski, B.; Lelièvre, J.; Ouji, M.; Ben Haddou, T.; Ariey, F.; Robert, A.; Augereau, J.M.; Ménard, D.; Meunier, B.; et al. Endoperoxide-based compounds: Cross-resistance with artemisinins and selection of a Plasmodium falciparum lineage with a K13 non-synonymous polymorphism. *J. Antimicrob. Chemother.* **2018**, *73*, 395–403. [CrossRef]
144. Schirmer, R.H.; Coulibaly, B.; Stich, A.; Scheiwein, M.; Merkle, H.; Eubel, J.; Becker, K.; Becher, H.; Müller, O.; Zich, T.; et al. Methylene blue as an antimalarial agent. *Redox Rep.* **2003**, *8*, 272–275. [CrossRef] [PubMed]
145. McNamara, C.W.; Lee, M.C.S.; Lim, C.S.; Lim, S.H.; Roland, J.; Nagle, A.; Simon, O.; Yeung, B.K.S.; Chatterjee, A.K.; McCormack, S.L.; et al. Targeting Plasmodium PI(4)K to eliminate malaria. *Nature* **2013**, *504*, 248–253. [CrossRef] [PubMed]
146. Tiwari, N.K.; Reynolds, P.J.; Calderón, A.I. Preliminary LC-MS Based Screening for Inhibitors of Plasmodium falciparum Thioredoxin Reductase (PfTrxR) among a Set of Antimalarials from the Malaria Box. *Molecules* **2016**, *21*, 424. [CrossRef] [PubMed]
147. Vallières, C.; Avery, S.V. The candidate antimalarial drug mmv665909 causes oxygen-dependent mRNA mistranslation and synergizes with quinoline-derived antimalarials. *Antimicrob. Agents Chemother.* **2017**, *61*, e00459-17. [CrossRef]

antioxidants

MDPI

Review

Evolutionary Adaptations of Parasitic Flatworms to Different Oxygen Tensions

José de Jesús Martínez-González, Alberto Guevara-Flores and Irene Patricia del Arenal Mena *

Departamento de Bioquímica, Facultad de Medicina, Universidad Nacional Autónoma de México, Ciudad de Mexico 04510, Mexico; jjmtz@bq.unam.mx (J.d.J.M.-G.); guevarafa@bq.unam.mx (A.G.-F.)
* Correspondence: darenal@bq.unam.mx

Abstract: During the evolution of the Earth, the increase in the atmospheric concentration of oxygen gave rise to the development of organisms with aerobic metabolism, which utilized this molecule as the ultimate electron acceptor, whereas other organisms maintained an anaerobic metabolism. Platyhelminthes exhibit both aerobic and anaerobic metabolism depending on the availability of oxygen in their environment and/or due to differential oxygen tensions during certain stages of their life cycle. As these organisms do not have a circulatory system, gas exchange occurs by the passive diffusion through their body wall. Consequently, the flatworms developed several adaptations related to the oxygen gradient that is established between the aerobic tegument and the cellular parenchyma that is mostly anaerobic. Because of the aerobic metabolism, hydrogen peroxide (H_2O_2) is produced in abundance. Catalase usually scavenges H_2O_2 in mammals; however, this enzyme is absent in parasitic platyhelminths. Thus, the architecture of the antioxidant systems is different, depending primarily on the superoxide dismutase, glutathione peroxidase, and peroxiredoxin enzymes represented mainly in the tegument. Here, we discuss the adaptations that parasitic flatworms have developed to be able to transit from the different metabolic conditions to those they are exposed to during their life cycle.

Keywords: platyhelminthes; Cestoda; oxygen tension; anaerobic metabolism; tegument; mitochondria

Citation: Martínez-González, J.d.J.; Guevara-Flores, A.; del Arenal Mena, I.P. Evolutionary Adaptations of Parasitic Flatworms to Different Oxygen Tensions. *Antioxidants* **2022**, *11*, 1102. https://doi.org/10.3390/antiox11061102

Academic Editor: Serge Ankri

Received: 29 March 2022
Accepted: 29 May 2022
Published: 31 May 2022

Publisher's Note: MDPI stays neutral with regard to jurisdictional claims in published maps and institutional affiliations.

1. Introduction

A crucial moment in the evolution of the planet was the change from an anoxic primordial atmosphere to one rich in oxygen (O_2). Currently, it is accepted that this change began with the origin of cyanobacteria and the development of oxygenic photosynthesis [1,2] and continued later with the emergence and diversification of photosynthetic pigments in different types of algae and plants [3,4]. The latter allowed the atmosphere to accumulate oxygen over millions of years and, after some fluctuations in the Carboniferous period, to reach its current level (around 21%) [5,6]. In turn, this process influenced the evolution of life on Earth, since geological and fossil evidence has allowed us to infer that the increase and accumulation of O_2 in the atmosphere gave rise to the establishment of specific ecological niches. Hence, some of the organisms adapted and developed in the increasingly aerobic conditions, whereas others established themselves in a microaerophilic environment and still others in sites where fully anaerobic conditions predominated [7,8]. There is even a proposal that considers that the first organisms to emerge were anaerobes, which allowed them to adapt and eventually live in hypoxic conditions [9,10]. In fact, it is recognized that anaerobic glycolysis is an ancestral metabolic pathway as it is present in these first living beings, which in turn allowed them to produce ATP at substrate-level phosphorylation in the absence of O_2 [11,12]. Moreover, the late oxygen accumulation in the atmosphere caused the emergence of aerobic-type energy metabolism, in which the now available O_2 is the final electron acceptor, enabling the ability to obtain a greater amount of ATP from a glucose molecule. This would favor the appearance and diversification of new metabolic pathways [13], which in turn enabled organisms to evolve into more complex forms.

Oxygen is one of the two main products derived from oxygenic photosynthesis. This molecule is chemically interesting since it has two unpaired electrons in the anti-union orbitals (with the same spin), which makes it difficult for it to oxidize another molecule and accept two electrons simultaneously [14]. This restriction in molecular oxygen (that is, the dioxygen di-radical) significantly decreases its reactivity; however, exposure to physical factors such as high temperatures or some source of radiation can cause a change in the spin, thereby decreasing said restriction, which favors the acceptance of one electron at a time (that is, oxygen becomes more reactive), making reactions very slow [15]. Due to this, an incomplete reduction of O_2 can occur, generating a series of molecules that, when accumulated, can cause adverse effects in the cell. These molecules are generically called reactive oxygen species (ROS); they include the superoxide radical anion ($O_2^{\bullet -}$), hydrogen peroxide (H_2O_2), and the hydroxyl radical ($HO^{\bullet}$), among others.

The purpose of this review is to analyze some of the basic adaptations presented by parasitic flatworms, a specific group of organisms that face changes in the concentration of oxygen (and related molecules) throughout their life cycle.

2. Parasitic Flatworms and Oxygen Availability

2.1. General Information about Flatworms

The Platyhelminthes, also known as flatworms, are dorsoventrally flattened organisms with bilateral symmetry [16] that morphologically constitute a heterogeneous group. Throughout history, the phylum Platyhelminthes has been the subject of multiple controversies, especially regarding its phylogeny since morphological data of some species can be troublesome [17]. We currently have nuclear and mitochondrial genomes and transcriptomic analyses that allow us to reach a consensus. The phylum Platyhelminthes is currently included within the Lophotrochozoa supergroup, which includes other invertebrates, as well as annelids and mollusks [18,19]. It is interesting to note that another phylum, which was traditionally thought to be close to flatworms, the phylum Nematoda, is included, together with arthropods and related groups, in the supergroup Ecdysozoa [18]. This is highly important to consider to understand the morphophysiological differences between flatworms and nematodes.

Already within the phylum Platyhelminthes is the paraphyletic group Turbellaria, which includes all free-living flatworms. These are found in aqueous environments, both in salt and fresh water, although there are some adapted to terrestrial environments with high humidity. An important characteristic that is present in the members of this group is that their external surface has a simple epidermis, composed of a single layer of columnar epithelium located on top of a basement membrane and several layers of muscle. This epithelium usually has cilia, which are used by these worms to swim in waterbodies or to glide over the substrate [20].

On the other hand, we have the monophyletic group Neodermata, whose innovation is replacement of the simple epidermis of the turbellarians by the presence of a syncytial-type tegument that is formed by extensions of cells that are below the basement membrane and that fuse together in the tegument creating a syncytium. This characteristic is considered an adaptation to parasitic life and is so important to understand their physiology that it is a criterion for defining this group, which is composed exclusively of parasitic organisms. Three types of flatworms with a clearer phylogenetic relationship are currently recognized [21]. On the one hand, we have the group of ectoparasitic flatworms of the class Monogenea, whose representatives are characterized by having a single host throughout their life cycle [22]; they usually live on the gills or skin of aquatic vertebrate animals. On the other hand, we have the flatworms that are endoparasites and are grouped within the class Cestoda and class Trematoda. These last two are considered the most successful parasites due to the great variety of vertebrates they infect; they are responsible for many diseases of livestock animals and humans [23–25]; thus, they are of great medical and economic relevance [26,27].

2.2. *The Complex Life Cycles of the Trematoda and Cestoda*

This group of organisms presents complex biological cycles in which some may have free-living stages or need one or more intermediate hosts of invertebrate or vertebrate origin to finally invade a definitive host in which they move until they find the tissue or organ, where they settle and reproduce sexually.

Trematodes, also known as flukes, have as their main characteristic the retention of the cecum, although they can also absorb nutrients and carry out gas exchange through their body wall (also called tegument) [28]. Their first intermediate hosts are generally mollusks, and adults can have a wide variety of diets, from blood to epithelia [29]. Some of the species in which pioneering studies on the biochemistry and immunology of trematodes have been made belong to this group: *Clonorchis sinensis*, *Fasciola hepatica*, and *Schistosoma mansoni*.

Cestodes are one of the most successful groups within the parasitic flatworms. This is due in part to their complete adaptation to parasitic life, including the total absence of an internal digestive system, the lack of an intermediate free-living form, the appearance of structures specific for attachment to the intestine of the definitive host, and serial repetition of a hermaphroditic reproductive complex [30]. This group has been problematic to study due to the difficulty in accessing and maintaining the biological material, the fragility of the specimens outside their hosts, and the contradictory information from the first studies [17]. Within this group, we have well-known species such as *Hymenolepis diminuta*, *Echinococcus granulosus*, and *Taenia solium*.

A representative life cycle of this group is found in *T. solium* (Figure 1). After the ingestion of feces contaminated with embryonated eggs (also called hexacanth larva) by a pig, the protective cover of such eggs is eliminated and the larval oncosphere form emerges. This oncosphere crosses the intestinal mucosa and migrates through the systemic circulation to lodge in various tissues, with a preference for the skeletal muscle and the nervous system. Already there, the larva develops to its metacestode form (also known as cysticercus), where it can stay for years, asymptomatically. Finally, when the definitive host (man) ingests pork meat contaminated with cysticerci, it carries out its last metamorphosis, which is a distinctive characteristic of cestodes [31]. It consists in the activation of the larva by means of pepsin and stomach acid, as well as the bile salts, of the definitive host, causing evagination of a fixing structure, the scolex, which will allow it to anchor itself to the intestinal epithelium. At this point, the adult tapeworm form rapidly begins to develop and matures sexually to generate a series of hermaphrodite structures called proglottids, each of which contains a complete set of male and female reproductive organs that mate with their other proglottid counterparts. Eventually they are filled with millions of fertile eggs (becoming gravid proglottids) that will detach and leave the host along with the feces to later be eaten by the pig to close the life cycle. Occasionally, man can accidentally ingest the eggs of *T. solium* that give rise to the development of the larva (metacestode) and that produces cysticercosis.

At this point, it is possible to assume that, during their free-living phase (in the case of trematodes) or the intermediate step between hosts, these endoparasitic flatworms are exposed to a higher O_2 tension that they can face inside the hosts cells, hence, their energy metabolism will preferably be aerobic [32]. During their transit and accommodation in the host, they are exposed to variable concentrations of O_2, so their anaerobic energy metabolism is expected [33–36] (Figure 2 and Table 1). For example, in their adult state, tapeworms settle in the intestine of their vertebrate host. In this organ, the partial pressure of O_2 (pO_2) can vary from 0 to 16 mmHg, with a three-times higher pressure in the mucosa than in the intestinal lumen, where it can reach zero [37]. Additionally, the presence of O_2 is also modified according to the postprandial state because an increase has been observed in O_2 when the digestion process begins as well as in this tissue's blood supply [38].

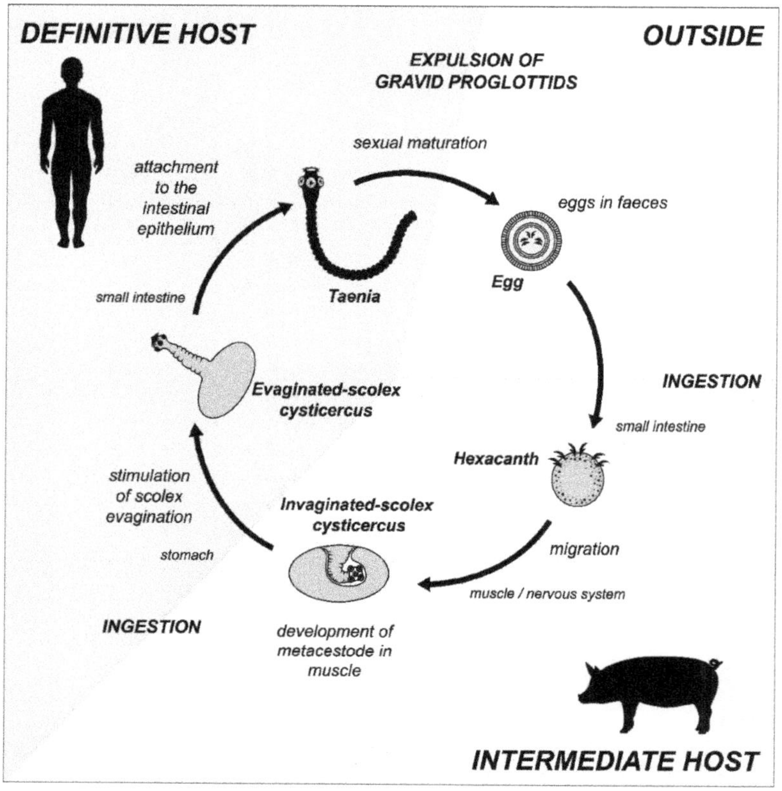

Figure 1. Life cycle of *Taenia solium*. Parasite stages and its migration through the interior of its intermediate host (e.g., pig, in green) and its definitive host (e.g., man, in purple), is illustrated.

Table 1. Oxygen concentration during the life cycle of *Taenia solium*.

Specie	Biologic Form	Host	Localization in Host	Oxygen Concentration		References
				pO_2	mmHg	
T. solium	Egg	Enviroment	NA	21.1	160	[33]
	Oncosphera	Pig	Duodenum	5.9	45	[34]
			Blood capillaries *	5.3–13.2	40–100	[33]
	Cysticerci		Muscle	4.9	37.5	[34]
			Brain	3.9	30	[34]
	Taenia	Human	Duodenum **	7.9	60	[35]
			Ileum **	1.3	10	[35]
			Cecum **	0	0	[35]
	Cysticerci		Muscle (in rest)	3.6–3.9	27–30	[36]
			Brain (grey matter)	2.1–5.3	16–40	[36]
			Brain (white matter)	3.2–4.4	24–33	[36]

* Human measurements. ** Mice measurements.

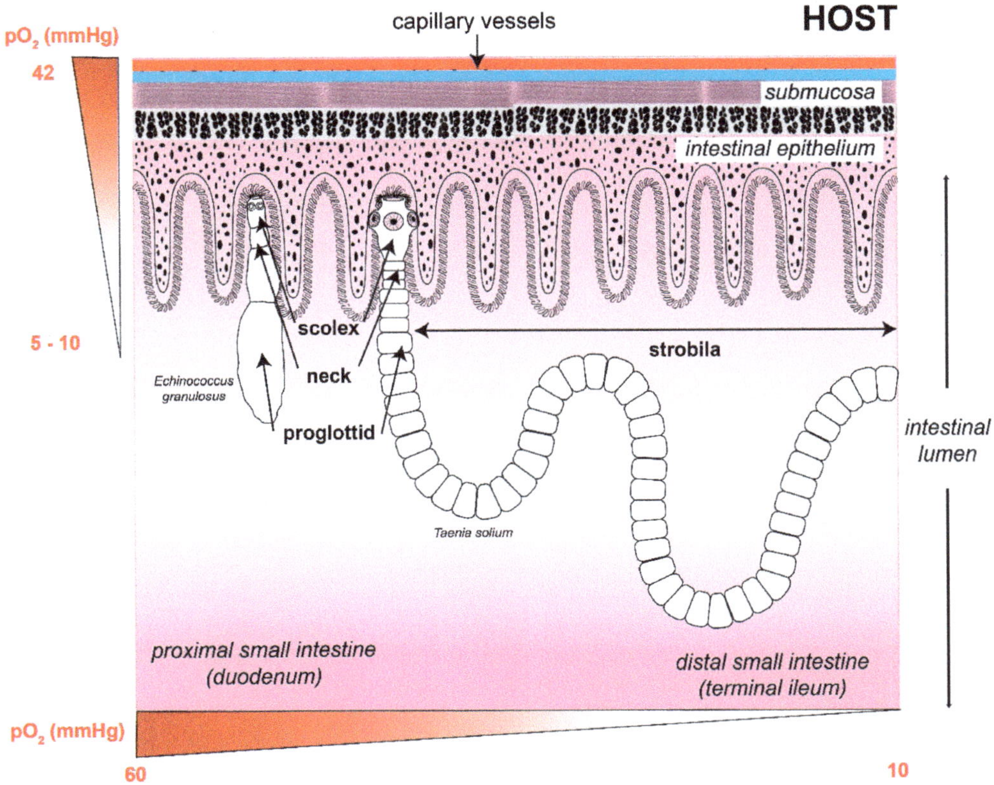

Figure 2. Adult form of the cestodes is exposed to the intestinal oxygen concentration. The image shows a structural drawing of the adult tapeworm form of *Echinococcus granulosus* (left; size range 2–7 mm) and *Taenia solium* (right; size range 2–7 m), attacking the intestinal epithelium of their definitive host. Oxygen tension in the intestinal tissue decreases the further away parasites are from the intestinal capillaries, while the oxygen concentration in the intestinal lumen decreases as parasites move towards the colon, where the environment is practically anaerobic. Oxygen concentrations were obtained from references [33–36]. The size of the parasites was obtained from the Laboratory Identification of Parasites of Public Health Concern website (https://www.cdc.gov/dpdx/ (accessed on 17 February 2022)).

The aforementioned indicates that the different habitats occupied by these parasites during their life cycle determine their energy metabolism and their transition from an aerobic to an anaerobic metabolism [39–41]. An example of this adaptation has been reported in trematodes such as *F. hepatica*. It was observed that its free-living larva has an aerobic metabolism, but when it invades the bile ducts of the vertebrate host, its metabolism is basically anaerobic [42,43] (see below).

3. Adaptations of Parasitic Flatworms to Changes in Oxygen Tension

3.1. Ultrastructural Adaptations

3.1.1. Body Wall (Tegument)

The parasitic flatworms of the Neodermata group have a glycocalyx rich in carbohydrates in their external part of the membrane that limits the tegument, which consists of a simple syncytium that covers the entire surface of these worms [44]. However, this tissue results from the fusion of cytoplasmic projections of parenchymal cells (also known as cytones) that are found below the basement membrane and whose function is to provide a constant flow of proteins and other molecules to the tegument [45] (Figure 3). Ultrastructural adaptations can be present such as microtriches in the case of cestodes, which increase the surface area of the parasite allowing a greater exchange between it and the host, as well as recognition mechanisms through the glycocalyx [46].

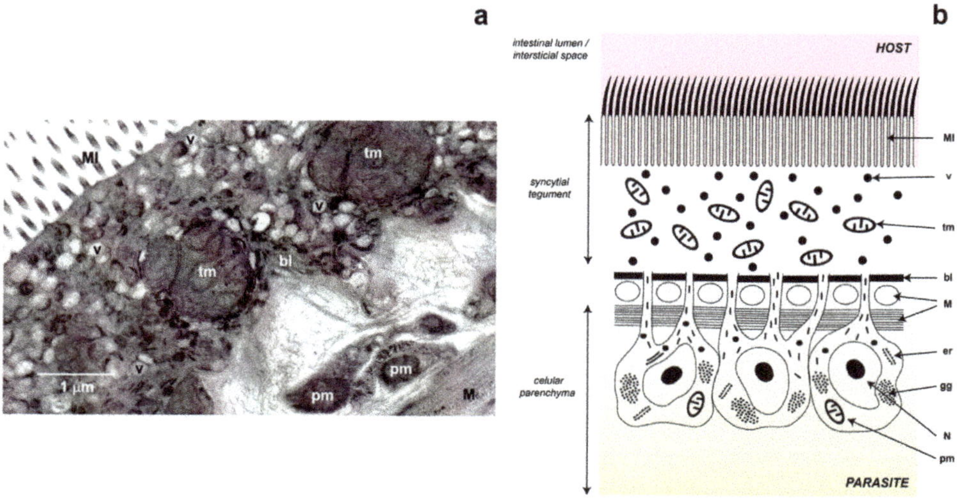

Figure 3. Tegument of the cestodes. Panel (**a**) represents a photograph of the tegument of *Taenia crassiceps* cysticercus obtained by transmission electron microscopy. Panel (**b**) is a schematic representation of the tegument of cestodes. *Abbreviations*: bl, basal lamina; er, endoplasmatic reticulum; gg, glycogen granules; M, muscle; MI, microtriches; N, nucleus; pm, parenchymal mitochondria (anaerobic mitochondria); tm, tegumental mitochondria (aerobic mitochondria); v, vesicles.

In the case of the phylum Platyhelminthes, due to their flattened morphology, gas exchange as well as nutrient uptake can take place through the body wall because these organisms lack a circulatory system. Naturally, the uptake of O_2 occurs by simple diffusion and is carried out through this structure. At this point, it is important to note that there is a gradient in the concentration of oxygen in the parasite, where the tegument, being the most exposed region, presents the highest pO_2, whereas the oxygen concentration decreases when entering the internal tissues of the parasite [47]. In addition to this, there is an important relationship with the size of the parasite, as in *F. hepatica* [48].

The tegument is essential for the success of these parasites and, in fact, it plays a key role in the evolution of parasitism in these animals due to the inseparable host-parasite bond that is generated [49]. In practice, it is a barrier that protects the parasite from the host's immune system [50,51] and from the hostile conditions in the digestive tract, blood, or other organs [46]. Additionally, it serves to house the molecular systems that will serve multiple purposes such as migration through the host's body, antioxidant defense, repair of damage caused by the attack of the immune system, and evasion and modulation of the immune system response [52]. We will deal with these points at the end of this review.

3.1.2. Diversity of Mitochondria

Parallel to the appearance and enrichment of O_2 in the atmosphere and, consequently, the diversification of living organisms, diverse types of mitochondria were also generated, from mitoplasts to aerobic mitochondria [53,54].

Palade in 1953 [55] recognized that variation in the size, shape, and internal organization of mitochondria seems to reflect their physiological and biochemical differences in different cells of an organism. It is now well known that tissues with a high demand in their energy metabolism contain several hundred mitochondria per cell and that they have many densely packed cristae, whereas in tissues with a lower energy demand, mitochondria with fewer cristae and smaller in size are present [56].

According to their metabolism, two large groups of mitochondria can be distinguished: aerobic and anaerobic. Aerobic mitochondria in the presence of O_2 carry out the Krebs cycle and oxidative phosphorylation. In contrast, anaerobic mitochondria are structurally similar to typical mitochondria but function in the absence of O_2; although their enzymatic repertoire is not very different from that of aerobic mitochondria [9,57], because many of these enzymes can catalyze the reverse reaction under certain conditions. Some enzymes of the Krebs cycle participate in these two metabolic pathways, such as fumarase and succinyl-CoA synthetase, whereas other enzymes participate in anaplerotic pathways such as phosphoenol-pyruvate kinase (PEPK) and mitochondrial malic enzyme (mME) (Figure 4). The above allows these two metabolisms, aerobic and anaerobic, to occur almost simultaneously or in different regions of a parasite, with pO_2 ultimately determining their prevalence [58].

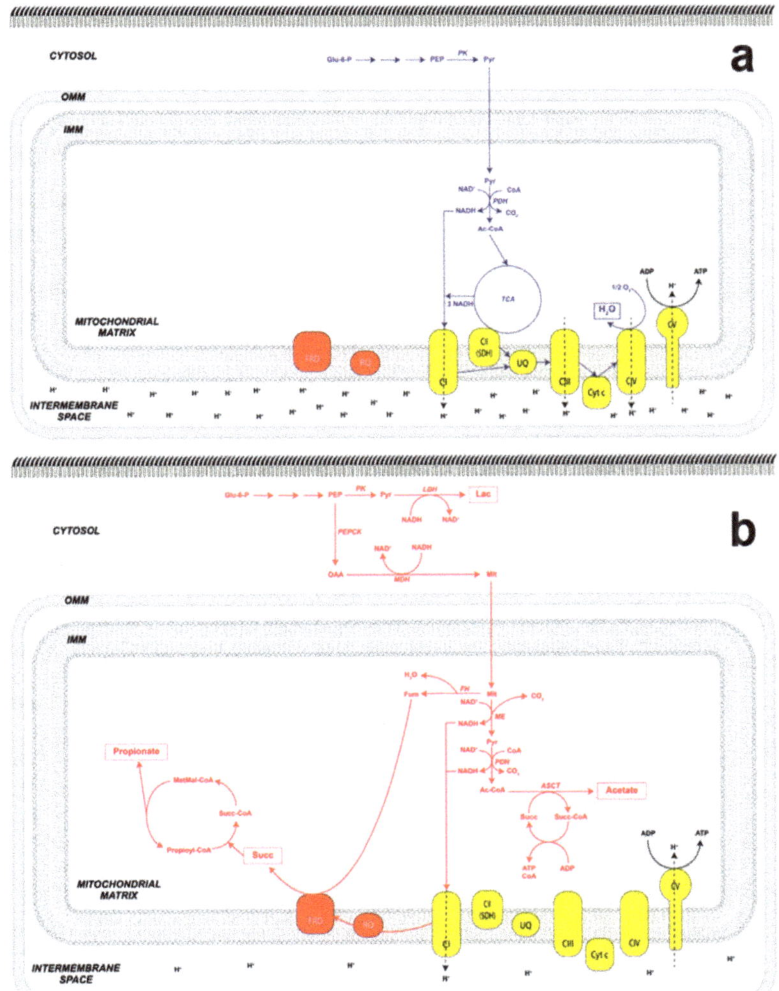

Figure 4. Aerobic and anaerobic energy metabolism in parasitic flatworms. Representation of the electron flow in aerobic metabolism: Panel (**a**), while Panel (**b**), represents the electron flow corresponding

to anaerobic metabolism. Abbreviations: OMM, outer mitochondria membrane; IMM, inner mitochondria membrane; Ac-CoA, Acetyl coenzyme A; ASCT, acetate succinate-CoA transferase; cytc, cytochrome *c*; Fum, fumarate; FH, fumarate hydratase (fumarase); FRD, fumarate reductase; Glu-6-P, glucose 6-phosphate; Lac, lactate; LDH, lactate dehydrogenase; MDH, malate dehydrogenase; ME, mitochondrial malic enzyme; MetMal-CoA, methylmalonyl coenzyme A; Mlt, malate; PEP, phosphoenol pyruvate; PEPCK, phosphoenol pyruvate carboxykinase; PDH, pyruvate dehydrogenase complex; PK, pyruvate kinase; Pyr, pyruvate; OAA, oxaloacetate; RQ, rhodoquinone; Succ, succinate; Succ-CoA, succinyl-coenzime A; SDH, succinate dehydrogenase; TCA, tricarboxylic acid cycle; UQ, ubiquinone.

Regarding the pO_2 to which they are exposed, structurally and metabolically different mitochondria have been described in the same organism related to the gradient that is established when O_2 diffuses from the tegument towards the cell parenchyma [59]; thus, it is expected that the tegument presents a higher concentration of O_2 than the parenchyma [60]. In 1967, Lumsden [61] described the presence of a heterogeneous population of mitochondria in the cestode *Lacistorhynchus tenuis*, where he reports that the parenchyma cells are larger despite occurring in smaller numbers and having fewer cristae compared to mitochondria of the tegument. This differential distribution of the types of mitochondria in the tissues of the cestodes was subsequently corroborated in the *Taenia crassiceps* metacestode [62], where we determined the aerobic metabolism of the mitochondria present in the tegument of the cysticercus and were able to observe that in addition to being very numerous, they have highly developed cristae (Figure 3). The presence of aerobic mitochondria in the tegument is not exclusive to cestodes; Takamiya observed the presence of several types of mitochondria in the trematode *Paragonimus westermani* [63]. On the one hand, in the tegument, he described numerous mitochondria with a larger number of cristae and a greater amount of cytochrome *c* oxidase activity (a marker of aerobic metabolism) compared to that of the parenchyma cells, where this author reported the presence of two types of mitochondria similar in size but one being completely anaerobic and the other one only partially so.

3.2. Metabolic Adaptations

3.2.1. Oxygen Carriers and Storage

Due to its active metabolism and energy needs that require a high production of eggs, a determining factor in the biology of parasitic flatworms is the availability of oxygen, both in trematodes and in cestodes.

As a consequence of the low solubility of O_2 in aqueous medium [64], some animals have developed a whole repertoire of respiratory pigments, which are metalloproteins and whose metallic element allows them to bind to O_2 temporarily to transport it through the circulatory system and to distribute it in body tissues [65]. Thus, for example, mammals depend on myoglobin (Mb), a monomeric protein associated mainly with cardiac and skeletal muscle, and on hemoglobin (Hb), a tetrameric protein contained in erythrocytes; both present a coordinated iron atom in a tetrapyrrolic chemical structure [66].

Several homologues of myoglobins have been reported in vertebrates: androglobin (Adgb), neuroglobin (Ngb), globin X (GbX), myoglobin (Mb), and cytoglobin (Cygb) [67]. However, these present an unequal distribution among the groups of flatworms. In the case of the trematodes, *F. hepatica* and *P. westermani*, an Mb with a high affinity for O_2 and a low K_d, was initially described [68], similar to that reported in the nematode *Ascaris suum* [69,70]. In the case of cestodes, until 2002, Mb could be co-purified in the cysticercus of *T. solium*, possibly associated with muscle fibers of the subtegumental area [71]. Currently, through genomic and phylogenetic analyses, the presence of globins belonging to each of the GbX, Ngb, and Mb subfamilies was deduced in trematodes, whereas the cestode species had an Ngb-like single protein, and in turbellarians it was not possible to detect this subfamily, but the GbX and Mb-like proteins [72] were identified.

It is widely accepted that myoglobins, due to their high affinity for O_2, function mainly as storage rather than for oxygen transport [73]. However, other additional functions have begun to be proposed for this type of molecule, such as reserves of the heme group (associated with egg production) or to serve as a ROS detoxifier [72]. For example, in the adult trematode *C. sinensis*, five different types of globins (CsMb 1–5) were identified, of which only CsMB-1 was found in the subtegumental area (parenchyma) and was the only one to respond to stimuli by exogenous O_2, whereas the other globins were located exclusively in sexual organs and intrauterine eggs, which supports the participation of some globins in other specialized non-respiratory tasks.

Nevertheless, a high pO_2 at the parasite's site does not necessarily imply that aerobic metabolism is present. An example of this is observed in *S. mansoni*, whose adult form lodges in the portal vein and whose metabolism has been reported to be preferably anaerobic, although this is not exclusive [74,75]. In contrast, the exceptionally high affinity for oxygen presented by the Hb of the fluke *Ophisthorchis viverrini* allows it to live in a practically anaerobic environment, such as the bile ducts [76].

3.2.2. HIF and the Detection of Oxygen in the Environment

As mentioned above, in most mammalian tissues there is between 2% and 9% O_2 (an average of 40 mm Hg). In this sense, hypoxia is usually defined as $\leq$2% O_2, whereas severe hypoxia (or anoxia) is defined as $\leq$0.02% O_2 [77]. Therefore, it is important to have mechanisms that detect pO_2 in real time to allow the cell to respond appropriately. The change in oxygenation is sensed by HIF (hypoxia-inducible factor), a transfer factor that in hypoxic conditions induces the change from aerobiosis to anaerobiosis. This, in turn, brings about a cascade of events such as the overexpression and activation of enzymes like lactate dehydrogenase (LDH), which is necessary for lactate formation and, simultaneously, activating pyruvate dehydrogenase kinase (PDHK) to prevent pyruvate production [58].

Apparently, the mechanism mediated by HIF and related proteins is highly conserved in animals, making it possible to detect cnidarians and sponges even in basal groups of metazoans [78]. In its active form, HIF is a heterodimer consisting of HIF-1α and HIF-1β. In the case of parasitic flatworms, it is expected that, due to their exposure to different concentrations of O_2, during their life cycles, HIF plays a relevant role. In fact, the HIF-1α and HIF-1β proteins have already been characterized in the parasitic nematode *A. suum* [79]. However, it was not until 2019 that a HIF-1α homologue (as well as other associated genes) could be isolated and characterized in the trematode *C. sinensis* (CsHIF-1α) [80]. As expected, CsHIF-1α was highly induced in adults under hypoxic conditions in vitro. Interestingly, CsHIF-1α was sensitive to changes in nitrite and nitric oxide, hence, the authors suggest that these molecules, together with O_2, participate in the induction of the response to hypoxia in this organism.

3.2.3. Aerobic Metabolism

In principle, we must recognize that glucose is the main source of energy used by parasitic flatworms (both in adult and larval forms [81]), while glycogen formation is their main energy conservation strategy [82–87]. In fact, in the case of glycogen, it has been previously identified through histochemical techniques and later through transmission electron microscopy, where many glycogen granules are observed that can serve in cestodes as a source of energy in fasting situations as reported in *T. solium* tapeworms [88] and the metacestode (traditionally also known as *Cysticercus cellulosae*) [89].

Therefore, under aerobic conditions, these organisms will use the traditional pathways to obtain reducing power through the oxidation of glucose (glycolysis and the Krebs cycle) and the subsequent production of energy coupled to oxidative phosphorylation (OXPHOS) (Figure 4) [90]. In both trematodes [91] and cestodes [92], it has been possible to identify the genes encoding enzymes of each of these pathways at the genome level. However, transcriptomic analyses have shown a differential expression of these enzymes in the tissues of the parasite, as demonstrated in the metacestode of *Echinococcus granulosus* in which the expression of enzymes from fermentative pathways associated with the germ layer and the gluconeogenic pathway associated with both the germinal layer and the protoscolex were detected [83]. Additionally, the expression can be influenced by experimental conditions. Fraga et al. [93] were able to successfully detect all the metabolites associated with the Krebs cycle in the metacestode of *T. crassiceps* under in vivo conditions; thus, it was inferred that this pathway is complete in the cysticercus. In this same parasite, we characterized the aerobic metabolism in the mitochondria of the tegument [62,94].

In the case of the metabolic pathways of lipids and proteins, there are great changes, which may be due to the adaptation to the conditions of a parasite related to what the host provides. In 2013, Tsai and a large team of collaborators reported the massive sequencing and comparison of four cestode genomes (*E. granulosus*, *Echinococcus multilocularis*, *Hymenolepis microstoma*, and *T. solium*) [92]. Basically, they reported a significant reduction in the metabolic capacity of these organisms, as well as the presence of specialized elements in the uptake of nutrients.

3.2.4. Anaerobic Metabolism

Glycolysis can be considered a universal pathway by which many organisms can obtain energy. Its final product, pyruvate (Pyr), can be used in other alternative pathways known as fermentative pathways; they occur in the absence of O_2 and allow the NADH generated during glycolysis to be oxidized to NAD^+, a necessary substrate for, so that this path can continue (Figure 4).

A classic adaptation of anaerobic metabolism is lactate fermentation, in which pyruvate is reduced to lactate (Lac) by the enzyme lactate dehydrogenase (LDH) using electrons from NADH. It is now known that this pathway is also used in parasitic helminths. Direct evidence of its presence is the secretion of Lac into the medium, as has been reported in cestodes such as *Moniezia expansa* [95,96], *E. granulosus* [83], and *E. multilocularis* [86].

In addition to Lac secretion, the secretion of other reduced compounds such as succinate (Succ), acetate, and propionate (PPO) has been reported. This has been reported in *M. expansa* [32], and confirmed in *E. granulosus* [83], *T. crassiceps* [84], and *E. multilocularis* [86], where the main secreted product was succinate; these products are the result of a pathway known as malate dismutation (Figure 4).

Malate dismutation is the main anaerobic pathway present in parasitic platyhelminths [43,97–99], and has as its final products a reduced molecule and an oxidized (as occurs in the reactions called dismutation). During this process, phosphoenolpyruvate (PEP) produced during glycolysis is carboxylated to oxaloacetate (OAA) by PEP carboxykinase (PEPCK), producing ATP by substrate-level phosphorylation. OAA is reduced to malate (Mlt) through the cytosolic malate dehydrogenase (cMDH), which has NADH produced during glycolysis as another substrate. Subsequently, Mlt enters the mitochondria and, on the one pathway, through fumarase (also named fumarate hydratase, FH), it produces fumarate (Fum) and, by another, the mitochondrial malic enzyme (mME) oxidatively decarboxylates it to Pyr that can later generate acetate (Figure 4).

The fumarate produced is a substrate for the enzyme fumarate reductase (FDR) that reduces it to Succ; this is the main product of electron secretion in cestodes [95,96,100,101]. To not accumulate this metabolite and to maintain its redox balance, Succ is secreted into the surrounding medium, as reported in the culture medium of *T. crassiceps*, as well as in the cysticerci of *T. solium* removed from pig brains [84,96]. Additionally, it has been reported that Succ can generate PPO. Recently, two alternative pathways for propionate formation have been reported: (a) from succinyl-CoA to methylmalonyl-CoA that is decarboxylated to generate ATP and propionyl-CoA, which, in the presence of Succ, releases PPO and acetylates succinate; this process appears to occur under prolonged anaerobic conditions [102]; and (b) via Lac accumulation and its transformation to propionyl-CoA releasing PPO and regenerating CoA [103]. This contrasts with what Ritler et al. reported, after they could not detect propionate as a secretion product in *E. multilocularis* [86].

The other malate dismutation reaction is the one that produces acetate where, as mentioned, the pyruvate generated in the mitochondrial matrix, through the mME (and in addition NAD(P)H is generated) [98,101] and through the pyruvate dehydrogenase (PDH) complex, is oxidatively decarboxylated to generate acetyl-CoA and NADH. Finally, through the enzyme acetate-succinate-CoA transferase (ASCT), CoA is transferred to Succ, producing acetate and succinyl-CoA [104]. A search of available genomic databases indicates the presence of ASCT genes in the flukes *S. mansoni*, *P. westermani*, *C. sinensis*, and *F. hepatica* [105], as well as in the nematodes *Ostertagia ostertagi*, *Anisakis simplex*, and *Brugia malayi*. Although the presence of the gene encoding for ASCT in cestodes has not been reported, it is possible to suggest the presence of the enzyme (or an analogous pathway) since the presence of acetate has been reported as an end-product of anaerobic respiration in both *T. solium* [84] and *E. multilocularis* [86].

The NADH generated in the previous reactions transfers its electrons to the mitochondrial complex I NADH-rhodoquinone oxidoreductase which, contrary to what happens in aerobic conditions, reduces the rhodoquinone (RQ) instead of reducing the ubiquinone (UQ). This transfer of electrons is favorable because RQ has a redox potential of $E_m{}' = -63$ mV, which is lower than that of UQ ($E_m{}' = +110$ mV) [106,107]. The RQ donates electrons to the FDR to generate succinate from fumarate [108]. The measurement of FDR activity [39,109] is indicative of anaerobic metabolism [47,99], whereas the measurement of cytochrome *c* oxidase activity, of SDH, as well as the sensitivity of the electron transport chain (ETC) to different inhibitors (such as cyanide), are indicative of aerobic metabolism [62,110]. One point to highlight is that when NADH-rhodoquinone oxidoreductase participates, protons are translocated from the mitochondrial matrix to the intermembrane space, which, in turn, maintains a chemiosmotic gradient and generates ATP even in the absence of O_2 (Figure 4).

In anaerobic metabolism [39], FRD performs the reverse reaction of succinate dehydrogenase (SDH) [40]. Both enzymes, SDH and FRD, are heterotetramers that share: (a) subunit 1 (Fp) that contains flavin-adenine dinucleotide (FAD); (b) subunit 2 (Ip) with three Fe-S centers; and (c) two subunits CybL and CybS, which maintain, on the one hand, binding to the inner mitochondrial membrane and, on the other, binding to the corresponding quinone [111,112]. In *A. suum*, there are isoforms in two of the four subunits as well; Fp and CybS are different between the aerobic larva and the anaerobic adult; no isoforms have been reported for the Ip and CybL subunits [113].

Considering the above, we can note that anaerobiosis-specific reactions are those catalyzed by FRD and ASCT. However, it is the presence of RQ that appears to be the only real difference between aerobic and anaerobic energy metabolism [114], as FRD expression has been described in cancer cells [115,116], while ASCT is an enzyme homologous to other transferases [105,117,118].

To recapitulate, in the cytosol of muscle cells under hypoxic conditions, lactate is produced by lactic acid fermentation. Unlike this, malate dismutation or malic fermentation has the following relevant aspects:

- It occurs in two cell compartments: cytosol and mitochondrion
- It is coupled to the generation of the proton gradient produced in complex I of the ETC, when the NADH coming from the formation of Acetyl-CoA and the one resulting from the activity of mME are oxidized (Figure 1).
- ATP is produced in the absence of O_2 since the redox potential difference between NAD^+/NADH ($E_m{}' = -320$ mV) and fumarate/succinate ($E_m{}' = +30$ mV) is sufficient for its synthesis [57,63]
- At the end of the pathway, several reduced compounds are obtained, including succinate, acetate, and propionate [119].

However, both aerobic and anaerobic metabolisms have the following aspects in common:

- The need for a final electron acceptor molecule
- Maintenance of redox balance

- Both are carried out in the mitochondrial compartments, which allows the formation of a proton gradient and therefore the synthesis of ATP.

Regardless of the type of energy metabolism, the redox balance is maintained. To keep it, organisms recycle their electron transporting coenzymes; thus, the number of reactions that produce NADH is equal to the reactions that consume it, or else, electrons are excreted in form to water, in aerobic organisms, and through succinate mainly in anaerobes [9].

3.3. Molecular Adaptations

The parasite-host relationship is, in most cases, a reciprocal interaction, in which the behavior of the parasite causes feedback in the host and vice versa. Because of this, analyzing this type of feedback mechanism is essential to understand the complex connections between animal behavior, ecology, and parasite evolution [120]. Up to this point, we can reflect on the convenience of using fermentative pathways in the maintenance of parasitic platyhelminths when aerobic respiration (consequently, settling in a place where an abundant supply of oxygen is available) would result in a greater supply of energy and, possibly, a higher metabolic rate. However, when dealing with parasitic forms, it is more important to establish the parasite in a strategic ecological niche that ensures a constant supply of substrates coming from the host (such as liver tissues in the case of flukes or the duodenum in the case of cestodes). Consequently, by having the resources secured, the parasites will concentrate on managing them [121].

One consequence of the foregoing is the parasite's need to interact with its host and be able to maintain its ecological niche by engaging in chemical communication with it. It is not surprising then that the parasite has developed and fine-tuned mechanisms to evade the immune response. For example, it is known that some adult schistosomes can live from three to 10 years in humans, despite the harsh intravascular environment and their constant exposure to the immune system [46]. In fact, it has been possible to verify a registry of patients infected with *S. mansoni* for more than 30 years and the case of a patient infected with *E. granulosus* for 53 years [121].

The feedback between parasitic flatworms and their hosts has been studied using model organisms. For example, the murine experimental model of cysticercosis has made it possible to evaluate the interaction of the host (mice) with the cysticercus of *T. crassiceps* during its proliferation in the peritoneal cavity. This made it possible to describe some of these complex interactions [122], like the importance of the genetic factors of the host (the murine strain used) in the establishment and proliferation of the parasite. Another interesting observation is the importance of the sex of the host as it has been observed that cysticerci grow preferentially in female mice, regardless of the strain of *T. crassiceps*. Apparently, this sexual dimorphism is mediated by hormonal factors, since estrogens favor and androgens hinder the asexual reproduction of cysticerci [123]. Interestingly, when such cysticerci are inoculated into male mice, a feminization phenomenon can be observed in which testosterone levels decrease and estradiol levels increase [124]. The last section of this review will discuss parasite-host feedback in greater detail.

3.3.1. Immune Response and Oxidative Stress in Parasitic Flatworms

Sources of Exposure to Reactive Oxygen Species

During establishment of the infection, the parasites induce a rapid immune response in the host, although it is nonspecific [125]. In general, this involves the activation of eosinophils, neutrophils, and macrophages, as well as the release of cytokines and the production of antibodies (IgE) [126]. These cells can produce large amounts of ROS and reactive nitrogen species (RNS), capable of directly destroying parasite cells. For example, liver flukes, such as *Opisthorchis viverrini*, induce chronic inflammation of the hepatobiliary system, exposing themselves to large amounts of ROS/RNS released by activated inflammatory cells [127]. Similarly, when an infection by the cestode *Taenia hydatigena* occurs in the peritoneum, an increase in the infiltration of small peritoneal macrophages responsible for a high production of nitric oxide (NO) can be observed, which harms the parasite and modulates the immune response [128]. However, the presence of immune cells induced by parasitic flatworms may be due to their participation in other processes such as wound repair caused by the migration of *F. hepatica* through the liver parenchyma [129].

The production of ROS/RNS due to the immune response has already been discussed in detail previously [130]. In general, the precursor of all ROS is the superoxide anion radical ($O_2^{\bullet-}$), which is generated in leukocytes through the integral membrane enzyme NADPH oxidase (NOX), and by transferring an electron from NADPH to O_2. $O_2^{\bullet-}$ can undergo a spontaneous dismutation reaction generating hydrogen peroxide (H_2O_2) and O_2. H_2O_2 can serve as a substrate for the enzyme myeloperoxidase (MPO) to generate the microbicidal compound hypochlorous acid (HClO). In the presence of transition elements such as ferrous (Fe^{2+}) or copper (Cu^+) ions, H_2O_2 can be reduced by the Fenton reaction, which produces the hydroxyl anion (HO^-) and the hydroxyl radical ($HO^\bullet$). The $HO^\bullet$ radical is highly reactive, so it can subtract electrons from other biomolecules, like proteins, changing their properties and biological activities, with DNA generating mutations and membrane lipids initiating the lipid peroxidation process. This damage can lead to altered metabolism and eventually cell death (Figure 5).

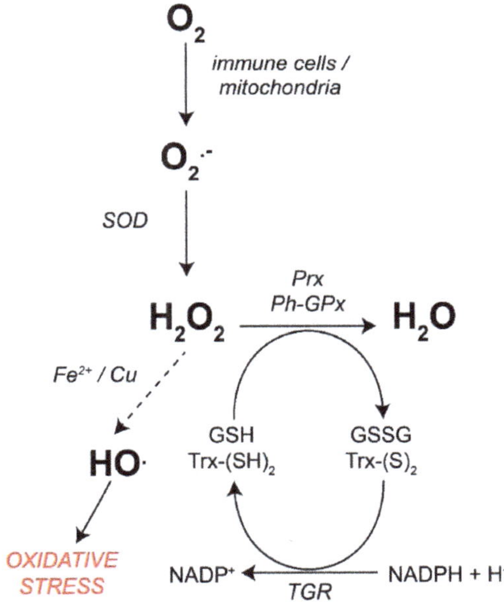

Figure 5. The antioxidant system of parasitic flatworms. *Abbreviations:* GSH reduced glutathione; GSSG, oxidized glutathione; H_2O_2, hydrogen peroxide; $HO^\bullet$, hydroxyl radical; O_2, molecular oxygen; $O_2^{\bullet-}$ superoxide anion radical; Ph-GPx, glutathione phospholipid peroxidase; Prx, peroxiredoxin; SOD, superoxide dismutase; TGR, thioredoxin-glutathione reductase; Trx-$(SH)_2$, reduced thioredoxin; Trx-$(S)_2$, oxidized thioredoxin.

It is important to clarify that another source of ROS is the metabolism of the parasites themselves, possibly due to their accelerated metabolism and the use of the malate dismutation pathway to obtain energy. It is in this pathway where the mitochondrial complex I continues to work, becoming an important place of ROS generation [131]. In mitochondria isolated from the tegument of *T. crassiceps*, a high production of H_2O_2 was recorded, unlike that observed in rat liver mitochondria. The high production of H_2O_2 associated with tegumental aerobic mitochondria has been observed with confocal microscopy in the cysticercus of *T. crassiceps* [94].

An important characteristic of these ROS and RNS is that they are short-lived intermediate products enzymatically synthesized by aerobic organisms and their clearance is regulated by enzymatic or non-enzymatic antioxidants. In this sense, there is a major co-evolutionary arms race competition between ROS production by the host and ROS scavenger by parasites; both closely related. For example, the production of H_2O_2 by hemocytes of the snail *Biomphalaria glabrata* when infected with *S. mansoni* sporocysts varies from population to population and in those snails with a high natural resistance against *S. mansoni* a higher production of H_2O_2 is observed, being preferentially infected by sporocysts with high levels of expression of their antioxidant systems [132].

Functioning and Localization of Enzymatic Antioxidant Systems in Parasitic Flatworms

(1) Superoxide Dismutase and Peroxidases

As mentioned before, $HO^{\bullet}$ is chemically very reactive, so the first line of defense of parasitic flatworms is to prevent its production, the controlled reduction of H_2O_2 to H_2O being vital. The classic enzyme that carries out this reaction in aerobic organisms is catalase (CAT), which is absent in the genomes of parasitic flatworms [92,133] but is conserved in their free-living counterparts [134]. Hence, these parasites depend on enzymes glutathione peroxidase (GPx) and peroxiredoxin (Prx) in addition to the superoxide dismutase (SOD) necessary for $O_2^{\bullet -}$ scavenging.

SOD is a family of metalloenzymes specialized in carrying out the dismutation of $O_2^{\bullet -}$, generating O_2 and H_2O_2. In animals, it is possible to identify two isoforms: (a) cytoplasmic SOD dependent on one atom of copper and another of Zn (Cu/Zn SOD); (b) mitochondrial SOD, dependent on a manganese atom (Mn SOD). Experimentally, it has been possible to isolate the enzyme [71] or clone the gene [135] from some parasitic flatworms, such as *T. solium*; thus, it was not surprising to confirm that, in genomes/transcriptomes, both trematode [125] and cestode [92,136] genes are present in both isoforms. Regarding their expression, their presence has been determined in all stages of the life cycle associated mainly with the tegument of these organisms [31], although a greater expression has been observed in the adult forms with respect to the larval [137].

GPx comprises isoenzymes that carry out the reduction of H_2O_2 to H_2O, requiring electrons from two molecules of the tripeptide glutathione (GSH), taking it to its oxidized disulfide form (GSSG). Although this family has many representatives in mammals [138], in trematodes [125] and cestodes it has only been possible to identify a single gene [92] corresponding to a selenium-dependent GPx. This GPx type turned out to be membrane integral and it is exclusive for the reduction of lipid hydroperoxides (Ph-GPx); thus, it makes sense that it has been located in the tegument [137] of the different stages of trematodes *S. mansoni* [125] and *Fasciola gigantica* [139], and of the cestode *E. granulosus* [140]. Like SOD, this enzyme has a differential expression that is a function of the pO_2 to which the parasite is exposed, finding its highest expression in the aerobic life cycle stages and its lowest in the anaerobic [141]. However, the overexpression of this enzyme responds to the exposure of exogenous oxidants, as shown in *C. sinensis* under in vitro conditions [142]. Cai et al., suggesting that Ph-GPx activity could be more focused on egg production than on the maintenance of the redox status [142].

Prxs are homodimeric proteins that catalyze the reduction of H_2O_2 and alkyl hydroperoxides to water and alcohol, respectively. The electrons that translocate frequently come from the 12 kDa protein thioredoxin (Trx), which is why these enzymes are also called thioredoxin peroxidases (TPx), although some isoforms can also obtain electrons from GSH. Despite the abundance of Prxs, their catalytic efficiency is lower than that of CAT or GPxs by one to three orders of magnitude [143]. However, this family of enzymes seems to be the most important in the H_2O_2 degradation process, both in trematodes and cestodes. In fact, in the genomes of *E. granulosus*, *E. multilocularis*, *H. microstoma*, and *T. solium*, it has been possible to identify three different genes encoding Prx 1–3 [92], which is consistent with the identification of three Prxs in *S. mansoni* [85]. Wang et al. reported that GPx activity in echinococcal cysts is practically undetectable, suggesting the relevance of Prxs in this parasite [144]. Using western blot analysis, it was found that in the trematode *O. viverrini*, OvTPx-1 is expressed in all stages of development; even if its location is different, depending on the isoenzyme [127]. For example, in adult flukes of *Schistosoma japonicum* it has been possible to locate Prx-1 in the tegument whereas Prx-2 has been found associated with the parenchyma, vitelline glands, and gastric epithelium [145]. In *H. diminuta*, a peroxidase-like activity was described in 1968 [146].

(2) Thioredoxin-Glutathione Reductase

To carry out peroxide hydrogen reduction, both the GPx and Prx (TPx) need to take electrons from GSH and Trx and then generate their oxidized forms, the substrates (i.e., GSSG and Trx-S_2, respectively). Reductases responsible for reducing these substrates again and helping to maintain the homeostatic redox cycle are significantly necessary. In mammals, there are two enzymes called glutathione reductase (GR) and thioredoxin reductase (TrxR), both are NADPH-dependent for the reduction of GSSG and Trx-S_2, respectively [147]. However, in 2001, Gladyshev et al. identified an enzyme capable of reducing both Trx-S_2 and GSSG in the mouse testis. This enzyme was called thioredoxin-glutathione reductase (TGR) [148].

Although this enzyme has been identified in vertebrate organisms, including humans (hTGR) [149], it is in the group of parasitic flatworms where its study has gained relevance. Early on, it was isolated and characterized in the trematode *S. mansoni* (SmTGR) [150], as well as in the cestodes *E. granulosus* (EgTGR) [151] and *T. crassiceps* (TcTGR) [152]. Later on, it was possible to deduce its presence in other parasitic flatworms thanks to genomic advances [92]. Unlike their vertebrate hosts and their free-living counterparts [153], these organisms depend exclusively on this enzyme to carry out the reduction of GSSG and Trx-S_2. Although it has been proposed that having an enzyme capable of reducing substrates belonging to two independent redox systems represents an evolutionary advantage [148], it is also possible to note that the dependence of parasitic flatworms on this enzyme makes it an excellent pharmacological target [154–161].

The genomes of trematodes and cestodes [76,87,92,129,162], as well as of the monogenean *Gyrodactylus salaris* [21], have corroborated the classic GRs and TrxRs in this group of parasites and have revealed that the TGR is encoded by a single gene; therefore, the cytosolic and mitochondrial forms must be generated by alternative splicing [163]. Despite being exactly of the same sequence, the environment in which it is located (either the cytosol or the mitochondrial matrix) affects its kinetic constants [164]. This enzyme is expressed in all stages of the life cycle, as reported in *S. mansoni* [125]. We previously reported that the in vivo inhibition of the TGR in *T. crassiceps* cysticerci is sufficient to compromise the viability of the parasite by altering its redox state and glutathione metabolism [165], which agrees with observations made when incubating schistosomula in the presence of anti TGR iRNA [166]. Due to its importance, TGR has been crystallized [167], promoting the search for drugs capable of inhibiting it [154–161].

(3) Glutathione-S-Transferase and Other Antioxidant Molecules

Glutathione-S-transferase (GST) is a family of highly conserved detoxifying enzymes that participate in the metabolism of many xenobiotics, although, in mammals, it has been found that it can participate in other relevant physiological processes such as the synthesis of leukotrienes, prostaglandins, and steroid hormones, as well as in amino acid catabolism and modulation of signaling processes [168]. GST can conjugate a wide range of substrates of an electrophilic nature, like the glutathione thiolate anion (GSH), to generate glutathionylated compounds, which are less reactive and more soluble and can be eliminated more efficiently by the cell. The enzyme can also detoxify by non-covalently binding to a series of hydrophobic ligands [169]. In the case of parasitic flatworms, it has been proposed that GSTs are essential for survival because they eliminate toxic and xenobiotic compounds derived endogenously or exogenously (generated or administered by the host) [170].

There are three families of GSTs with distinctive structural characteristics and different evolutionary origins: (a) cytosolic GSTs, (b) microsomal GSTs or also called MAPEG (Membrane associated proteins involved in eicosanoid and glutathione metabolism), and

(c) kappa-class mitochondrial GSTs [171]. However, cytosolic GSTs have been the most studied, identifying up to seven different classes in mammals: alpha, mu, pi, sigma, theta, zeta, and omega [172]. Parkinson et al. reported the presence of two sigma-type cytosolic GSTs in the larval form of *E. granulosus*; one mu class and one microsomal GST [83]. However, an in-depth analysis of the *C. sinensis* trematode genome suggests the presence of 12 cytosolic GSTs distributed in the mu, sigma, zeta, and omega classes; in addition to mitochondrial GST and microsomal GST [173]. This agrees with findings in cestode genomes, where an important presence of 10 genes for cytosolic GSTs of the mu class have been reported, in addition to two GST genes of the sigma class and one of the MAPEG class [92]. Interestingly, Nguyen et al. reported the presence of a new type of cytosolic sigma GST in the metacestode of *T. solium* (TsMsGST), specifically expressed in the cytosol of the scolex tegument and susceptible to praziquantel (a drug used against neurocysticercosis, and which does not normally interact with sigma GSTs) [170]. Finally, Iriarte et al. analyzed the genomic information available from several flatworm representatives and discovered the potential absence of omega-class GST in cestodes, contrary to observations in other species of *Schistosoma* as the planarian *Schmidtea mediterranea* [174].

Regarding their location and expression in parasitic flatworms, GSTs present complex patterns and depend on both the species and the stage of the life cycle in which they are found. For example, Mei and Loverde reported that, regardless of the parasite stage, when analyzing the transcript levels of different antioxidant enzymes in *S. mansoni*, GST transcripts were 100-times more abundant than GPx transcripts and 10-times more than SOD isoforms. However, when enzymes were localized by immunofluorescence in the adult fluke, GST isoforms were restricted to a reduced subpopulation of parenchymal cells as well as to immature germ cells in both males and females. In contrast, SOD and GPX isoforms were localized in the tegument [137].

Another example is observed in the mRNA expression of the omega class GSTs of *C. sinensis* (CsGSTo 1 and 2). It begins with a growing pattern in juveniles of two to four weeks of age, but there is no expression in the metacercaria form and, in contrast, they are overexpressed in eggs [175]. This was contrasted with immunodetection techniques, locating CsGSTo in the egg, vitelline follicles, seminal receptacles, and testes. Because the expression of CsGSTo remains at high levels, regardless of environmental stimuli, the authors propose that the expression of these GSTs is conditioned by sexual reproduction within the host and that the abundance of CsGSTo in the egg is a preparation for the hostile conditions that the parasite will face when expelled from the definitive host.

Other Detoxifying Proteins That Have Been Reported in Parasitic Flatworms:

(a) *Cytochrome p450*. This cytochrome has monooxygenase activity, which allows it to oxidize multiple exogenous molecules and contributes to their detoxification. Both in the flukes of *S. mansoni* [176] and *Opisthorchis felineus* [177], as well as in the genomes of the cestodes, only one copy of the gene has been found [92].

(b) *Phytochelatin synthase (PCS)*. This enzyme works together with GST in the detoxification of xenobiotics and in the uptake of potentially harmful transition metals [178] through the formation of glutathione biopolymers [179]. Originally reported in plants, the presence of a functional PCS was reported in *S. mansoni* [180] and its presence was later confirmed in the genomes of cestodes [92] as well as in the parasitic nematode *Ancylostoma ceylanicum* [181]. Previously, we reported the presence of three unknown thiols in an extract of low molecular weight thiols obtained from the cysticercus of *T. crassiceps* [165] and whose retention patterns coincide with those found in *S. mansoni* and are associated with phytochelatins of different sizes [180]. Significantly, both phytochelatins and the PSC gene are absent in the mammalian hosts, suggesting that it is an adaptation to parasitic life [182], although its specific function is still under discussion [183].

(c) *Myoglobin (Mb)*. We have previously talked about the capacity of Mb to store O_2. Other activities that Mb presents are peroxidase/dioxygenase, having the ability to interact with O_2 molecules such as NO, CO, and H_2O_2 [184]. Ren et al. reported that a globin from *C. sinensis* (CsMb) showed peroxidase activity and that it may be important for ROS detoxification because of its overexpression after incubation with exogenous H_2O_2 [185]. This was later corroborated by Kim et al., who showed that incubation of *C. sinensis* flukes under aerobic conditions or in the presence of nitric oxide or nitrite is sufficient to induce the expression of the gene encoding CsMb [72]. Interestingly, overexpression of CsMb was also observed when flukes were co-incubated with human cholangiocytes (bile epithelial cells).

(d) *Other enzymes.* Under conditions of oxidative stress, hydroxyl groups can be non-specifically oxidized to their aldehyde form. Similarly, reactions with radicals can lead to the formation of reactive carbonyls. As part of the characterization of the response of *E. granulosus* protoscolex to oxidative stress by exogenous H_2O_2, Cancela et al. reported high levels of a type of aldo-keto reductase (AKR), estradiol-17-beta dehydrogenase, and the enzyme carbonyl reductase 1 (CBR) [31]. AKRs are NADPH-dependent enzymes that can reduce aldehydes to alcohols [186]. On the other hand, CBR is an enzyme necessary to detoxify reactive carbonyls [187].

(4) Complexity of the Antioxidant Response

The different antioxidant enzymes do not work in isolation because to successfully face oxidative challenges, all systems must work together and simultaneously to avoid, as much as possible, the generation of highly toxic ROS such as the HO• radical. The latter maintains the peroxidases functioning by regenerating their electron source and repairing the damage caused during oxidative stress. In addition to this, there may be other factors that influence the antioxidant response, such as:

- *Time elapsed since the establishment of the infection.* Skrzycki et al. compared oxidative stress markers and the presence of antioxidant enzymes in two populations of the adult *H. diminuta* cestode, one with a short experimental infection time and another with a well-consolidated infection [126]. They found a high activity of the enzymes SOD, Ph-GPx, and Prx in the anterior end (close to the intestinal epithelium) comparable to that of both tapeworms. However, in older tapeworms they found higher GST activity and lower GSH concentration, which suggests that adults also face a constant detoxification process. As the tapeworm size increases and occupies the ileum, oxidant indicators increase with a progressive decrease in antioxidant enzymes (except GST); however, at the posterior end of the parasite, where the proglottids are sexually mature, antioxidant enzymes increase again. This suggests that the production and storage of eggs, which occurs in the mature and gravid proglottids located at the posterior end of tapeworms, requires the participation of antioxidant systems. In the case of old tapeworms, a similar pattern of antioxidant enzyme activity is observed, but contrary to expectations, oxidative stress markers always remained below the levels reached by their young counterparts. This suggests that by consolidating the infection, old tapeworms have managed to modulate the immune response, which leads to less exposure to ROS. Finally, the only enzyme that does not significantly reduce its activity is GST, which implies that the parasite is always ready to purge toxic metabolites.
- *Sexual dimorphism of the parasite.* Oliveira et al. compared the contribution of nutrients and gender of unpaired adults of *S. mansoni* on O_2 consumption pathways and susceptibility to oxidative stress [85]. In general, they found a greater contribution of glutamine to respiration in females, which contrasts with a greater contribution of glucose in the case of males. The O_2 consumption rate was higher in males compared to females, regardless of the respiratory substrate. In contrast, the rate of ROS production and the expression of antioxidant enzymes was higher in females than in males. This suggests that the physiological process of egg production is related to an increase in endogenous ROS. Finally, females were more tolerant to exogenous oxidative stress than males, possibly due to basal overexpression of their antioxidant systems.

3.3.2. Parasite-Host Relationships

Parasitic platyhelminths establish a chemical dialogue with the host by taking elements from it and by sending molecules from the parasite, having a different impact in their relationship with the host. In addition to oxygen diffusion, this class of parasites must ensure access to the nutrients they need for their development, so the presence of a sophisticated recognition and acquisition system through the use of specific transporters (similar to those of the host) is not surprising [30,188]. However, the parasite-host relationship is not limited to that, as the parasite can use the metabolism of the host cells to its advantage [93],

Through proteomic analysis, the presence of intact and functional host proteins has been confirmed in the hydatid fluid of *E. granulosus* and in the vesicular fluid of various species of the genus *Taenia* [50]. Although the ratio of parasite/host proteins is specific to each organism, the composition of these fluids against the composition of the serum of the respective host has been analyzed [50,189–192]. Some of the most abundant host proteins reported are serum albumin and immunoglobulins [87,190]. In the case of the former, the parasite can use it to maintain internal osmotic pressure, whereas the latter could help to prevent antigen exposure of the immune system [192,193]. Surprisingly, it has been reported that these organisms can incorporate various host antioxidant proteins to their antioxidant repertoire, such as the SOD, Prxs, and CAT isoforms [81,87].

In parasitic flatworms, the presence of various families of transporters specialized in the removal of metabolites and drugs has been reported. Although this representation is not homogeneous in flatworms, its participation in detoxification processes has been demonstrated [194–201]. In a previous experiment, we inhibited the TGR enzyme activity of *T. crassiceps* cysticerci under in vitro conditions and observed the appearance of GSSG in the culture medium, which led to the proposition that the cestode expels GSSG excess as a mechanism to avoid the change of the redox environment inside [165]. By searching for a transporter capable of carrying out the translocation of this oxidized species, we were able to identify some multidrug resistance (MDR) transporters in the genome of *T. solium*, which may potentially be responsible for carrying out this function [165].

4. Conclusions

Oxygen has a dual function in organisms. In aerobic organisms it works mainly as a final electron acceptor during respiration, which results in greater energy production through the catabolic pathways, and its presence is related to the generation of ROS resulting in the expression of antioxidant systems involved in redox homeostasis maintenance.

Among flatworms, trematodes and cestodes have life cycles that develop in environments with different oxygen tension, which determine the development of special characteristics that have allowed them to adapt to varied conditions, such as:

- An energy metabolism that transits between aerobiosis and anaerobiosis depending on the availability of oxygen.
- This is possible because they have an enzymatic repertoire with common metabolic pathways involving enzymes that catalyze reversible reactions.
- In anaerobiosis, in addition to lactic fermentation, they have another fermentation pathway known as malate dismutation that allows them to obtain a greater amount of energy even in the absence of oxygen. Additionally, this pathway allows them to maintain their redox balance by eliminating the electrons in molecules that are secreted into the medium, mainly succinate, acetate, and propionate.
- Due to the absence of a circulatory system, they developed a tegument through which O_2 diffuses.
- The diffusion of oxygen generates the formation of a concentration gradient, its presence being greater in the tegument than in the parenchyma.
- Two populations of mitochondria, aerobic and anaerobic, have been described; the first located mainly in the tegument.

- Finally, the exposure of the tegument to a higher concentration of O_2 implies a greater production of ROS in it, as indirectly demonstrated by a significant presence of antioxidant enzymes in this region (SOD, GPx, Prx).

These overall data provide more information on the type of metabolism that is performed in the parasite in relation to pO_2. However, as Boyunaga comments [82], "one must be cautious when trying to relate this O_2 tension where these parasites develop" with the type of energy metabolism they carry out, since the reports in the literature can be controversial. Thus, an important aspect that must be considered is the presence of both types of metabolism, aerobic and anaerobic, in the same organism and its relation to the size of the parasite, the stage of the life cycle, and the degree of purity of the mitochondrion (at least two mitochondrial types in these organisms). Having pure populations of mitochondria would make it possible to determine with greater certainty what type of energy metabolism occurs at what time in the life cycle and in what region of the parasite.

Funding: This work was supported by the research grant IN217920 from Dirección General de Asuntos del Personal Académico (DGAPA), at Universidad Nacional Autónoma de México (UNAM).

Acknowledgments: Authors want to thank Ingrid Helena Mascher Gramlich, for proofreading the English-language version of this manuscript. Authors want to thank Fashion Designer Juan Antonio Rocha-Santiago for redesign and elaboration of the figures used in this paper.

Conflicts of Interest: The authors declare no conflict of interest.

References

1. Hamilton, T.L.; Bryant, D.A.; Macalady, J.L. The role of biology in planetary evolution: Cyanobacterial primary production in low-oxygen Proterozoic oceans. *Environ. Microbiol.* **2016**, *18*, 325–340. [CrossRef] [PubMed]
2. Fisher, W.W.; Hemp, J.; Valentine, J.S. How did life survive earth's great oxygenation? *Curr. Opin. Chem. Biol.* **2016**, *31*, 166–178. [CrossRef] [PubMed]
3. Lenton, T.M.; Dahl, T.W.; Daines, S.J.; Mills, B.J.W.; Ozaki, K.; Saltzman, M.R.; Porada, P. Earliest land plants created modern levels of atmospheric oxygen. *Proc. Natl. Acad. Sci. USA* **2016**, *113*, 9704–9709. [CrossRef] [PubMed]
4. Morris, J.L.; Puttick, M.N.; Clark, J.D.; Donoghue, P.C.J. The timescale of early land plant evolution. *Proc. Natl. Acad. Sci. USA* **2018**, *115*, E2274–E2283. [CrossRef] [PubMed]
5. Lyons, T.W.; Reinhard, C.T.; Planavsky, N.J. The rise of oxygen in Earth´s ocean and atmosphere. *Nature* **2014**, *506*, 307–315. [CrossRef] [PubMed]
6. Zimorski, V.; Mental, M.; Tielens, A.G.M.; Martin, W.F. Energy metabolism in anaerobic eukaryotes and Earth's late oxygenation. *Free Radic. Biol. Med.* **2019**, *140*, 279–294. [CrossRef]
7. Mentel, M.; Martin, W. Energy metabolism among eukaryotic anaerobes in light of Proterozoic ocean chemistry. *Phil. Trans. R. Soc. B Biol. Sci.* **2008**, *363*, 2717–2729. [CrossRef]
8. Mentel, M.; Rottger, M.; Leys, S.; Tielens, A.G.M.; Martin, W.F. Of early animals, anaerobic mitochondria, and a modern sponge. *Bioassays* **2014**, *36*, 924–932. [CrossRef]
9. Muller, M. Biochemistry and evolution of anaerobic energy metabolism in eukaryotes. *Microbiol. Mol. Biol. Rev.* **2012**, *76*, 453–474. [CrossRef]
10. Tielens, A.G.M.; Rotte, C.; van Hellemond, J.J.; Martin, W. Mitochondria as we don't know them. *Trends Biochem. Sci.* **2002**, *27*, 564–572. [CrossRef]
11. Meléndez-Hevia, E.; Montero-Gómez, N.; Montero, F. From prebiotic chemistry to cellular metabolism-The chemical evolution of metabolism before Darwinian natural selection. *J. Theor. Biol.* **2008**, *252*, 505–519. [CrossRef] [PubMed]
12. Romano, A.H.; Conway, T. Evolution of carbohydrate metabolic pathways. *Res. Microbiol.* **1996**, *147*, 448–455. [CrossRef]
13. Van Der Giezen, M.; Lenton, T.M. The rise of oxygen and complex life. *J. Eukaryot. Microbiol.* **2012**, *59*, 111–113. [CrossRef]
14. Halliwell, B.; Gutteridge, J.M. Oxygen toxicity, oxygen radicals, transition metals and disease. *Biochem. J.* **1984**, *219*, 1–14. [CrossRef] [PubMed]
15. Sawyer, D.T.; Valentine, J.S. How Super is Superoxide? *Acc. Chem. Res.* **1981**, *14*, 393–400. [CrossRef]
16. Collins, J.J., 3rd. Platyhelminthes. *Curr. Biol.* **2017**, *27*, R252–R256. [CrossRef]
17. Bobes, R.J.; Fragoso, G.; Fleury, A.; García-Varela, M.; Sciutto, E.; Larralde, C.; Laclette, J.P. Evolution, molecular epidemiology and perspectives on the research of taeniid parasites with special emphasis on *Taenia solium*. *Infect. Genet. Evol.* **2014**, *23*, 150–160. [CrossRef]
18. Egger, B.; Steinke, D.; Tarui, H.; De Mulder, K.; Arendt, D.; Borgonie, G.; Funayama, N.; Gschwentner, R.; Hartenstein, V.; Hobmayer, B.; et al. To be or not to be a flatworm: The acoel controversy. *PLoS ONE* **2009**, *4*, e5502. [CrossRef]

19. Riutort, M.; Álvarez-Presas, M.; Lázaro, E.; Solà, E.; Paps, J. Evolutionary history of the Tricladida and the Platyhelminthes: An up-to-date phylogenetic and systematic account. *Int. J. Dev. Biol.* **2012**, *56*, 5–17. [CrossRef]
20. Cheng, L.C.; Tu, K.C.; Seidel, C.W.; Robb, S.; Guo, F.; Sánchez Alvarado, A. Cellular, ultrastructural and molecular analyses of epidermal cell development in the planarian Schmidtea mediterranea. *Dev. Biol.* **2018**, *433*, 357–373. [CrossRef]
21. Hahn, C.; Fromm, B.; Bachmann, L. Comparative genomics of flatworms (platyhelminthes) reveals shared genomic features of ecto- and endoparastic neodermata. *Genome Biol. Evol.* **2014**, *6*, 1105–1117. [CrossRef] [PubMed]
22. Perkins, E.M.; Donnellan, S.C.; Bertozzi, T.; Whittington, I.D. Closing the mitochondrial circle on paraphyly of the Monogenea (Platyhelminthes) infers evolution in the diet of parasitic flatworms. *Int. J. Parasitol.* **2010**, *40*, 1237–1245. [CrossRef] [PubMed]
23. Harrington, D.; Lamberton, P.; McGregor, A. Human liver flukes. *Lancet Gastroenterol. Hepatol.* **2017**, *2*, 680–689. [CrossRef]
24. Toledo, A.; Osorio, R.; Matus, C.; Martinez Lopez, Y.; Ramirez Cruz, N.; Sciutto, E.; Fragoso, G.; Arauz, A.; Carrillo-Mezo, R.; Fleury, A. Human Extraparenchymal Neurocysticercosis: The Control of Inflammation Favors the Host … but Also the Parasite. *Front. Immunol.* **2018**, *9*, 2652. [CrossRef] [PubMed]
25. Hayward, A.D.; Skuce, P.J.; McNeilly, T.N. The influence of liver fluke infection on production in sheep and cattle: A meta-analysis. *Int. J. Parasitol.* **2021**, *51*, 913–924. [CrossRef]
26. Cwiklinski, K.; O'Neill, S.M.; Donnelly, S.; Dalton, J.P. A prospective view of animal and human Fasciolosis. *Parasite Immunol.* **2016**, *38*, 558–568. [CrossRef] [PubMed]
27. Trevisan, C.; Devleesschauwer, B.; Schmidt, V.; Winkler, A.S.; Harrison, W.; Johansen, M.V. The societal cost of *Taenia solium* cysticercosis in Tanzania. *Acta Trop.* **2017**, *165*, 141–154. [CrossRef]
28. Neves, L.X.; Wilson, R.A.; Brownridge, P.; Harman, V.M.; Holman, S.W.; Beynon, R.J.; Eyers, C.E.; DeMarco, R.; Castro-Borges, W. Quantitative Proteomics of Enriched Esophageal and Gut Tissues from the Human Blood Fluke *Schistosoma mansoni* Pinpoints Secreted Proteins for Vaccine Development. *J. Proteome Res.* **2020**, *19*, 314–326. [CrossRef]
29. Morley, N.J. Ecology of free-living metacercariae (Trematoda). *Adv. Parasitol.* **2015**, *89*, 1–78.
30. Li, W.H.; Yang, Y.; Zhang, N.Z.; Wang, J.K.; Liu, Y.J.; Li, L.; Yan, H.B.; Jia, W.Z.; Fu, B. Comparative Transcriptome Analyses of the Developmental Stages of *Taenia multiceps*. *Front. Vet. Sci.* **2021**, *8*, 677045. [CrossRef]
31. Cancela, M.; Paes, J.A.; Moura, H.; Barr, J.R.; Zaha, A.; Ferreira, H.B. Unraveling oxidative stress response in the cestode parasite Echinococcus granulosus. *Sci. Rep.* **2019**, *9*, 15876. [CrossRef] [PubMed]
32. Bryant, C.; Behm, C.A. *Biochemistry of Parasites and Host Parasite Relationships*; Van den Bossche, H., Ed.; North Holland: Amsterdam, The Netherland, 1976; pp. 89–94.
33. Carreau, A.; El Hafny-Rahbi, B.; Matejuk, A.; Grillon, C.; Kieda, C. Why is the partial oxygen pressure of human tissues a crucial parameter? Small molecules and hypoxia. *J. Cell Mol. Med.* **2011**, *15*, 1239–1253. [CrossRef] [PubMed]
34. De Santis, V.; Singer, M. Tissue oxygen tension monitoring of organ perfusion: Rationale, methodologies, and literature review. *Br. J. Anaesth.* **2015**, *115*, 357–365. [CrossRef] [PubMed]
35. Friedman, E.S.; Bittinger, K.; Esipova, T.V.; Hou, L.; Chau, L.; Jiang, J.; Mesaros, C.; Lund, P.J.; Liang, X.; FitzGerald, G.A.; et al. Microbes vs. chemistry in the origin of the anaerobic gut lumen. *Proc. Natl. Acad. Sci. USA* **2018**, *115*, 4170–4175. [CrossRef]
36. Mori, M.P.; Penjweini, R.; Knutson, J.R.; Wang, P.; Hwang, P.M. Mitochondria and oxygen homeostasis. *FEBS J.* **2021**. [CrossRef]
37. Komuniecki, R.; Tielens, A.G.M. Carbohydrate and energy metabolism helminths. In *Molecular Medical Parasitology*; Marr, J.J., Nilsen, T.W., Komuniecki, R.W., Eds.; Academic Press: London, UK, 2003; pp. 339–358.
38. Ward, J.B.J.; Keely, S.J.; Keely, S.J. Oxygen in the regulation of intestinal epithelial transport. *J. Phisiol.* **2014**, *592*, 2473–2489. [CrossRef]
39. Harada, S.; Inaoka, D.K.; Ohmori, J.; Kita, K. Diversity of parasite complex II. *BBA* **2013**, *1827*, 658–667. [CrossRef]
40. Van Hellemond, J.J.; Tielens, A.G.M. Expression and functional properties of fumarate reductase. *Biochem. J.* **1994**, *304*, 321–333. [CrossRef]
41. Kita, K.; Nihei, C.; Tomitsuka, E. Parasite Mitochondria as Drug Target: Diversity and Dynamic Changes During the Life Cycle. *Curr. Med. Chem.* **2003**, *10*, 2535–2548. [CrossRef]
42. Tielens, A.G.M.; van den Heuvel, J.M.; van den Berg, S.G. Differences in intermediary energy metabolism between juvenile and adult *Fasciola hepatica*. *Mol. Biochem. Parasitol.* **1987**, *24*, 273–281. [CrossRef]
43. Tielens, A.G.M. The carbohydrate metabolism of *Fasciola hepatica*, an example of biochemical adaptations in parasitic helminths. *Acta Parasitol.* **2000**, *45*, 59–66.
44. Poddubnaya, L.G.; Scholz, T.; Kuchta, R.; Levron, C.; Brunanská, M. Ultrastructure of the proglottid tegument (neodermis) of the cestode *Echinophallus wageneri* (Pseudophyllidea: Echinophallidae), a parasite of the bathypelagic fish *Centrolophus niger*. *Parasitol. Res.* **2007**, *101*, 373–383. [CrossRef] [PubMed]
45. Wendt, G.R.; Collins, J.N.; Pei, J.; Pearson, M.S.; Bennett, H.M.; Loukas, A.; Berriman, M.; Grishin, N.V.; Collins, J.J., 3rd. Flatworm-specific transcriptional regulators promote the specification of tegumental progenitors in *Schistosoma mansoni*. *Elife* **2018**, *7*, e33221. [CrossRef] [PubMed]
46. Sotillo, J.; Pearson, M.; Becker, L.; Mulvenna, J.; Loukas, A. A quantitative proteomic analysis of the tegumental proteins from *Schistosoma mansoni* schistosomula reveals novel potential therapeutic targets. *Int. J. Parasitol.* **2015**, *45*, 505–516. [CrossRef] [PubMed]

47. Takamiya, S.; Fukuda, K.; Nakamura, T.; Aoki, T.; Sugiyama, H. *Paragonimus westermani* possesses aerobic and anaerobic mitochondria in different tissues, adapting to fluctuating oxygen tension in microaerobic habitats. *Int. J. Parasitol.* **2010**, *40*, 1651–1658. [CrossRef]
48. Tielens, A.G.M.; van den Heuvel, J.M.; van den Bergh, S.G. The energy metabolism of *Fasciola hepatica* during its development in the final host. *Mol. Biochem. Parasitol.* **1984**, *13*, 301–307. [CrossRef]
49. Starling, J.A. Tegumental carbohydrate transport in intestinal helminths: Correlation between mechanisms of membrane transport and the biochemical environment of absorptive surfaces. *Trends Am. Microsc. Soc.* **1975**, *94*, 508–523. [CrossRef]
50. Brehm, K.; Koziol, U. Echinococcus-Host Interactions at Cellular and Molecular Levels. *Adv. Parasitol.* **2017**, *95*, 147–212.
51. Van Hellemond, J.J.; Retra, K.; Brouwers, J.F.H.M.; van Balkom, M.Y.; Shoemarker, C.B.; Tielens, A.G.T. Functions of the tegument of schistosomes: Clues from the proteome and lipidome. *Int. J. Parasitol.* **2006**, *36*, 691–699. [CrossRef]
52. Leow, C.Y.; Willis, C.; Hofmann, A.; Jones, M.K. Structure-function analysis of apical membrane-associated molecules of the tegument of schistosome parasites of humans: Prospects for identification of novel targets for parasite control. *Br. J. Pharmacol.* **2015**, *172*, 1653–1663. [CrossRef]
53. Tovar, J.; Fischer, A.; Clark, C.G. The mitosome, a novel organelle related to mitochondria in the amitochondrial parasite *Entamoeba histolytica*. *Mol. Microbiol.* **1999**, *32*, 1013–1021. [CrossRef]
54. Makiuchi, T.; Nozaki, T. Highly divergent mitochondrion-related organelles in anaerobic parasitic protozoa. *Biochimie* **2014**, *100*, 3–17. [CrossRef] [PubMed]
55. Palade, G.E. An electron microscope study of the mitochondrial structure. *J. Histochem. Chem. Cytochem.* **1953**, *1*, 188–211. [CrossRef] [PubMed]
56. Scheffler, I.E. Structure and morphology. Integration into the cell. In *Mitochondria*, 2nd ed.; John Wiley & Sons: Hoboken, NJ, USA, 2011; pp. 18–59.
57. Kita, K.; Takamiya, S. Electron-transfer complexes in *Ascaris* mitochondria. *Adv. Parasitol.* **2002**, *51*, 95–131. [PubMed]
58. Semenza, G.L. Hypoxia-inducible factors in physiology and medicine. *Cell* **2012**, *148*, 399–408. [CrossRef] [PubMed]
59. Thompson, D.P.; Geary, T.G. The Structure and Function of Helminth Surfaces: Structural. In *Biochemistry and Molecular Biology of Parasites*; Marr, J.J., Müller, M., Eds.; Academic Press: San Diego, CA, USA, 2003; pp. 203–232.
60. Tielens, A.G.M. Energy generation in parasitic helminths. *Parasitol. Today* **1994**, *10*, 346–352. [CrossRef]
61. Lumsden, R.D. Ultrastructure of mitochondria in a cestode, *Lacistorrhynchus tenuis* (V. Benden, 1858). *J. Parasitol.* **1967**, *53*, 65–77. [CrossRef] [PubMed]
62. Del Arenal, M.I.P.; Cea, B.A.; Moreno-Sanchez, R.; Escamilla, J.E. A method for the isolation of tegument syncytium mitochondria from *Taenia crassiceps* cysticerci and partial characterization of their aerobic metabolism. *J. Parasitol.* **1998**, *84*, 461–468. [CrossRef]
63. Takamiya, S.; Wang, H.; Hiraishi, A.; Yu, Y.; Hamajima, F. Respiratory chain of the lung *Paragonimus westermani*: Facultative anaerobic mitochondria. *Arch. Biochem. Biophys.* **1994**, *312*, 142–150. [CrossRef]
64. Roppongi, T.; Mizuno, N.; Miyagawa, Y.; Kobayashi, T.; Nakagawa, K.; Adachi, S. Solubility and mass transfer coefficient of oxygen through gas- and water-lipid interfaces. *J. Food Sci.* **2021**, *86*, 867–873. [CrossRef]
65. Terwilliger, N.B. Functional adaptations of oxygen-transport proteins. *J. Exp. Biol.* **1998**, *201*, 1085–1098. [CrossRef] [PubMed]
66. Gell, D.A. Structure and function of haemoglobins. *Blood Cells Mol. Dis.* **2018**, *70*, 13–42. [CrossRef] [PubMed]
67. Storz, J.F.; Opazo, J.C.; Hoffmann, F.G. Gene duplication, genome duplication, and the functional diversification of vertebrate globins. *Mol. Phylogenet. Evol.* **2013**, *66*, 469–478. [CrossRef] [PubMed]
68. De Guzman, J.V.; Yu, H.S.; Jeong, H.J.; Hong, Y.C.; Kim, J.; Kong, H.H.; Chung, D.I. Molecular characterization of two myoglobins of *Paragonimus westermani*. *J. Parasitol.* **2007**, *93*, 97–103. [CrossRef]
69. Goldberg, D.E. The enigmatic oxygen-avid hemoglobin of *Ascaris*. *Bioessays* **1995**, *17*, 177–182. [CrossRef]
70. Kiger, L.; Rashid, A.K.; Griffon, N.; Haque, M.; Moens, L.; Gibson, Q.H.; Poyart, C.; Marden, M.C. Trematode hemoglobins show exceptionally high oxygen affinity. *Biophys. J.* **1998**, *75*, 990–998. [CrossRef]
71. González, R.; Mendoza-Hernández, G.; Plancarte, A. Purification of *Taenia solium* cysticerci superoxide dismutase and myoglobin copurification. *Parasitol. Res.* **2002**, *88*, 881–887.
72. Kim, S.H.; Yang, D.; Bae, Y.A. Hypoxic and nitrosative stress conditions modulate expression of myoglobin genes in a carcinogenic hepatobiliary trematode, *Clonorchis sinensis*. *PLoS Negl. Trop. Dis.* **2021**, *15*, e0009811. [CrossRef]
73. Burmester, T.; Hankeln, T. Function and evolution of vertebrate globins. *Acta Physiol.* **2014**, *211*, 501–514. [CrossRef]
74. McManus, D.P. Intermediary metabolism in parasitic helminths. *Int. J. Parasitol.* **1987**, *17*, 79–95. [CrossRef]
75. Tielens, A.G.M.; van de Pas, F.A.; van den Heuvel, J.M.; van den Bergh, S.G. The aerobic energy metabolism of *Schistosoma mansoni* miracidia. *Mol. Biochem. Parasitol.* **1991**, *46*, 181–184. [CrossRef]
76. Young, N.D.; Nagarajan, N.; Lin, S.J.; Korhonen, P.K.; Jex, A.R.; Hall, R.S.; Safavi-Hemami, H.; Kaewkong, W.; Bertrand, D.; Gao, S.; et al. The *Opisthorchis viverrini* genome provides insights into life in the bile duct. *Nat. Commun.* **2014**, *5*, 4378. [CrossRef] [PubMed]
77. Bertout, J.A.; Patel, S.A.; Simon, M.C. The impact of O2 availability on human cancer. *Nat. Rev. Cancer.* **2008**, *8*, 967–975. [CrossRef] [PubMed]
78. Rytkonen, K.T.; Storz, J.F. Evolutionary origins of oxygen sensing in animals. *EMBO Rep.* **2011**, *12*, 2–4. [CrossRef]

79. Goto, M.; Amino, H.; Nakajima, M.; Tsuji, N.; Sakamoto, K.; Kita, K. Cloning and characterization of hypoxia-inducible factor-1 subunits from Ascaris suum-a parasitic nematode highly adapted to changes of oxygen conditions during its life cycle. *Gene* **2013**, *516*, 39–47. [CrossRef] [PubMed]
80. Kim, S.H.; Oh, G.S.; Sohn, W.M.; Lee, K.; Yang, H.J.; Bae, Y.A. Molecular characteristics and induction profiles of hypoxia-inducible factor-1α and other basic helix-loop-helix and Per-Arnt-Sim domain-containing proteins identified in a carcinogenic liver fluke *Clonorchis sinensis*. *Parasitology* **2019**, *146*, 176–186. [CrossRef]
81. Cui, S.J.; Xu, L.L.; Zhang, T.; Xu, M.; Yao, J.; Fang, C.Y.; Feng, Z.; Yang, P.Y.; Hu, W.; Liu, F. Proteomic characterization of larval and adult developmental stages in *Echinococcus granulosus* reveals novel insight into host-parasite interactions. *J. Proteom.* **2013**, *84*, 158–175. [CrossRef]
82. Boyunaga, H.; Schmitz, M.G.; Brouwers, J.F.; van Hellemond, J.J.; Tielens, A.G.M. *Fasciola hepatica* miracidia are dependent on respiration and endogenous glycogen degradations for their energy generation. *Parasitology* **2001**, *122*, 169–173. [CrossRef]
83. Parkinson, J.; Wasmuth, J.D.; Salinas, G.; Bizarro, C.V.; Sanford, C.; Berriman, M.; Ferreira, H.B.; Zaha, A.; Blaxter, M.L.; Maizels, R.M.; et al. A transcriptomic analysis of *Echinococcus granulosus* larval stages: Implications for parasite biology and host adaptation. *PLoS Negl. Trop. Dis.* **2012**, *6*, e1897. [CrossRef]
84. De Almeida Leandro, L.; Fraga, C.M.; de Souza Lino, R., Jr.; Vinaud, M.C. Partial reverse of the TCA cycle is enhanced in *Taenia crassiceps* experimental neurocysticercosis after in vivo treatment with anthelminthic drugs. *Parasitol. Res.* **2014**, *113*, 1313–1317. [CrossRef]
85. Oliveira, M.P.; Correa Soares, J.B.; Oliveira, M.F. Sexual Preferences in Nutrient Utilization Regulate Oxygen Consumption and Reactive Oxygen Species Generation in *Schistosoma mansoni*: Potential Implications for Parasite Redox Biology. *PLoS ONE* **2016**, *11*, e0158429. [CrossRef] [PubMed]
86. Ritler, D.; Rufener, R.; Li, J.V.; Kämpfer, U.; Müller, J.; Bühr, C.; Schürch, S.; Lundström-Stadelmann, B. In vitro metabolomic footprint of the *Echinococcus multilocularis* metacestode. *Sci. Rep.* **2019**, *9*, 19438. [CrossRef] [PubMed]
87. Zhang, S. Comparative Transcriptomic Analysis of the Larval and Adult Stages of *Taenia pisiformis*. *Genes* **2019**, *10*, 507. [CrossRef] [PubMed]
88. Willms, K.; Robert, L.; Caro, J.A. Ultrastructure of smooth muscle, gap junctions and glycogen distribution in *Taenia solium* tapeworms from experimentally infected hamsters. *Parasitol. Res.* **2003**, *89*, 308–316. [CrossRef]
89. Valkounová, J.; Zdárská, Z.; Slais, J. Histochemistry of the racemose form of *Cysticercus cellulosae*. *Folia Parasitol.* **1992**, *39*, 207–226.
90. Chekulayev, V.; Mado, K.; Shevchuk, I.; Koit, A.; Kaldma, A.; Klepinin, A.; Timohhina, N.; Tepp, K.; Kandashvili, M.; Ounpuu, L.; et al. Metabolic remodeling in human colorectal cancer and surrounding tissues: Alterations in regulation of mitochondrial respiration and metabolic fluxes. *Biochem. Biophys. Rep.* **2015**, *4*, 111–125. [CrossRef]
91. Skelly, P.J.; Shoemaker, C.B. A molecular genetic study of the variations in metabolic function during schistosome development. *Mem. Ins. Oswaldo Cruz.* **1995**, *90*, 281–284. [CrossRef]
92. Tsai, I.J.; Zarowiecki, M.; Holroyd, N.; Garciarrubio, A.; Sánchez-Flores, A.; Brooks, K.L.; Tracey, A.; Bobes, R.J.; Fragoso, G.; Sciutto, E.; et al. The genomes of four tapeworm species reveal adaptations to parasitism. *Nature* **2013**, *496*, 57–63. [CrossRef]
93. Fraga, C.M.; Costa, T.L.; Bezerra, J.C.; de Souza Lino, R., Jr.; Vinaud, M.C. *Taenia crassiceps*: Host treatment alters glycolisis and tricarboxilic acid cycle in cysticerci. *Exp. Parasitol.* **2012**, *130*, 146–151. [CrossRef]
94. Del Arenal, I.P.; Rubio, M.E.; Ramírez, J.; Rendón, J.L.; Escamilla, J.E. Cyanide-resistant respiration in *Taenia crassiceps* metacestode (cysticerci) is explained by the H_2O_2-producing side-reaction of respiratory complex I with O_2. *Parasitol. Int.* **2005**, *54*, 185–193. [CrossRef]
95. Bennet, E.M.; Behm, C.A.; Bryant, C. The role of the host in the regulation of end-product formation in two strains of the rat tapeworm, *Hymenolepis diminuta*. *Int. J. Parasitol.* **1990**, *20*, 841–848. [CrossRef]
96. Bryant, C. Organic acid excretion by helminths. *Parasitol. Today* **1993**, *9*, 58–60. [CrossRef]
97. Campbell, T.; Rubin, N.; Komuniecki, R. Succinate-dependent energy generation in *Ascaris suum* mitochondria. *Mol. Biochem. Parasitol.* **1989**, *33*, 1–12. [CrossRef]
98. Tielens, A.G.M.; van Hellemond, J.J. The electron transport chain in anaerobically functioning eukaryotes. *Biochim. Biophys. Acta* **1998**, *1365*, 71–78. [CrossRef]
99. Matsumoto, J.; Sakamoto, K.; Shinjyo, N.; Kido, Y.; Yamamoto, N.; Yagi, K.; Miyoshi, H.; Nonaka, N.; Katakura, K.; Kita, K.; et al. Anaerobic NADH-fumarate reductase system is predominant in the respiratory chain of *Echinococcus multilocularis*, providing a novel target for the chemotherapy of Alveolar Echinococcosis. *Antimicrob. Agents Chemother.* **2008**, *52*, 164–170. [CrossRef]
100. Ovington, K.S.; Bryant, C. The role of carbon dioxide in the formation of end-products *Hymenolepis diminuta*. *Int. J. Parasitol.* **1981**, *11*, 221–228. [CrossRef]
101. Fioravanti, C.F.; Vandock, K.P. Transhydrogenase and the anaerobic mitochondrial metabolism of adult *Hymenolepis diminuta*. *Parasitology* **2010**, *137*, 395–410. [CrossRef]
102. Van Hellemond, J.J.; van der Klei, A.; van Weelden, S.W.H.; Tielens, A.G.M. Biochemical and evolutionary aspects of anaerobically functioning mitochondria. *Philos. Trans. R. Soc. B Biol. Sci.* **2003**, *358*, 213–215. [CrossRef]
103. Lima, N.F.; Picanço, G.A.; Costa, T.L.; de Souza Lino Junior, R.; Vinaud, M.C. In Vivo Treatment with the Combination of Nitazoxanide and Flubendazole Induces Gluconeogenesis and Protein Catabolism in *Taenia crassiceps* cysticerci. *Acta Parasitol.* **2021**, *66*, 98–103. [CrossRef]

104. Tielens, A.G.M.; van Grinsven, K.; Henze, K.; van Hellemond, J.J.; Martin, W. Acetate formation in the energy metabolism of parasitic helminths and protists. *Int. J. Parasitol.* **2010**, *40*, 387–397. [CrossRef]
105. Van Grinsven, K.W.A.; van Hellemond, J.J.; Tielens, A.G.M. Acetate: Succinate CoA-transferase in the anaerobic mitochondria of *Fasciola hepatica*. *Mol. Biochem. Parasitol.* **2009**, *164*, 74–79. [CrossRef] [PubMed]
106. Moore, H.W.; Folkers, K.; Coenzyme, Q. LXII. Structure and Synthesis of Rhodoquinone, a Natural Aminoquinone of the Coenzyme Q Group. *J. Am. Chem. Soc.* **1965**, *87*, 1409–1410. [CrossRef] [PubMed]
107. Van Hellemond, J.J.; Klockiewicz, M.; Gaasenbeek, C.P.; Roos, M.H.; Tielens, A.G.M. Rhodoquinone and complex II of the electron transport chain in anaerobically functioning eukaryotes. *J. Biol. Chem.* **1995**, *270*, 31065–31070. [CrossRef] [PubMed]
108. Fioravanti, C.F.; Kim, Y. Rhodoquinone requirement of the *Hymenolepis diminuta* mitochondrial electron transport system. *Mol. Biochem. Parasitol.* **1988**, *28*, 129–134. [CrossRef]
109. Boveris, A.; Hertig, C.; Turrens, J. Fumarate reductase and other mitochondrial activities in *trypanosoma cruzi*. *Mol. Biochem. Parasitol.* **1986**, *19*, 163–169. [CrossRef]
110. Arrigoni, O.; Singer, T.P. Limitations of the phenazine methosulfate assay for succinic and related dehydrogenases. *Nature* **1992**, *193*, 1256–1258. [CrossRef]
111. Kuramochi, T.; Hirawake, H.; Kojima, S.; Takamiya, S.; Furushima, R.; Aoki, T.; Komuniecki, R.; Kita, K. Sequence comparison between the flavoprotein subunit of the fumarate reductase (complex II) of the anaerobic parasitic nematode, *Ascaris suum* and the succinate dehydrogenase of the aerobic, free-living nematode, *Caenorhabditis elegans*. *Mol. Biol. Parasitol.* **1994**, *68*, 177–187. [CrossRef]
112. Amino, H.; Osanai, A.; Miyadera, H.; Shinjyo, N.; Tomitsuka, E.; Taka, H.; Mineki, R.; Murayama, K.; Takamiya, S.; Aoki, T.; et al. Isolation and characterization of the stage-specific cytochrome b small subunit (CybS) of *Ascaris suum* complex II from the aerobic respiratory chain of larval mitochondria. *Mol. Biochem. Parasitol.* **2003**, *128*, 175–186. [CrossRef]
113. Amino, H.; Wang, H.; Hirawake, H.; Saruta, F.; Mizuchi, D.; Mineki, R.; Shindo, N.; Murayama, K.; Takamiya, S.; Aoki, T.; et al. Stage specific isoforms of *Ascaris suum* complex II: The fumarate reductase of the parasitic adult and the succinate dehydrogenase of free-living larvae share a common iron-sulfur subunit. *Mol. Biochem. Parasitol.* **2000**, *106*, 63–76. [CrossRef]
114. Salinas, G.; Langelaan, D.N.; Shepherd, J.N. Rhodoquinone in bacteria and animals: Two distinct pathways for biosynthesis of this key electron transporter used in anaerobic bioenergetics. *BBA Bioenerg.* **2020**, *1861*, 1–14. [CrossRef]
115. Sakai, C.; Tomitsuka, E.; Esumi, H.; Harada, S.; Kita, K. Mitochondrial fumarate reductase as a target of chemotherapy: From parasites to cancer cells. *BBA* **2012**, *1820*, 643–651. [CrossRef] [PubMed]
116. Tomitsuka, E.; Kita, K.; Esumi, H. An anticancer agent, pyrvinium pamoate inhibits the NADH-fumarate reductase system-a unique mitochondrial energy metabolism in tumor microenvironments. *J. Biochem.* **2012**, *152*, 171–183. [CrossRef] [PubMed]
117. Saz, H.J.; deBruyn, B.; de Mata, Z. Acyl-CoA transferase activities in homogenates of *Fasciola hepatica* adults. *J. Parasitol.* **1996**, *82*, 694–696. [CrossRef]
118. Rivière, L.; van Weelden, S.W.; Glass, P.; Vegh, P.; Coustou, V.; Biran, M.; van Hellemond, J.J.; Bringaud, F.; Tielens, A.G.M.; Boshart, M. Acetyl: Succinate CoA-transferase in procyclic *Trypanosoma brucei*. Gene identification and role in carbohydrate metabolism. *J. Biol. Chem.* **2004**, *279*, 45337–45346. [CrossRef] [PubMed]
119. Fraga, C.M.; De Castro, A.M.; Reynoso-Ducoing, O.; Ambrosio, J.; Hernández-Campos, A.; Castillo, R.; Vinaud, M.C. Alternative energy production pathways in *Taenia crassiceps* in vitro exposed to a benzimidazole derivative (RCB20). *Parasitology* **2016**, *143*, 88–493. [CrossRef]
120. Ezenwa, V.O.; Archie, E.A.; Craft, M.E.; Hawley, D.M.; Martin, L.B.; Moore, J.; White, L. Host behaviour-parasite feedback: An essential link between animal behaviour and disease ecology. *Proc. Biol. Sci.* **2016**, *283*, 20153078. [CrossRef]
121. Zarowiecki, M.; Berriman, M. What helminth genomes have taught us about parasite evolution. *Parasitology* **2015**, *142*, S85–S97. [CrossRef]
122. Fragoso, G.; Bobes, R.J.; Espinoza, B.; Martínez, M.L.; Pérez-Morales, D.; Rosas, G.; Sciutto, E.; Laclette, J.P. Changes in cyst's nuclear chromatin resulting after experimental manipulation of *Taenia crassiceps* mice infections: Biological implications. *Exp. Parasitol.* **2012**, *130*, 423–429. [CrossRef]
123. Escobedo, G.; Larralde, C.; Chavarria, A.; Cerbón, M.A.; Morales-Montor, J. Molecular mechanisms involved in the differential effects of sex steroids on the reproduction and infectivity of *Taenia crassiceps*. *J. Parasitol.* **2004**, *90*, 1235–1244. [CrossRef]
124. Larralde, C.; Morales, J.; Terrazas, I.; Govezensky, T.; Romano, M.C. Sex hormone changes induced by the parasite lead to feminization of the male host in murine *Taenia crassiceps* cysticercosis. *J. Steroid Biochem. Mol. Biol.* **1995**, *52*, 575–580. [CrossRef]
125. Mourão, M.; Dinguirard, N.; Franco, G.R.; Yoshino, T.P. Role of the endogenous antioxidant system in the protection of *Schistosoma mansoni* primary sporocysts against exogenous oxidative stress. *PLoS Negl. Trop. Dis.* **2009**, *3*, e550. [CrossRef] [PubMed]
126. Skrzycki, M.; Majewska, M.; Podsiad, M.; Czeczot, H.; Salamatin, R.; Twarowska, J.; Grytner-Zięcina, B. *Hymenolepis diminuta*: Experimental studies on the antioxidant system with short and long term infection periods in the rats. *Exp. Parasitol.* **2011**, *129*, 158–163. [CrossRef] [PubMed]
127. Suttiprapa, S.; Sotillo, J.; Smout, M.; Suyapoh, W.; Chaiyadet, S.; Tripathi, T.; Laha, T.; Loukas, A. Opisthorchis viverrini Proteome and Host-Parasite Interactions. *Adv. Parasitol.* **2018**, *102*, 45–72. [PubMed]
128. Zheng, Y. Proteomic analysis of *Taenia hydatigena* cyst fluid reveals unique internal microenvironment. *Acta Trop.* **2017**, *176*, 224–227. [CrossRef] [PubMed]

129. Dorey, A.; Cwiklinski, K.; Rooney, J.; De Marco Verissimo, C.; López Corrales, J.; Jewhurst, H.; Fazekas, B.; Calvani, N.; Hamon, S.; Gaughan, S.; et al. Autonomous Non Antioxidant Roles for *Fasciola hepatica* Secreted Thioredoxin-1 and Peroxiredoxin-1. *Front. Cell Infect. Microbiol.* **2021**, *11*, 667272. [CrossRef]
130. Al-Shehri, S.S. Reactive oxygen and nitrogen species and innate immune response. *Biochimie* **2021**, *181*, 52–64. [CrossRef]
131. Vinogradov, A.D.; Grivennikova, V.G. Oxidation of NADH and ROS production by respiratory complex I. *Biochim. Biophys. Acta* **2016**, *1857*, 863–871. [CrossRef]
132. Moné, Y.; Ribou, A.C.; Cosseau, C.; Duval, D.; Théron, A.; Mitta, G.; Gourbal, B. An example of molecular co-evolution: Reactive oxygen species (ROS) and ROS scavenger levels in *Schistosoma mansoni/Biomphalaria glabrata* interactions. *Int. J. Parasitol.* **2011**, *41*, 721–730. [CrossRef]
133. Berriman, M.; Haas, B.J.; LoVerde, P.T.; Wilson, R.A.; Dillon, G.P.; Cerqueira, G.C.; Mashiyama, S.T.; Al-Lazikani, B.; Andrade, L.F.; Ashton, P.D.; et al. The genome of the blood fluke *Schistosoma mansoni*. *Nature* **2009**, *460*, 352–358. [CrossRef]
134. Zhang, H.C.; Ma, K.X.; Yang, Y.J.; Shi, C.Y.; Chen, G.W.; Liu, D.Z. Molecular cloning, characterization, expression and enzyme activity of catalase from planarian *Dugesia japonica* in response to environmental pollutants. *Ecotoxicol. Environ. Saf.* **2018**, *165*, 88–95. [CrossRef]
135. Hernández-Santoyo, A.; Landa, A.; González-Mondragón, E.; Pedraza-Escalona, M.; Parra-Unda, R.; Rodríguez-Romero, A. Crystal structure of Cu/Zn superoxide dismutase from *Taenia solium* reveals metal-mediated self-assembly. *FEBS J.* **2011**, *278*, 3308–3318. [CrossRef] [PubMed]
136. Yang, D.; Fu, Y.; Wu, X.; Xie, Y.; Nie, H.; Chen, L.; Nong, X.; Gu, X.; Wang, S.; Peng, X.; et al. Annotation of the transcriptome from Taenia pisiformis and its comparative analysis with three Taeniidae species. *PLoS ONE* **2012**, *7*, e32283. [CrossRef] [PubMed]
137. Mei, H.; LoVerde, P.T. *Schistosoma mansoni*: The developmental regulation and immunolocalization of antioxidant enzymes. *Exp. Parasitol.* **1997**, *86*, 69–78. [CrossRef] [PubMed]
138. Toppo, S.; Vanin, S.; Bosello, V.; Tosatto, S.C. Evolutionary and structural insights into the multifaceted glutathione peroxidase (Gpx) superfamily. *Antioxid. Redox Signal.* **2008**, *10*, 1501–1514. [CrossRef]
139. Changklungmoa, N.; Chaithirayanon, K.; Cheukamud, W.; Chaiwichien, A.; Osotprasit, S.; Samrit, T.; Sobhon, P.; Kueakhai, P. Expression and characterization of glutathione peroxidase of the liver fluke. *Parasitol. Res.* **2018**, *117*, 3487–3495. [CrossRef]
140. Fan, J.; Wu, H.; Li, K.; Liu, X.; Tan, Q.; Cao, W.; Liang, B.; Ye, B. Transcriptomic Features of *Echinococcus granulosus* Protoscolex during the Encystation Process. *Korean J. Parasitol.* **2020**, *58*, 287–299. [CrossRef]
141. Zelck, U.E.; Von Janowsky, B. Antioxidant enzymes in intramolluscan *Schistosoma mansoni* and ROS-induced changes in expression. *Parasitology* **2004**, *128*, 493–501. [CrossRef]
142. Cai, G.B.; Bae, Y.A.; Kim, S.H.; Sohn, W.M.; Lee, Y.S.; Jiang, M.S.; Kim, T.S.; Kong, Y. Vitellocyte-specific expression of phospholipid hydroperoxide glutathione peroxidases in *Clonorchis sinensis*. *Int. J. Parasitol.* **2008**, *38*, 1613–1623. [CrossRef]
143. Low, F.M.; Hampton, M.B.; Winterbourn, C.C. Peroxiredoxin 2 and peroxide metabolism in the erythrocyte. *Antioxid. Redox Signal.* **2008**, *10*, 1621–1630. [CrossRef]
144. Wang, H.; Li, J.; Zhang, C.; Guo, B.; Wei, Q.; Li, L.; Yang, N.; Peter McManus, D.; Gao, X.; Zhang, W.; et al. *Echinococcus granulosus* sensu stricto: Silencing of thioredoxin peroxidase impairs the differentiation of protoscoleces into metacestodes. *Parasite* **2018**, *25*, 57. [CrossRef]
145. Kumagai, T.; Osada, Y.; Kanazawa, T. 2-Cys peroxiredoxins from Schistosoma japonicum: The expression profile and localization in the life cycle. *Mol. Biochem. Parasitol.* **2006**, *149*, 135–143. [CrossRef]
146. Threadgold, L.T.; Arme, C.; Read, C.P. Ultrastructure localization of a peroxidase in the tapeworm, *Hymenolepis diminuta*. *J. Parasitol.* **1968**, *54*, 802–807. [CrossRef] [PubMed]
147. Circu, M.L.; Aw, T.Y. Redox biology of the intestine. *Free Radic. Res.* **2011**, *45*, 1245–1266. [CrossRef]
148. Sun, Q.A.; Kirnarsky, L.; Sherman, S.; Gladyshev, V.N. Selenoprotein oxidoreductase with specificity for thioredoxin and glutathione systems. *Proc. Natl. Acad. Sci. USA* **2001**, *98*, 3673–3678. [CrossRef] [PubMed]
149. Su, D.; Novoselov, S.V.; Sun, Q.A.; Moustafa, M.E.; Zhou, Y.; Oko, R.; Hatfield, D.L.; Gladyshev, V.N. Mammalian selenoprotein thioredoxin-glutathione reductase. Roles in disulfide bond formation and sperm maturation. *J. Biol. Chem.* **2005**, *280*, 26491–26498. [CrossRef] [PubMed]
150. Alger, H.M.; Williams, D.L. The disulfide redox system of *Schistosoma mansoni* and the importance of a multifunctional enzyme, thioredoxin glutathione reductase. *Mol. Biochem. Parasitol.* **2002**, *121*, 129–139. [CrossRef]
151. Agorio, A.; Chalar, C.; Cardozo, S.; Salinas, G. Alternative mRNAs arising from trans-splicing code for mitochondrial and cytosolic variants of *Echinococcus granulosus* thioredoxin Glutathione reductase. *J. Biol. Chem.* **2003**, *278*, 12920–12928. [CrossRef]
152. Rendón, J.L.; del Arenal, I.P.; Guevara-Flores, A.; Uribe, A.; Plancarte, A.; Mendoza-Hernández, G. Purification, characterization and kinetic properties of the multifunctional thioredoxin-glutathione reductase from *Taenia crassiceps* metacestode (cysticerci). *Mol. Biochem. Parasitol.* **2004**, *133*, 61–69. [CrossRef]
153. Otero, L.; Bonilla, M.; Protasio, A.V.; Fernández, C.; Gladyshev, V.N.; Salinas, G. Thioredoxin and glutathione systems differ in parasitic and free-living platyhelminths. *BMC Genom.* **2010**, *11*, 237. [CrossRef]
154. Martínez-González, J.J.; Guevara-Flores, A.; Alvarez, G.; Rendón-Gómez, J.L.; Del Arenal, I.P. In vitro killing action of auranofin on *Taenia crassiceps* metacestode (cysticerci) and inactivation of thioredoxin-glutathione reductase (TGR). *Parasitol. Res.* **2010**, *107*, 227–231. [CrossRef]

155. Prast-Nielsen, S.; Huang, H.H.; Williams, D.L. Thioredoxin glutathione reductase: Its role in redox biology and potential as a target for drugs against neglected diseases. *Biochim. Biophys. Acta* **2011**, *1810*, 1262–1271. [CrossRef] [PubMed]
156. Song, L.; Li, J.; Xie, S.; Qian, C.; Wang, J.; Zhang, W.; Yin, X.; Hua, Z.; Yu, C. Thioredoxin glutathione reductase as a novel drug target: Evidence from Schistosoma japonicum. *PLoS ONE* **2012**, *7*, e31456. [CrossRef] [PubMed]
157. Eweas, A.F.; Allam, G. Targeting thioredoxin glutathione reductase as a potential antischistosomal drug target. *Mol. Biochem. Parasitol.* **2018**, *225*, 94–102. [CrossRef] [PubMed]
158. Shukla, R.; Shukla, H.; Kalita, P.; Tripathi, T. Structural insights into natural compounds as inhibitors of Fasciola gigantica thioredoxin glutathione reductase. *J. Cell Biochem.* **2018**, *119*, 3067–3080. [CrossRef] [PubMed]
159. Guevara-Flores, A.; Martínez-González, J.J.; Herrera-Juárez, Á.M.; Rendón, J.L.; González-Andrade, M.; Torres Durán, P.V.; Enríquez-Habib, R.G.; Del Arenal Mena, I.P. Effect of curcuminoids and curcumin derivate products on thioredoxin-glutathione reductase from *Taenia crassiceps* cysticerci. Evidence suggesting a curcumin oxidation product as a suitable inhibitor. *PLoS ONE* **2019**, *14*, e0220098.
160. Lyu, H.; Petukhov, P.A.; Banta, P.R.; Jadhav, A.; Lea, W.A.; Cheng, Q.; Arnér, E.; Simeonov, A.; Thatcher, G.; Angelucci, F.; et al. Characterization of Lead Compounds Targeting the Selenoprotein Thioredoxin Glutathione Reductase for Treatment of Schistosomiasis. *ACS Infec. Dis.* **2020**, *6*, 393–405. [CrossRef] [PubMed]
161. Faixová, D.; Hrčková, G.; Mačák Kubašková, T.; Mudroňová, D. Antiparasitic Effects of Selected Isoflavones on Flatworms. *Helminthologia* **2021**, *58*, 1–16. [CrossRef] [PubMed]
162. Cwiklinski, K.; Dalton, J.P.; Dufresne, P.J.; La Course, J.; Williams, D.J.; Hodgkinson, J.; Paterson, S. The *Fasciola hepatica* genome: Gene duplication and polymorphism reveals adaptation to the host environment and the capacity for rapid evolution. *Genome Biol.* **2015**, *16*, 71. [CrossRef]
163. Bonilla, M.; Denicola, A.; Novoselov, S.V.; Turanov, A.A.; Protasio, A.; Izmendi, D.; Gladyshev, V.N.; Salinas, G. Platyhelminth mitochondrial and cytosolic redox homeostasis is controlled by a single thioredoxin glutathione reductase and dependent on selenium and glutathione. *J. Biol. Chem.* **2008**, *283*, 17898–17907. [CrossRef]
164. Guevara-Flores, A.; Del Arenal, I.P.; Mendoza-Hernández, G.; Pardo, J.P.; Flores-Herrera, O.; Rendón, J.L. Mitochondrial Thioredoxin-Glutathione Reductase from Larval *Taenia crassiceps* (Cysticerci). *J. Parasitol. Res.* **2010**, *2010*, 719856. [CrossRef]
165. Martínez-González, J.J.; Guevara-Flores, A.; Rendón, J.L.; Arenal, I. Auranofin-induced oxidative stress causes redistribution of the glutathione pool in *Taenia crassiceps* cysticerci. *Mol. Biochem. Parasitol.* **2015**, *201*, 16–25. [CrossRef] [PubMed]
166. Kuntz, A.N.; Davioud-Charvet, E.; Sayed, A.A.; Califf, L.L.; Dessolin, J.; Arnér, E.S.; Williams, D.L. Thioredoxin glutathione reductase from *Schistosoma mansoni*: An essential parasite enzyme and a key drug target. *PLoS Med.* **2007**, *4*, e206.
167. Angelucci, F.; Miele, A.E.; Boumis, G.; Dimastrogiovanni, D.; Brunori, M.; Bellelli, A. Glutathione reductase and thioredoxin reductase at the crossroad: The structure of *Schistosoma mansoni* thioredoxin glutathione reductase. *Proteins* **2008**, *72*, 936–945. [CrossRef] [PubMed]
168. Hayes, J.D.; Flanagan, J.U.; Jowsey, I.R. Glutathione transferases. *Annu. Rev. Pharmacol. Toxicol.* **2005**, *45*, 51–88. [CrossRef] [PubMed]
169. Wolkoff, A.W. The glutathione S-transferases: Their role in the transport of organic anions from blood to bile. *Int. Rev. Physiol.* **1980**, *21*, 150–169.
170. Nguyen, H.A.; Bae, Y.A.; Lee, E.G.; Kim, S.H.; Diaz-Camacho, S.P.; Nawa, Y.; Kang, I.; Kong, Y. A novel sigma-like glutathione transferase of *Taenia solium* metacestode. *Int. J. Parasitol.* **2010**, *40*, 1097–1106. [CrossRef]
171. Pearson, W.R. Phylogenies of glutathione transferase families. *Methods Enzymol.* **2005**, *401*, 186–204.
172. Wu, B.; Dong, D. Human cytosolic glutathione transferases: Structure, function, and drug discovery. *Trends Pharmacol. Sci.* **2012**, *33*, 656–668. [CrossRef]
173. Bae, Y.A.; Kim, J.G.; Kong, Y. Phylogenetic characterization of *Clonorchis sinensis* proteins homologous to the sigma-class glutathione transferase and their differential expression profiles. *Mol. Biochem. Parasitol.* **2016**, *206*, 46–55. [CrossRef]
174. Iriarte, A.; Arbildi, P.; La-Rocca, S.; Musto, H.; Fernández, V. Identification of novel glutathione transferases in *Echinococcus granulosus*. An evolutionary perspective. *Acta Trop.* **2012**, *123*, 208–216. [CrossRef]
175. Kim, J.G.; Ahn, C.S.; Kim, S.H.; Bae, Y.A.; Kwon, N.Y.; Kang, I.; Yang, H.J.; Sohn, W.M.; Kong, Y. *Clonorchis sinensis* omega-class glutathione transferases play major roles in the protection of the reproductive system during maturation and the response to oxidative stress. *Parasites Vectors* **2016**, *9*, 337. [CrossRef] [PubMed]
176. Ziniel, P.D.; Karumudi, B.; Barnard, A.H.; Fisher, E.M.; Thatcher, G.R.; Podust, L.M.; Williams, D.L. The *Schistosoma mansoni* Cytochrome P450 (CYP3050A1) Is Essential for Worm Survival and Egg Development. *PLoS Negl. Trop. Dis.* **2015**, *9*, e0004279. [CrossRef] [PubMed]
177. Pakharukova, M.Y.; Vavilin, V.A.; Sripa, B.; Laha, T.; Brindley, P.J.; Mordvinov, V.A. Functional Analysis of the Unique Cytochrome P450 of the Liver Fluke Opisthorchis felineus. *PLoS Negl. Trop. Dis.* **2015**, *9*, e0004258. [CrossRef]
178. Pal, R.; Rai, J.P. Phytochelatins: Peptides involved in heavy metal detoxification. *Appl. Biochem. Biotechnol.* **2010**, *160*, 945–963. [CrossRef] [PubMed]
179. Rea, P.A. Phytochelatin synthase: Of a protease a peptide polymerase made. *Physiol. Plant.* **2012**, *145*, 154–164. [CrossRef]
180. Ray, D.; Williams, D.L. Characterization of the phytochelatin synthase of *Schistosoma mansoni*. *PLoS Negl. Trop. Dis.* **2011**, *5*, e1168. [CrossRef] [PubMed]

181. Rigouin, C.; Vermeire, J.J.; Nylin, E.; Williams, D.L. Characterization of the phytochelatin synthase from the human parasitic nematode Ancylostoma ceylanicum. *Mol. Biochem. Parasitol.* **2013**, *191*, 1–6. [CrossRef]
182. Williams, D.L.; Bonilla, M.; Gladyshev, V.N.; Salinas, G. Thioredoxin glutathione reductase-dependent redox networks in platyhelminth parasites. *Antioxid. Redox Signal.* **2013**, *19*, 735–745. [CrossRef]
183. Bundy, J.G.; Kille, P.; Liebeke, M.; Spurgeon, D.J. Metallothioneins may not be enough–the role of phytochelatins in invertebrate metal detoxification. *Environ. Sci. Technol.* **2014**, *48*, 885–886. [CrossRef]
184. Mannino, M.H.; Patel, R.S.; Eccardt, A.M.; Perez Magnelli, R.A.; Robinson, C.; Janowiak, B.E.; Warren, D.E.; Fisher, J.S. Myoglobin as a versatile peroxidase: Implications for a more important role for vertebrate striated muscle in antioxidant defense. *Comp. Biochem. Physiol. B Biochem. Mol. Biol.* **2019**, *234*, 9–17. [CrossRef]
185. Ren, M.; He, L.; Huang, Y.; Mao, Q.; Li, S.; Qu, H.; Bian, M.; Liang, P.; Chen, X.; Ling, J.; et al. Molecular characterization of *Clonorchis sinensis* secretory myoglobin: Delineating its role in anti-oxidative survival. *Parasites Vectors* **2014**, *7*, 250. [CrossRef] [PubMed]
186. Penning, T.M. The aldo-keto reductases (AKRs): Overview. *Chem. Biol. Interact.* **2015**, *234*, 236–246. [CrossRef] [PubMed]
187. Forrest, G.L.; González, B. Carbonyl reductase. *Chem. Biol. Interact.* **2000**, *129*, 21–40. [CrossRef]
188. Escobedo, G.; Romano, M.C.; Morales-Montor, J. Differential in vitro effects of insulin on *Taenia crassiceps* and *Taenia solium* cysticerci. *J. Helminthol.* **2009**, *83*, 403–412. [CrossRef]
189. Adalid-Peralta, L.; Rosas, G.; Arce-Sillas, A.; Bobes, R.J.; Cárdenas, G.; Hernández, M.; Trejo, C.; Meneses, G.; Hernández, B.; Estrada, K.; et al. Effect of Transforming Growth Factor-β upon *Taenia solium* and *Taenia crassiceps* Cysticerci. *Sci. Rep.* **2017**, *7*, 12345. [CrossRef]
190. Navarrete-Perea, J.; Moguel, B.; Mendoza-Hernández, G.; Fragoso, G.; Sciutto, E.; Bobes, R.J.; Laclette, J.P. Identification and quantification of host proteins in the vesicular fluid of porcine *Taenia solium* cysticerci. *Exp. Parasitol.* **2014**, *143*, 11–17. [CrossRef]
191. Ahn, C.S.; Han, X.; Bae, Y.A.; Ma, X.; Kim, J.T.; Cai, H.; Yang, H.J.; Kang, I.; Wang, H.; Kong, Y. Alteration of immunoproteome profile of *Echinococcus granulosus* hydatid fluid with progression of cystic echinococcosis. *Parasites Vectors* **2015**, *8*, 10. [CrossRef]
192. Ahn, C.S.; Kim, J.G.; Han, X.; Kang, I.; Kong, Y. Comparison of *Echinococcus multilocularis* and *Echinococcus granulosus* hydatid fluid proteome provides molecular strategies for specialized host-parasite interactions. *Oncotarget* **2017**, *8*, 97009–97024. [CrossRef]
193. Flores-Bautista, J.; Navarrete-Perea, J.; Fragoso, G.; Flisser, A.; Soberón, X.; Laclette, J.P. Fate of uptaken host proteins in *Taenia solium* and *Taenia crassiceps* cysticerci. *Biosci. Rep.* **2018**, *38*, BSR20180636. [CrossRef]
194. Ahn, C.S.; Kim, J.G.; Bae, Y.A.; Kim, S.H.; Shin, J.H.; Yang, Y.; Kang, I.; Kong, Y. Fasciclin-calcareous corpuscle binary complex mediated protein-protein interactions in *Taenia solium* metacestode. *Parasites Vectors* **2017**, *10*, 438. [CrossRef]
195. Loos, J.A.; Caparros, P.A.; Nicolao, M.C.; Denegri, G.M.; Cumino, A.C. Identification and pharmacological induction of autophagy in the larval stages of *Echinococcus granulosus*: An active catabolic process in calcareous corpuscles. *Int. J. Parasitol.* **2014**, *44*, 415–427. [CrossRef] [PubMed]
196. Cwiklinski, K.; Robinson, M.W.; Donnelly, S.; Dalton, J.P. Complementary transcriptomic and proteomic analyses reveal the cellular and molecular processes that drive growth and development of *Fasciola hepatica* in the host liver. *BMC Genom.* **2021**, *22*, 46. [CrossRef] [PubMed]
197. Becerro-Recio, D.; González-Miguel, J.; Ucero, A.; Sotillo, J.; Martínez-Moreno, Á.; Pérez-Arévalo, J.; Cwiklinski, K.; Dalton, J.P.; Siles-Lucas, M. Recognition Pattern of the *Fasciola hepatica* Excretome/Secretome during the Course of an Experimental Infection in Sheep by 2D Immunoproteomics. *Pathogens* **2021**, *10*, 725. [CrossRef] [PubMed]
198. Virginio, V.G.; Monteiro, K.M.; Drumond, F.; de Carvalho, M.O.; Vargas, D.M.; Zaha, A.; Ferreira, H.B. Excretory/secretory products from in vitro-cultured *Echinococcus granulosus* protoscoleces. *Mol. Biochem. Parasitol.* **2012**, *183*, 15–22. [CrossRef] [PubMed]
199. Wang, H.; Zhang, C.S.; Fang, B.B.; Li, Z.D.; Li, L.; Bi, X.J.; Li, W.D.; Zhang, N.; Lin, R.Y.; Wen, H. Thioredoxin peroxidase secreted by *Echinococcus granulosus* (sensu stricto) promotes the alternative activation of macrophages via PI3K/AKT/mTOR pathway. *Parasites Vectors* **2019**, *12*, 542. [CrossRef]
200. García-Montoya, G.M.; Mesa-Arango, J.A.; Isaza-Agudelo, J.P.; Agudelo-Lopez, S.P.; Cabarcas, F.; Barrera, L.F.; Alzate, J.F. Transcriptome profiling of the cysticercus stage of the laboratory model *Taenia crassiceps*, strain ORF. *Acta Trop.* **2016**, *154*, 50–62. [CrossRef]
201. Greenberg, R.M. ABC multidrug transporters in schistosomes and other parasitic flatworms. *Parasitol. Int.* **2013**, *62*, 647–653. [CrossRef]

antioxidants

Review

Ambivalent Roles of Oxidative Stress in Triangular Relationships among Arthropod Vectors, Pathogens and Hosts

Emmanuel Pacia Hernandez [1], Anisuzzaman [2], Md Abdul Alim [2], Hayato Kawada [1,3], Kofi Dadzie Kwofie [1,4], Danielle Ladzekpo [1,4,5], Yuki Koike [3], Takahiro Inoue [3], Sana Sasaki [3], Fusako Mikami [1], Makoto Matsubayashi [6], Tetsuya Tanaka [7], Naotoshi Tsuji [1,3] and Takeshi Hatta [1,3,*]

1 Department of Parasitology and Tropical Medicine, Kitasato University School of Medicine, 1-15-1 Kitasato, Minami, Sagamihara 252-0374, Kanagawa, Japan; ephernandez4@alum.up.edu.ph (E.P.H.); hkawada@med.kitasato-u.ac.jp (H.K.); kwofiek@gmail.com (K.D.K.); dladzekpo.vip@tmd.ac.jp (D.L.); mikami@kitasato-u.ac.jp (F.M.); tsujin@med.kitasato-u.ac.jp (N.T.)
2 Department of Parasitology, Faculty of Veterinary Science, Bangladesh Agricultural University, Mymensingh 2202, Bangladesh; zaman.a.bau@gmail.com (A.); aalimpara@bau.edu.bd (M.A.A.)
3 Department of Molecular and Cellular Parasitology, Kitasato University Graduate School of Medical Science, 1-15-1 Kitasato, Minami, Sagamihara 252-0374, Kanagawa, Japan; koike.yuki@st.kitasato-u.ac.jp (Y.K.); dm21005@st.kitasato-u.ac.jp (T.I.); mm21025@st.kitasato-u.ac.jp (S.S.)
4 Department of Parasitology, Noguchi Memorial Institute for Medical Research, College of Health Sciences, University of Ghana, Legon, Accra P.O. Box LG 581, Ghana
5 Department of Environmental Parasitology, Tokyo Medical and Dental University, 1-5-45 Yushima, Bunkyo-Ku, Tokyo 113-8510, Japan
6 Department of Veterinary Immunology, Graduate School of Veterinary Sciences, Osaka Metropolitan University, Izumisano 598-8531, Osaka, Japan; matsubayashi@omu.ac.jp
7 Laboratory of Infectious Diseases, Joint Faculty of Veterinary Medicine, Kagoshima University, 1-21-24 Korimoto, Kagoshima 890-0065, Japan; k6199431@kadai.jp
* Correspondence: htakeshi@med.kitasato-u.ac.jp

Citation: Hernandez, E.P.; Anisuzzaman; Alim, M.A.; Kawada, H.; Kwofie, K.D.; Ladzekpo, D.; Koike, Y.; Inoue, T.; Sasaki, S.; Mikami, F.; et al. Ambivalent Roles of Oxidative Stress in Triangular Relationships among Arthropod Vectors, Pathogens and Hosts. *Antioxidants* **2022**, *11*, 1254. https://doi.org/10.3390/antiox11071254

Academic Editor: Serge Ankri

Received: 10 May 2022
Accepted: 21 June 2022
Published: 25 June 2022

Abstract: Blood-feeding arthropods, particularly ticks and mosquitoes are considered the most important vectors of arthropod-borne diseases affecting humans and animals. While feeding on blood meals, arthropods are exposed to high levels of reactive oxygen species (ROS) since heme and other blood components can induce oxidative stress. Different ROS have important roles in interactions among the pathogens, vectors, and hosts. ROS influence various metabolic processes of the arthropods and some have detrimental effects. In this review, we investigate the various roles of ROS in these arthropods, including their innate immunity and the homeostasis of their microbiomes, that is, how ROS are utilized to maintain the balance between the natural microbiota and potential pathogens. We elucidate the mechanism of how ROS are utilized to fight off invading pathogens and how the arthropod-borne pathogens use the arthropods' antioxidant mechanism to defend against these ROS attacks and their possible impact on their vector potentials or their ability to acquire and transmit pathogens. In addition, we describe the possible roles of ROS in chemical insecticide/acaricide activity and/or in the development of resistance. Overall, this underscores the importance of the antioxidant system as a potential target for the control of arthropod and arthropod-borne pathogens.

Keywords: ticks; mosquitoes; oxidative stress; ROS; microbiome; acaricide; insecticide resistance

1. Introduction

Every year more than 700,000 individuals die from diseases transmitted by hematophagous arthropods, of which malaria, dengue, human African trypanosomiasis, leishmaniasis, Chagas disease, yellow fever, Japanese encephalitis, onchocerciasis, and tick-borne encephalitis are particularly detrimental [1]. Hematophagous arthropods such as mosquitoes, triatomine bed bugs, sandflies, blackflies, midges, and ticks derive their nutritional requirement from blood feeding. While some of them are facultative blood feeders such as the female mosquitoes that need blood to trigger oogenesis; others, such as ticks, are obligate blood feeders that solely depend on blood meals from their hosts for their propagations, molting, development, and survival [2–7]. Most of them ingest huge amounts of blood in a single meal ranging from three to ten times their body weight to up to several hundred times their unfed body weight [8,9]. Due to this unique feeding behavior, these arthropods have become efficient bridging agents between vector-borne pathogens and their hosts. Therefore, the control of these vectors could also lead to the control of these diseases. Among blood-feeding arthropods, mosquitoes and ticks are recognized as the most important vectors of pathogens affecting animals or humans. Of the arthropod-borne diseases, only the mosquito-borne pathogens cause high morbidity and mortality globally and the collective global burden of mosquito-borne diseases is not less than that of AIDS. Malaria, a mosquito-borne protozoan disease, alone attributes to 45 million disability-adjusted life years (DALYs) per year and is considered one of the "big three" along with AIDS and tuberculosis [10]. On the other hand, ticks are only second to mosquitoes in terms of their potential to transmit a wide array of devastating pathogens, posing a global threat to both human and animal health. Therefore, the control of these vectors is of public health and veterinary concern [11].

Blood is an excellent source of nutrients for hematophagous arthropods; however, its components including heme, iron, amino acids, and even water could be deleterious for the attacking vectors themselves. Heme and free iron lead to the formation of reactive oxygen species (ROS), resulting in oxidative stress. Previous works of literature have already shed some light into the mechanisms of how vectors are engaged in the neutralization of oxidative stresses for their survival [8,12–14]. In this review, we describe the roles of ROS in mosquitoes and ticks, particularly in vector competency and the development of insecticide/acaricide resistance.

2. ROS Generation and Oxidative Stress in Arthropods

ROS are free radicals that are highly unstable and reactive and contain one or more unpaired electrons. They are naturally produced in all cells and organisms and are crucial in several metabolic processes including cell signaling, cellular proliferation, and transcription regulation [15,16]. A tight balance between ROS production and its neutralization supports healthy life and vector potentials and ameliorates the effects of insecticides/acaricides. However, an imbalance between ROS production and antioxidant activities results in cellular damage, a phenomenon known as oxidative stress, which is highly detrimental to the life of any organism, including hematophagous vectors.

2.1. *Blood Meal and the Fate of Heme within Arthropods*

Blood is an excellent source of nutrients, rich in proteins; it also contains the necessary carbohydrates, salts, and lipids that can supply the needs of the arthropod as well as its embryo development [17–20]. Hemoglobin and albumin, the most abundant proteins in the blood, make up almost 80% of the total proteins of blood and are greatly utilized by hematophagous arthropods and blood-dwelling parasites [21]. When hemoglobin is broken down during hematophagy, it results in the liberation of large amounts of heme. Heme itself is capable of promoting the production of ROS [14,22]. In vivo study demonstrates that most of the heme ingested is not absorbed but readily excreted in the feces through several mechanisms. In hemipterans, the lipid membranes in the gut epithelium catalyze the formation of crystalline heme aggregates known as hemozoin. In the mosquitoes, the heme is trapped in the peritrophic matrix of the gut. Ticks, on the other hand, digest blood intracellularly [23–26], wherein the liberated heme is trapped in hemosomes, specialized organelles in tick-gut cells designated to trap the liberated heme [8,22]. Although the gut of hematophagous arthropods efficiently tends to thwart the absorption of heme, a significant amount of heme is retained within and is capable of inducing oxidative damage at the cellular level and causes lipid peroxidation [12,14]. A mechanism to degrade heme and reduce its toxic effects is also present within vectors. Although seemingly beneficial, it can cause another problem by releasing highly reactive free Fe^{2+}, which may trigger the Fenton reaction, ultimately resulting in accelerated ROS production, and thus, eventually exacerbating oxidative stress [12]. However, this phenomenon is not observed in ticks since they lack the heme oxygenase gene [27]. In sandflies and *Aedes aegypti*, hemoglobin digestion intermediates also induce oxidative stress upon contact with the gut epithelium [16,28]. Iron is also acquired from the serum from the hosts' ferritin, transferrin, and other iron-binding proteins found in the host blood [13].

Hard ticks are pool feeders, and they attach for a long period that can range for 3–12 days depending on the species and developmental stage of life cycle. Interestingly, the attachment period may extend up to 60 days when they attach to a reptile host [3,5,6,29–31]. During feeding, ROS are also generated from the hosts' eosinophils, neutrophils, and macrophages through the lesion brought about by blood feeding [5,30,32,33]. In blood-feeding arthropods such as mosquitoes and ticks, the control of oxidative stress from heme and iron toxicity, which includes postprandial downregulation of ROS production in insect cells, matrix peritrophins, heme catabolism, and iron chaperones and shuttling, has already been thoroughly discussed in previous works [12,17,22].

2.2. Other Biological Sources of ROS

A variety of cellular developmental and metabolic activities and other biological events can lead to the generation of ROS (Table 1). Increased malondialdehyde (MDA) concentrations, an indicator of oxidative stress, were also observed during embryonic development as well as larval maturation of the ticks such as *Rhipicephalus microplus* and *Haemaphysalis longicornis* [34,35]. In mosquitoes, an increase in metabolic activities, including oogenesis, also results in the generation of ROS and these ROS accumulate over time and can affect the fecundity of mosquitoes [36]. Since mosquitoes are flying insects, ROS generation is further accelerated by the increased mitochondrial activity of the flight muscles [37]. Although it is not proven in blood-feeding arthropods, studies in insect models such as *Drosophila* indicate that a high production of ROS in the insect nervous system results in the ROS-mediated decline of neuron survival [16].

Table 1. Biological sources of ROS aside from blood meals.

Biological Activity	Reference
Activity for reproduction and growth	
Oogenesis	[36]
Embryonic development	[34,35]
Hatching and molting	[38]
Larval development	[34]
Normal homeostasis	
Flight activity	[37]
Nervous system activity	[16]
Cellular activity	[38]
Pathogen infection	
Microbial killing by ROS	[39]
Melanocytic encapsulation	[40]
Immune signaling	[41,42]
Arthropod microbiome	
Enterobacter production	[43]
Arthropod control	
Metabolism of acaricide/insecticide	[16]
Insecticide resistance	[44]
Phytochemical control	[45]

3. Vector Competency and Oxidative Stress

Vector competence (also termed vector potential) refers to the ability of arthropods to transmit pathogens, which is greatly influenced by the genetic and/or other intrinsic factors of arthropod vectors [46]. Additionally, it is also governed by the factors exerted by hosts themselves during pathogen inoculation, development, and propagation in particular hosts. During an infection, ROS have pivotal roles in the triangular relationship among vectors, pathogens, and hosts and may influence the triad either positively or negatively. A pluripotent molecule isolated from the salivary glands of *H. longicornis*, called longistatin [47], plays a central role in the feeding and development of ticks [3,4,9,48] and has been elegantly shown to ameliorate cellular ROS production in human endothelial cells [5], making it a key molecule in the survival of hard ticks. On the other hand, the acquisition of pathogens into a vector also induces modification of the normal ROS production resulting in oxidative stress to arthropod cells, which ultimately is being utilized by hematophagous arthropods to eliminate invading pathogens. Therefore, vector competence largely depends on a smart balance of the ROS that ensures the entrance, survival, and proliferation of pathogens into a vector. At the same time, maintenance of an optimum level of ROS is essential for the assurance of survival of the arthropods themselves to allow the feeding behavior that eventually ensures pathogen transmission [32,49–51].

3.1. ROS and Arthropod's Innate Immunity

To ensure their own survival and existence, arthropods use ROS to eliminate invading pathogens as well as to mount a better immune response during infection (Figure 1A) [52–54]. *Anopheles gambiae*, for example, can survive better at higher levels of systemic ROS when challenged with *Micrococcus* and *Escherichia*. Furthermore, the supplementation of antioxidants in the diet results in significantly higher mortality during bacterial infection [55], indicating that ROS and oxidative stress play a critical role in the arthropods' survival during the acquisition and transmission of infections. On the other hand, a *Plasmodium* refractory strain of *Ano. gambiae* was observed to be in a chronic state of oxidative stress, while the same parasite would survive if antioxidants were provided in the diet [8,40]. The same effects of oxidative stress are observed in the *Rh. microplus* (BME26) cell line. During infection with *Rickettsia ricketsii* or exposure to heat-killed microorganisms, upregulation of genes encoding for ROS production was observed, while antioxidant genes were downregulated [56]. Oxidative burst by macrophages efficiently eliminates pathogens basically by ROS, which are either toxic to the pathogen or work together with hydrolases, reactive nitrogen species, and the NADPH oxidase system (NOX). Supporting the above notion, bacterial infections have been shown to increase ROS in the ticks' hemocytes [39]. ROS play a role to block pathogen transmission by melanotic encapsulation, where invading pathogens are encapsulated to help the prevention of transmission. Melanotic encapsulation of *Plasmodium* has been shown in *Ano. gambiae.* The melanocytic capsule of the refractory strains of *Ano. gambiae* constructed around *Plasmodium* can block parasite development in the mosquitoes' midgut and the strain was observed to have higher levels of ROS [40]. On the other hand, in mosquitoes, ROS also act as a signaling molecule for the mitogen-activated protein kinase (MAPK)–dependent cascade and phosphatidylinositol 3-kinase (PI3K)/Akt-dependent pathway, which has been shown to regulate innate immunity and affects the physiology and development of the malarial parasite [41,42].

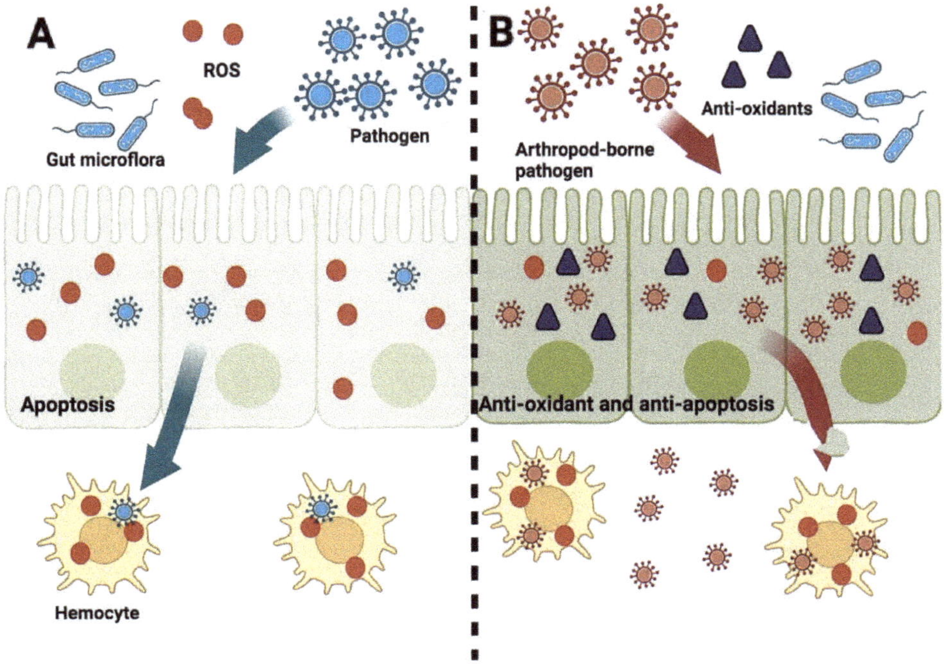

Figure 1. Schematic diagram of the life cycle of the pathogen (**A**) after infection of the arthropod versus the arthropod-borne pathogen's (**B**) life cycle and its interaction with ROS. Created with Biorender.com (accessed on 10 May 2022).

3.2. *ROS after the Establishment of Infection in Hosts*

While ROS are an important component of defense by the host against the pathogen, ROS are also generated during the establishment of a pathogen within a host and continue to be produced throughout the progression of the disease, for example, flaviviruses are known to induce the production of ROS, which are linked to apoptosis and are, thus, involved in the killing of infected cells together with the pathogen and eventually support the survival of the remaining non-infected cells. Through the production of ROS, infected individuals battle against pathogens to eliminate invading microbes in the early stage of invasion to prevent the progression of damage to the adjacent cells caused by the pathogens themselves and by the triggered inflammatory insults as well (Figure 1A) [57,58]. Therefore, this cellular response could affect the vectorial capacity of the arthropods. However, the coevolution of the arthropods with the pathogens carried by them has led to their coadaptation with each other's immune responses (Figure 1B). Pathogens have devised various protective shields to evade host responses to ensure their transmission, colonization, and survival in a hostile environment within vertebrate hosts, until either the recovery or the death of the host [49,50,59].

During dengue virus (DENV) infection, apoptosis is the usual outcome [60]. Apoptosis is usually brought about by the production of viral proteins, which disrupts the function of the endoplasmic reticulum (ER), resulting in the accumulation of misfolded and unfolded proteins. The presence of these misfolded and unfolded proteins activates the unfolded protein response (UPR). Even with the UPR, the mitigation of the effect of ER stress may not be addressed within a specific time and will still result in apoptosis [61,62]. However, in a mosquito cell line, it was found that mosquito cells were neither severely damaged nor subjected to apoptosis, rather the infection persisted in this setting, and ROS were detected. Interestingly, a p53 paralogue was upregulated during infection. The p53 is a transcription factor that selectively transcribes the catalase gene, which alleviates ROS accumulation within the cells, therefore reducing the rate of apoptosis (Figure 1B). In experiments that reduced the expression of the *p53* gene, ROS accumulated in the infected cells [60]. In *Ae. aegypti*, knockdown of the catalase gene also resulted in reduced oviposition and lifespan with H_2O_2 challenge and reduced virus titer in the midgut upon infection with DENV [63]. Aside from the catalase activity, glutathione S-transferase (GST) activity was also higher in DENV-infected cells. GST suppression resulted in an earlier release of superoxide ions and higher cell mortality. Interestingly, this upregulation was not observed in mammalian cells infected with the same virus, indicating that this phenomenon may be limited to only mosquito cells [64]. Besides GST activity, an additive anti-apoptotic activity was observed due to the upregulation of the inhibitor of apoptosis (IAP) [61]. ROS and oxidative stress are also believed to be controlled by the proper refolding of misfolded proteins, and this is usually achieved through the production of the X-box binding protein 1 (XBP1), which is presumed to be a critical transcription factor for various chaperones, including the *BiP/GRP 78* mRNA [62]. In contrast, West Nile virus, another flavivirus, also induces ROS production. However, the exact mechanism of this ROS production is still unknown. Mosquito cells infected with this virus-induced upregulation of Nrf2- and NRF1-mediated antioxidant genes, eventually result in elevated reduced glutathione (GSH) levels. This ultimately increased the oxidative capacity of the cells to withstand the oxidative stress elicited by the virus infection [65].

In contrast, transfection with the nonstructural protein 1 of the flavivirus, e.g., tick-borne encephalitis virus, induces oxidative stress in HEK293T cells and activates the antioxidant defense of these cells [66]. Moreover, in tick cells, the knockdown of the antioxidant *GST* molecule with subsequent infection of the Langat virus, another member of Flaviviridae, resulted in increased mortality, decreased proliferation, and decreased viral titer [67]. Furthermore, infection of LGTV in tick cells indicates a possible correction of the protein folding as seen by the upregulation of chaperone proteins, specifically heat shock proteins (HSPs) 90 and 70 [68,69]. These HSPs help in the refolding of misfolded proteins or are related to the degradation of terminally misfolded proteins to prevent protein aggregation, thereby creating an anti-apoptotic environment within cellular niches [58,70]. This corrective response was also observed in *Anaplasma phagocytophilum* infection in ticks and tick cells, wherein HSP20, HSP70, and HSP90 expressions have been upregulated [58,71,72]. Metabolomics also indicates that terminally misfolded proteins tend to prevent ER stress and apoptosis. Accordingly, *HSP70* knockdown decreases *Ana. phagocytophilum* titer in ticks [71].

Furthermore, *Ana. marginale* infection upregulated genes closely related to the generation of antioxidants. Simultaneous knockdown of catalase, glutathione peroxidase, and thioredoxin together with oxidative resistance 1 gene favored the colonization of *Ana. marginale* in BME26 cells, strongly supporting that the oxidant response is involved in the control of infection and the maintenance of cell survival [56]. Additionally, mitochondrial ROS production also increases in response to *Ana. phagocytophilum* to control the parasite. Conversely, to maintain the parasite's fitness and maintain the infection, other alternative ROS production and apoptosis pathways are also inhibited [73].

Another group of molecules that are also upregulated during infection are selenoproteins. They are a group of proteins that both catalyze and regulate several redox reactions [38]. It has been proposed that pathogens can induce the production of selenoproteins that not only allow their proliferation and transmission but also play a key role in pathogen acquisition. In *Borrelia burgdorferi* infection, Salp25D, a tick selenoprotein, is utilized against the oxidative stress from the inflammatory process at the biting site [32]. Knockdown of selenoprotein M reduces the titer of *Ana. marginale* and inhibits the development of the infective stage of *Ana. marginale* [74]. Selenoprotein P (SelP) is upregulated in the salivary glands of the *Ri. parkeri*–infected ticks to ameliorate oxidative stress during feeding. Furthermore, the knockdown of *SelP* genes also reduced the transovarial transmission of pathogens [52]. In addition to the direct control of ROS through antioxidant enzymes, the generation of ROS is also controlled by regulating free cations such as iron, which augments ROS productions. Ferritins that sequester free iron are upregulated in *Dermacentor variabilis* during *Escherichia coli* infection [13]. Bacterial iron-binding proteins could also sequester iron in the blood meal, which is expressed in the infective stage of *Ana. phagocytophilum* [75].

3.3. ROS and Arthropod Microbiome

In this review, microbiome refers to the overall genome of microorganisms in a certain niche, which has been shown to shape the phenome [76,77]. Therefore, attaining a balance between the natural microbiota and potential pathogens is very crucial. One way to maintain this balance is through the dual oxidase (Duox)–dependent ROS generation system [16]. Duox is a protein that mainly functions in the generation of ROS; however, the Duox-ROS pathway remains inactive unless proliferation occurs and bacteria come in contact with the mucosal barrier. In this manner, ROS produced from the Duox pathways attack invading pathogens through the mucosal barrier, particularly by H_2O_2. These attacks can disrupt the tyrosine phosphorylation network of invading pathogens and, thereby, reduce their fitness. Enterobacteria dominate in the midgut by maintaining gut homeostasis [78–80], and these bacterial species are adapted to survive within blood-feeding arthropods (e.g., ticks) as they are resistant to ROS killing. Since enterobacterial species are ROS-generating bacteria, during blood feeding, the gut environment would favor the growth of these bacterial species, overcoming the possible proliferation of *Plasmodium* and other pathogenic organisms [43,81–83]. Challenge with pathogens, therefore, accelerates the expansion of bacterial populations during blood feeding and they escape the constraints of the Duox system. One way for the natural microbiota to escape the Duox system is by avoiding contact with the gut epithelium. To achieve this, bacteria are encaged in a blood bolus during blood feeding. The formation of a dityrosine network (DTN) on the luminal surface of the gut epithelium also makes it difficult for the soluble immune mediators to penetrate the blood bolus; thus, the microbiota avoids activation of the immune responses [84]. The protective effects of this DTN are not only beneficial for the microbiota but also for the *Plasmodium* parasite [85]. The same DTN is also found in ticks and has been proved to maintain *B. burgdorferi* infection.

Furthermore, in ticks, some pathogenic organisms have already adapted to this strategy by altering their transcription mechanism and ameliorating the antioxidant mechanisms, including selenoproteins, to maintain the favorable levels of ROS that allow for their survival and growth. The knockdown of the *SelP* gene increases oxidative stress, thus, decreasing *Ri. parkeri* loads and increasing levels of *Francisella*-like symbionts such as *Candidatus* Midichloria mitochondrii and other bacteria [43,52].

4. ROS and Chemical Control of Arthropods

The metabolism of xenobiotics including acaricides and insecticides also leads to the generation of ROS [16]. Enzymes involved in the detoxification of these acaricides are also known as antioxidants, including glutathione S-transferase, cytochrome P450 monooxygenases, and some esterases. Higher expression or activities of these enzymes are involved with some resistant strains in both mosquitoes and ticks [44,86,87]. Permethrin-resistant strains of *Ano. gambiae* were shown to have higher ROS production rates and the increased ROS production in turn resulted in increased *GST* and *catalase* gene expression [44]. The same observations were reported in DDT-resistant *Ano. arabiensis* and *Ano. funestus,* wherein increased catalase and glutathione peroxidase activity was observed, indicating that resistant strains of mosquitoes need a higher capacity for coping with oxidative stress [88]. During every single blood meal, a transient decrease of oxidative stress was observed in mosquitoes. Increased expression and activity of antioxidant enzymes after multiple bouts of blood feeding is essential to maintain the homeostasis of DDT- and permethrin-resistant phenotypes of mosquitoes [88]. Disruption of the oxidative stress, e.g., elevated oxidative stress or reduced capability of minimizing oxidative stress, increases the susceptibility of DDT-resistant *Ano. gambiae* to DDT [89]. Additionally, some phytochemicals extracted from plants can augment ROS production. When *Culex quinquefasciatus* is exposed to the extracts of the medicinal plant, *Stachytarpheta jamaicensis*, it can lead to the generation of a high amount of ROS, resulting in the death of their larvae [45]. Plant extracts can decrease the levels of antioxidative enzymes in mosquitoes, for example, phytochemicals reduce the levels of carboxylesterases and superoxide dismutases in *Ae. aegypti* larvae (Figure 2) [90,91]. In the ticks, a negative correlation between GST enzyme expression and different phytochemical concentrations was observed, indicating the ability of these phytochemicals to induce oxidative stress in ticks [92].

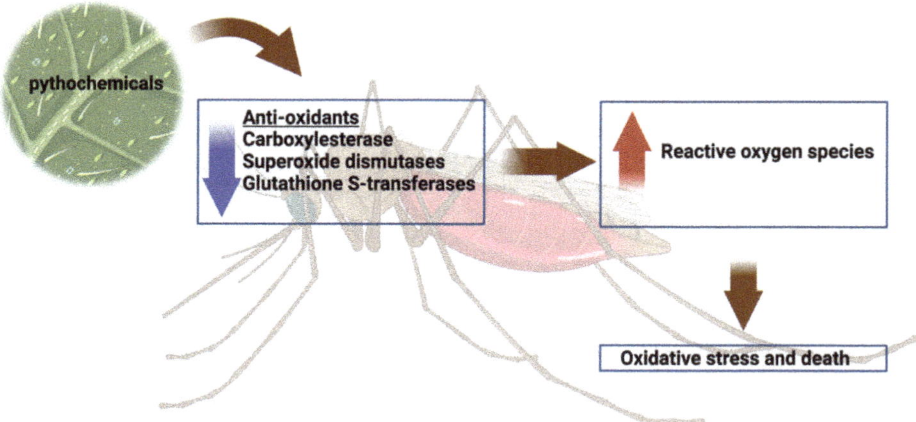

Figure 2. Schematic diagram on the activity of the phytochemicals and its possible mode of activity to cause mortality. Created with Biorender.com (accessed on 10 May 2022).

5. Conclusions

Mosquitoes and ticks are considered to be the most important vectors in the transmission of human and animal disease-causing pathogens. The vast array of pathogens includes viruses, bacteria, protozoa, and nematodes. Pathogens' acquisition and transmission largely depend on the ability of the pathogens to evade not only the arthropod's immune response but also their physiology. ROS are an integral part of the cellular physiology of any organism. In hematophagous arthropods, high amounts of ROS are produced by heme released during the digestion of blood meals. Moreover, several biological activities such as embryonic and larval development and flight activity augment ROS production. ROS have dual functions and are essential mediators of varieties of biological functions, including immune responses, cell signaling, and maintenance of the natural microbiome. ROS are also crucial in maintaining the natural microbiota in ticks. On the other hand, high amounts of ROS could result in redox imbalance, causing lipid peroxidation and DNA damage, and eventually death. ROS have pivotal roles in the effectiveness of mosquito resistance to insecticides. As pathogens have co-evolved with the arthropods that transmit them, the pathogens have devised ways to adapt to the physiological responses of the arthropods, even making use of their antioxidant response to evade ROS attacks of the host.

For the reasons mentioned above, it is important to consider the antioxidants or the antioxidant system as a potential target group in the development of drugs, vaccines, or other biological and chemical means to control the arthropods and their associated pathogens. In addition to the control aspects, this novel approach would address the vectorial capacity of these arthropods as well as their resistance to insecticides/acaricides. The use of phytochemicals has been a step in the right direction, but further understanding of the mechanism on how these phytochemicals work may be necessary to enable mass production globally.

Author Contributions: Conceptualization, E.P.H., T.T., N.T. and T.H.; writing, review, and editing, E.P.H., A. and M.A.A.; review and editing of final manuscript, H.K., K.D.K., D.L., Y.K., T.I., S.S., F.M. and M.M. All authors have read and agreed to the published version of the manuscript.

Funding: This work was supported by JSPS KAKENHI Grant Numbers 20F20093 and 20H03001.

Conflicts of Interest: The authors declare no conflict of interest.

References

1. Gubler, D.J. Vector-borne diseases. *Rev. Sci. Tech.* **2009**, *28*, 583–588. [CrossRef] [PubMed]
2. Ferreira, C.M.; Oliveira, M.P.; Paes, M.C.; Oliveira, M.F. Modulation of mitochondrial metabolism as a biochemical trait in blood feeding organisms: The redox vampire hypothesis redux. *Cell Biol. Int.* **2018**, *42*, 683–700. [CrossRef] [PubMed]
3. Anisuzzaman; Islam, M.K.; Alim, M.A.; Miyoshi, T.; Hatta, T.; Yamaji, K.; Matsumoto, Y.; Fujisaki, K.; Tsuji, N. Longistatin, a plasminogen activator, is key to the availability of blood-meals for ixodid ticks. *PLoS Pathog.* **2011**, *7*, e1001312. [CrossRef] [PubMed]
4. Anisuzzaman; Islam, M.K.; Alim, M.A.; Miyoshi, T.; Hatta, T.; Yamaji, K.; Matsumoto, Y.; Fujisaki, K.; Tsuji, N. Longistatin is an unconventional serine protease and induces protective immunity against tick infestation. *Mol. Biochem. Parasitol.* **2012**, *182*, 45–53. [CrossRef]
5. Anisuzzaman; Hatta, T.; Miyoshi, T.; Matsubayashi, M.; Islam, M.K.; Alim, M.A.; Anas, M.A.; Hasan, M.M.; Matsumoto, Y.; Yamamoto, Y.; et al. Longistatin in tick saliva blocks advanced glycation end-product receptor activation. *J. Clin. Investig.* **2014**, *124*, 4429–4444. [CrossRef] [PubMed]
6. Alim, M.A.; Tsuji, N.; Miyoshi, T.; Islam, M.K.; Hatta, T.; Fujisaki, K. Legumains from the hard tick *Haemaphysalis longicornis* play modulatory roles in blood feeding and gut cellular remodelling and impact on embryogenesis. *Int. J. Parasitol.* **2009**, *39*, 97–107. [CrossRef]
7. Hatta, T.; Tsuji, N.; Miyoshi, T.; Islam, M.K.; Alim, M.A.; Yamaji, K.; Anisuzzaman; Fujisaki, K. Leucine aminopeptidase, HlLAP, from the ixodid tick *Haemaphysalis longicornis*, plays vital roles in the development of oocytes. *Parasitol. Int.* **2010**, *59*, 286–289. [CrossRef]
8. Graça-Souza, A.V.; Maya-Monteiro, C.; Paiva-Silva, G.O.; Braz, G.R.C.; Paes, M.C.; Sorgine, M.H.F.; Oliveira, M.F.; Oliveira, P.L. Adaptations against heme toxicity in blood-feeding arthropods. *Insect Biochem. Mol. Biol.* **2006**, *36*, 322–335. [CrossRef]
9. Anisuzzaman; Islam, M.K.; Miyoshi, T.; Alim, M.A.; Hatta, T.; Yamaji, K.; Matsumoto, Y.; Fujisaki, K.; Tsuji, N. Longistatin, a novel EF-hand protein from the ixodid tick *Haemaphysalis longicornis*, is required for acquisition of host blood-meals. *Int. J. Parasitol.* **2010**, *40*, 721–729. [CrossRef]
10. GBD 2017 DALYs and HALE Collaborators Global. Regional, and national disability-adjusted life-years (DALYs) for 359 diseases and injuries and healthy life expectancy (HALE) for 195 countries and territories, 1990-2017: A systematic analysis for the Global Burden of Disease Study 2017. *Lancet* **2018**, *392*, 1859–1922. [CrossRef]
11. Dantas-Torres, F.; Otranto, D. Best Practices for Preventing Vector-Borne Diseases in Dogs and Humans. *Trends Parasitol.* **2016**, *32*, 43–55. [CrossRef] [PubMed]
12. Whiten, S.R.; Eggleston, H.; Adelman, Z.N. Ironing out the Details: Exploring the Role of Iron and Heme in Blood-Sucking Arthropods. *Front. Physiol.* **2017**, *8*, 1134. [CrossRef] [PubMed]
13. Galay, R.L.; Umemiya-Shirafuji, R.; Mochizuki, M.; Fujisaki, K.; Tanaka, T. Iron metabolism in hard ticks (Acari: Ixodidae): The antidote to their toxic diet. *Parasitol. Int.* **2015**, *64*, 182–189. [CrossRef] [PubMed]
14. Toh, S.Q.; Glanfield, A.; Gobert, G.N.; Jones, M.K. Heme and blood-feeding parasites: Friends or foes? *Parasit. Vectors* **2010**, *3*, 108. [CrossRef]
15. Hernandez, E.P.; Talactac, M.R.; Fujisaki, K.; Tanaka, T. The case for oxidative stress molecule involvement in the tick-pathogen interactions -an omics approach. *Dev. Comp. Immunol.* **2019**, *100*, 103409. [CrossRef]
16. Chaitanya, R.K.; Shashank, K.; Sridevi, P. Oxidative stress in invertebrate systems. In *Free Radicals and Diseases*; Ahmad, R., Ed.; InTech: London, UK, 2016.
17. Sabadin, G.A.; Xavier, M.A.; Vaz, I.D.S., Jr. Control of redox homeostasis in tick blood feeding. *Acta Sci. Vet.* **2019**, *47*, 1678. [CrossRef]
18. Fraga, A.; Moraes, J.; da Silva, J.R.; Costa, E.P.; Menezes, J.; da Silva Vaz, I.; Logullo, C.; da Fonseca, R.N.; Campos, E. Inorganic polyphosphates regulate hexokinase activity and reactive oxygen species generation in mitochondria of *Rhipicephalus* (*Boophilus*) *microplus* embryo. *Int. J. Biol. Sci.* **2013**, *9*, 842–852. [CrossRef]
19. Ibrahim, M.A. Traffic of the tick embryo basic protein during embryogenesis of the camel tick *Hyalomma dromedarii* (Acari: Ixodidae). *Exp. Appl. Acarol.* **1998**, *22*, 481–495. [CrossRef]
20. Santos, V.T.; Ribeiro, L.; Fraga, A.; de Barros, C.M.; Campos, E.; Moraes, J.; Fontenele, M.R.; Araújo, H.M.; Feitosa, N.M.; Logullo, C.; et al. The embryogenesis of the tick *Rhipicephalus* (*Boophilus*) *microplus*: The establishment of a new chelicerate model system. *Genesis* **2013**, *51*, 803–818. [CrossRef]
21. Frahm, S.; Anisuzzaman, A.; Prodjinotho, U.F.; Vejzagić, N.; Verschoor, A.; Prazeres da Costa, C. A novel cell-free method to culture *Schistosoma mansoni* from cercariae to juvenile worm stages for in vitro drug testing. *PLoS Negl. Trop. Dis.* **2019**, *13*, e0006590. [CrossRef]

22. Sterkel, M.; Oliveira, J.H.M.; Bottino-Rojas, V.; Paiva-Silva, G.O.; Oliveira, P.L. The Dose Makes the Poison: Nutritional Overload Determines the Life Traits of Blood-Feeding Arthropods. *Trends Parasitol.* **2017**, *33*, 633–644. [CrossRef] [PubMed]
23. Hatta, T.; Tsuji, N.; Miyoshi, T.; Alim, M.A.; Islam, M.K.; Fujisaki, K. Leucine aminopeptidase in the ixodid tick *Haemaphysalis longicornis*: Endogenous expression profiles in midgut. *J. Vet. Med. Sci.* **2009**, *71*, 589–594. [CrossRef] [PubMed]
24. Miyoshi, T.; Tsuji, N.; Islam, M.K.; Alim, M.A.; Hatta, T.; Huang, X.; Fujisaki, K. A set of serine proteinase paralogs are required for blood-digestion in the ixodid tick *Haemaphysalis longicornis*. *Parasitol. Int.* **2008**, *57*, 499–505. [CrossRef] [PubMed]
25. Alim, A.M.; Tsuji, N.; Miyoshi, T.; Khyrul Islam, M.; Huang, X.; Motobu, M.; Fujisaki, K. Characterization of asparaginyl endopeptidase, legumain induced by blood feeding in the ixodid tick *Haemaphysalis longicornis*. *Insect Biochem. Mol. Biol.* **2007**, *37*, 911–922. [CrossRef] [PubMed]
26. Alim, M.A.; Tsuji, N.; Miyoshi, T.; Islam, M.K.; Huang, X.; Hatta, T.; Fujisaki, K. HlLgm2, a member of asparaginyl endopeptidases/legumains in the midgut of the ixodid tick *Haemaphysalis longicornis*, is involved in blood-meal digestion. *J. Insect Physiol.* **2008**, *54*, 573–585. [CrossRef]
27. Perner, J.; Sobotka, R.; Sima, R.; Konvickova, J.; Sojka, D.; de Oliveira, P.L.; Hajdusek, O.; Kopacek, P. Acquisition of exogenous haem is essential for tick reproduction. *Elife* **2016**, *5*, e12318. [CrossRef]
28. de Araújo, A.P.O.; Telleria, E.L.; da Dutra, J.M.F.; Júlio, R.M.; Traub-Csekö, Y.M. Disruption of the peritrophic matrix by exogenous chitinase feeding reduces fecundity in *Lutzomyia longipalpis* females. *Mem. Inst. Oswaldo Cruz* **2012**, *107*, 543–545. [CrossRef]
29. Hatta, T.; Miyoshi, T.; Matsubayashi, M.; Islam, M.K.; Alim, M.A.; Anisuzzaman; Yamaji, K.; Fujisaki, K.; Tsuji, N. Semi-artificial mouse skin membrane feeding technique for adult tick, *Haemaphysalis longicornis*. *Parasit. Vectors* **2012**, *5*, 263. [CrossRef]
30. Sonenshine, D.E.; Anderson, J.M. Mouthparts and Digestive System: Anatomy and Molecular biology of Feeding and Digestion. In *Biology of Ticks*, 2nd ed.; Sonenshine, D.E., Roe, R.M., Eds.; Oxford University Press: New York, NY, USA, 2014; Volume 1.
31. Alim, M.A.; Tsuji, N.; Miyoshi, T.; Islam, M.K.; Hatta, T.; Yamaji, K.; Fujisaki, K. Developmental stage- and organ-specific expression profiles of asparaginyl endopeptidases/legumains in the ixodid tick *Haemaphysalis longicornis*. *J. Vet. Med. Sci.* **2008**, *70*, 1363–1366. [CrossRef]
32. Narasimhan, S.; Sukumaran, B.; Bozdogan, U.; Thomas, V.; Liang, X.; De Ponte, K.; Marcantonio, N.; Koski, R.A.; Anderson, J.F.; Kantor, F.; et al. A tick antioxidant facilitates the Lyme disease agent's successful migration from the mammalian host to the arthropod vector. *Cell Host Microbe* **2007**, *2*, 7–18. [CrossRef]
33. Narasimhan, S.; Fikrig, E. Tick microbiome: The force within. *Trends Parasitol.* **2015**, *31*, 315–323. [CrossRef] [PubMed]
34. Freitas, D.R.J.; Rosa, R.M.; Moraes, J.; Campos, E.; Logullo, C.; Da Silva Vaz, I.; Masuda, A. Relationship between glutathione S-transferase, catalase, oxygen consumption, lipid peroxidation and oxidative stress in eggs and larvae of *Boophilus microplus* (Acarina: Ixodidae). *Comp. Biochem. Physiol. Part A Mol. Integr. Physiol.* **2007**, *146*, 688–694. [CrossRef] [PubMed]
35. Hernandez, E.P.; Shimazaki, K.; Niihara, H.; Umemiya-Shirafuji, R.; Fujisaki, K.; Tanaka, T. Expression analysis of glutathione S-transferases and ferritins during the embryogenesis of the tick *Haemaphysalis longicornis*. *Heliyon* **2020**, *6*, e03644. [CrossRef] [PubMed]
36. DeJong, R.J.; Miller, L.M.; Molina-Cruz, A.; Gupta, L.; Kumar, S.; Barillas-Mury, C. Reactive oxygen species detoxification by catalase is a major determinant of fecundity in the mosquito *Anopheles gambiae*. *Proc. Natl. Acad. Sci. USA* **2007**, *104*, 2121–2126. [CrossRef] [PubMed]
37. Strode, C.; Steen, K.; Ortelli, F.; Ranson, H. Differential expression of the detoxification genes in the different life stages of the malaria vector *Anopheles gambiae*. *Insect Mol. Biol.* **2006**, *15*, 523–530. [CrossRef]
38. Reeves, M.A.; Hoffmann, P.R. The human selenoproteome: Recent insights into functions and regulation. *Cell Mol. Life Sci.* **2009**, *66*, 2457–2478. [CrossRef]
39. Pereira, L.S.; Oliveira, P.L.; Barja-Fidalgo, C.; Daffre, S. Production of reactive oxygen species by hemocytes from the cattle tick *Boophilus microplus*. *Exp. Parasitol.* **2001**, *99*, 66–72. [CrossRef]
40. Kumar, S.; Christophides, G.K.; Cantera, R.; Charles, B.; Han, Y.S.; Meister, S.; Dimopoulos, G.; Kafatos, F.C.; Barillas-Mury, C. The role of reactive oxygen species on *Plasmodium* melanotic encapsulation in *Anopheles gambiae*. *Proc. Natl. Acad. Sci. USA* **2003**, *100*, 14139–14144. [CrossRef]
41. Surachetpong, W.; Pakpour, N.; Cheung, K.W.; Luckhart, S. Reactive oxygen species-dependent cell signaling regulates the mosquito immune response to *Plasmodium falciparum*. *Antioxid. Redox Signal.* **2011**, *14*, 943–955. [CrossRef]
42. Corby-Harris, V.; Drexler, A.; Watkins de Jong, L.; Antonova, Y.; Pakpour, N.; Ziegler, R.; Ramberg, F.; Lewis, E.E.; Brown, J.M.; Luckhart, S.; et al. Activation of Akt signaling reduces the prevalence and intensity of malaria parasite infection and lifespan in *Anopheles stephensi* mosquitoes. *PLoS Pathog.* **2010**, *6*, e1001003. [CrossRef]
43. Budachetri, K.; Kumar, D.; Crispell, G.; Beck, C.; Dasch, G.; Karim, S. The tick endosymbiont *Candidatus* Midichloria mitochondrii and selenoproteins are essential for the growth of *Rickettsia parkeri* in the Gulf Coast tick vector. *Microbiome* **2018**, *6*, 141. [CrossRef] [PubMed]
44. Otali, D.; Novak, R.J.; Wan, W.; Bu, S.; Moellering, D.R.; De Luca, M. Increased production of mitochondrial reactive oxygen species and reduced adult life span in an insecticide-resistant strain of *Anopheles gambiae*. *Bull. Entomol. Res.* **2014**, *104*, 323–333. [CrossRef] [PubMed]
45. Rajan, V.; Puthur, R. Activation of the Oxidative Stress in *Culex quinquefasciatus* by the Augmented Production of Reactive Oxygen Species (ROS) in response to *Stachytarpheta jamaicensis* Exposure. *J. Commun. Dis.* **2021**, *53*, 43–51.

46. Beerntsen, B.T.; James, A.A.; Christensen, B.M. Genetics of mosquito vector competence. *Microbiol. Mol. Biol. Rev.* **2000**, *64*, 115–137. [CrossRef] [PubMed]
47. Anisuzzaman; Islam, M.K.; Alim, M.A.; Tsuji, N. Longistatin, an EF-hand Ca^{2+}-binding protein from vector tick: Identification, purification, and characterization. *Methods Mol. Biol.* **2013**, *963*, 127–146.
48. Anisuzzaman; Khyrul Islam, M.; Abdul Alim, M.; Miyoshi, T.; Hatta, T.; Yamaji, K.; Matsumoto, Y.; Fujisaki, K.; Tsuji, N. Longistatin, a novel plasminogen activator from vector ticks, is resistant to plasminogen activator inhibitor-1. *Biochem. Biophys. Res. Commun.* **2011**, *413*, 599–604. [CrossRef]
49. de la Fuente, J.; Antunes, S.; Bonnet, S.; Cabezas-Cruz, A.; Domingos, A.G.; Estrada-Peña, A.; Johnson, N.; Kocan, K.M.; Mansfield, K.L.; Nijhof, A.M.; et al. Tick-Pathogen Interactions and Vector Competence: Identification of Molecular Drivers for Tick-Borne Diseases. *Front. Cell Infect. Microbiol.* **2017**, *7*, 114. [CrossRef]
50. Hajdušek, O.; Síma, R.; Ayllón, N.; Jalovecká, M.; Perner, J.; de la Fuente, J.; Kopáček, P. Interaction of the tick immune system with transmitted pathogens. *Front. Cell Infect. Microbiol.* **2013**, *3*, 26. [CrossRef]
51. Tsuji, N.; Miyoshi, T.; Battsetseg, B.; Matsuo, T.; Xuan, X.; Fujisaki, K. A cysteine protease is critical for *Babesia* spp. transmission in *Haemaphysalis* ticks. *PLoS Pathog.* **2008**, *4*, e1000062. [CrossRef]
52. Budachetri, K.; Crispell, G.; Karim, S. *Amblyomma maculatum* SECIS binding protein 2 and putative selenoprotein P are indispensable for pathogen replication and tick fecundity. *Insect Biochem. Mol. Biol.* **2017**, *88*, 37–47. [CrossRef]
53. Ha, E.M.; Oh, C.T.; Ryu, J.H.; Bae, Y.S.; Kang, S.W.; Jang, I.H.; Brey, P.T.; Lee, W.J. An antioxidant system required for host protection against gut infection in *Drosophila*. *Dev. Cell* **2005**, *8*, 125–132. [CrossRef] [PubMed]
54. Muralidharan, S.; Mandrekar, P. Cellular stress response and innate immune signaling: Integrating pathways in host defense and inflammation. *J. Leukoc. Biol.* **2013**, *94*, 1167–1184. [CrossRef] [PubMed]
55. Molina-Cruz, A.; DeJong, R.J.; Charles, B.; Gupta, L.; Kumar, S.; Jaramillo-Gutierrez, G.; Barillas-Mury, C. Reactive oxygen species modulate *Anopheles gambiae* immunity against bacteria and *Plasmodium*. *J. Biol. Chem.* **2008**, *283*, 3217–3223. [CrossRef] [PubMed]
56. Kalil, S.P.; da Rosa, R.D.; Capelli-Peixoto, J.; Pohl, P.C.; de Oliveira, P.L.; Fogaça, A.C.; Daffre, S. Immune-related redox metabolism of embryonic cells of the tick *Rhipicephalus microplus* (BME26) in response to infection with *Anaplasma marginale*. *Parasit. Vectors* **2017**, *10*, 613. [CrossRef] [PubMed]
57. Ashida, H.; Mimuro, H.; Ogawa, M.; Kobayashi, T.; Sanada, T.; Kim, M.; Sasakawa, C. Cell death and infection: A double-edged sword for host and pathogen survival. *J. Cell Biol.* **2011**, *195*, 931–942. [CrossRef]
58. Villar, M.; Ayllón, N.; Alberdi, P.; Moreno, A.; Moreno, M.; Tobes, R.; Mateos-Hernández, L.; Weisheit, S.; Bell-Sakyi, L.; de la Fuente, J. Integrated Metabolomics, Transcriptomics and Proteomics Identifies Metabolic Pathways Affected by *Anaplasma phagocytophilum* Infection in Tick Cells. *Mol. Cell Proteom.* **2015**, *14*, 3154–3172. [CrossRef]
59. Villar, M.; Popara, M.; Bonzón-Kulichenko, E.; Ayllón, N.; Vázquez, J.; de la Fuente, J. Characterization of the tick-pathogen interface by quantitative proteomics. *Ticks Tick Borne Dis.* **2012**, *3*, 154–158. [CrossRef]
60. Chen, T.H.; Wu, Y.J.; Hou, J.N.; Chiang, Y.H.; Cheng, C.C.; Sifiyatun, E.; Chiu, C.H.; Wang, L.C.; Chen, W.J. A novel p53 paralogue mediates antioxidant defense of mosquito cells to survive dengue virus replication. *Virology* **2018**, *519*, 156–169. [CrossRef]
61. Chen, T.H.; Lo, Y.P.; Yang, C.F.; Chen, W.J. Additive protection by antioxidant and apoptosis-inhibiting effects on mosquito cells with dengue 2 virus infection. *PLoS Negl. Trop. Dis.* **2012**, *6*, e1613. [CrossRef]
62. Chen, T.H.; Chiang, Y.H.; Hou, J.N.; Cheng, C.C.; Sofiyatun, E.; Chiu, C.H.; Chen, W.J. XBP1-Mediated BiP/GRP78 Upregulation Copes with Oxidative Stress in Mosquito Cells during Dengue 2 Virus Infection. *Biomed Res. Int.* **2017**, *2017*, 3519158. [CrossRef]
63. Oliveira, J.H.M.; Talyuli, O.A.C.; Goncalves, R.L.S.; Paiva-Silva, G.O.; Sorgine, M.H.F.; Alvarenga, P.H.; Oliveira, P.L. Catalase protects *Aedes aegypti* from oxidative stress and increases midgut infection prevalence of dengue but not Zika. *PLoS Negl. Trop. Dis.* **2017**, *11*, e0005525. [CrossRef] [PubMed]
64. Chen, T.H.; Tang, P.; Yang, C.F.; Kao, L.H.; Lo, Y.P.; Chuang, C.K.; Shih, Y.T.; Chen, W.J. Antioxidant defense is one of the mechanisms by which mosquito cells survive dengue 2 viral infection. *Virology* **2011**, *410*, 410–417. [CrossRef]
65. Basu, M.; Courtney, S.C.; Brinton, M.A. Arsenite-induced stress granule formation is inhibited by elevated levels of reduced glutathione in West Nile virus-infected cells. *PLoS Pathog.* **2017**, *13*, e1006240. [CrossRef]
66. Kuzmenko, Y.V.; Smirnova, O.A.; Ivanov, A.V.; Starodubova, E.S.; Karpov, V.L. Nonstructural Protein 1 of Tick-Borne Encephalitis Virus Induces Oxidative Stress and Activates Antioxidant Defense by the Nrf2/ARE Pathway. *Intervirology* **2016**, *59*, 111–117. [CrossRef]
67. Hernandez, E.P.; Talactac, M.R.; Vitor, R.J.S.; Yoshii, K.; Tanaka, T. An *Ixodes scapularis* glutathione S-transferase plays a role in cell survival and viability during Langat virus infection of a tick cell line. *Acta Trop.* **2021**, *214*, 105763. [CrossRef] [PubMed]
68. Grabowski, J.M.; Perera, R.; Roumani, A.M.; Hedrick, V.E.; Inerowicz, H.D.; Hill, C.A.; Kuhn, R.J. Changes in the Proteome of Langat-Infected *Ixodes scapularis* ISE6 Cells: Metabolic Pathways Associated with Flavivirus Infection. *PLoS Negl. Trop. Dis.* **2016**, *10*, e0004180. [CrossRef] [PubMed]
69. Grabowski, J.M.; Gulia-Nuss, M.; Kuhn, R.J.; Hill, C.A. RNAi reveals proteins for metabolism and protein processing associated with Langat virus infection in *Ixodes scapularis* (black-legged tick) ISE6 cells. *Parasit. Vectors* **2017**, *10*, 24. [CrossRef] [PubMed]
70. Fulda, S.; Gorman, A.M.; Hori, O.; Samali, A. Cellular stress responses: Cell survival and cell death. *Int. J. Cell Biol.* **2010**, *2010*, 214074. [CrossRef]

71. Busby, A.T.; Ayllón, N.; Kocan, K.M.; Blouin, E.F.; de la Fuente, G.; Galindo, R.C.; Villar, M.; de la Fuente, J. Expression of heat shock proteins and subolesin affects stress responses, *Anaplasma phagocytophilum* infection and questing behaviour in the tick, *Ixodes scapularis*. *Med. Vet. Entomol.* **2012**, *26*, 92–102. [CrossRef]
72. Villar, M.; Ayllón, N.; Busby, A.T.; Galindo, R.C.; Blouin, E.F.; Kocan, K.M.; Bonzón-Kulichenko, E.; Zivkovic, Z.; Almazán, C.; Torina, A.; et al. Expression of Heat Shock and Other Stress Response Proteins in Ticks and Cultured Tick Cells in Response to *Anaplasma* spp. Infection and Heat Shock. *Int. J. Proteom.* **2010**, *2010*, 657261. [CrossRef]
73. Alberdi, P.; Cabezas-Cruz, A.; Prados, P.E.; Rayo, M.V.; Artigas-Jerónimo, S.; de la Fuente, J. The redox metabolic pathways function to limit *Anaplasma phagocytophilum* infection and multiplication while preserving fitness in tick vector cells. *Sci. Rep.* **2019**, *9*, 13236. [CrossRef] [PubMed]
74. Kocan, K.M.; Zivkovic, Z.; Blouin, E.F.; Naranjo, V.; Almazán, C.; Mitra, R.; de la Fuente, J. Silencing of genes involved in *Anaplasma marginale*-tick interactions affects the pathogen developmental cycle in *Dermacentor variabilis*. *BMC Dev. Biol.* **2009**, *9*, 42. [CrossRef] [PubMed]
75. Kahlon, A.; Ojogun, N.; Ragland, S.A.; Seidman, D.; Troese, M.J.; Ottens, A.K.; Mastronunzio, J.E.; Truchan, H.K.; Walker, N.J.; Borjesson, D.L.; et al. *Anaplasma phagocytophilum* Asp14 is an invasin that interacts with mammalian host cells via its C terminus to facilitate infection. *Infect. Immun.* **2013**, *81*, 65–79. [CrossRef] [PubMed]
76. Hooper, L.V.; Gordon, J.I. Commensal host-bacterial relationships in the gut. *Science* **2001**, *292*, 1115–1118. [CrossRef] [PubMed]
77. Kinross, J.M.; Darzi, A.W.; Nicholson, J.K. Gut microbiome-host interactions in health and disease. *Genome Med.* **2011**, *3*, 14. [CrossRef]
78. Budachetri, K.; Browning, R.E.; Adamson, S.W.; Dowd, S.E.; Chao, C.-C.; Ching, W.M.; Karim, S. An insight into the microbiome of the *Amblyomma maculatum* (Acari: Ixodidae). *J. Med. Entomol.* **2014**, *51*, 119–129. [CrossRef]
79. Cirimotich, C.M.; Ramirez, J.L.; Dimopoulos, G. Native microbiota shape insect vector competence for human pathogens. *Cell Host Microbe* **2011**, *10*, 307–310. [CrossRef]
80. Cirimotich, C.M.; Dong, Y.; Clayton, A.M.; Sandiford, S.L.; Souza-Neto, J.A.; Mulenga, M.; Dimopoulos, G. Natural microbe-mediated refractoriness to *Plasmodium* infection in *Anopheles gambiae*. *Science* **2011**, *332*, 855–858. [CrossRef]
81. Crispell, G.; Budachetri, K.; Karim, S. *Rickettsia parkeri* colonization in *Amblyomma maculatum*: The role of superoxide dismutases. *Parasit. Vectors* **2016**, *9*, 291. [CrossRef]
82. Kumar, D.; Budachetri, K.; Meyers, V.C.; Karim, S. Assessment of tick antioxidant responses to exogenous oxidative stressors and insight into the role of catalase in the reproductive fitness of the Gulf Coast tick, *Amblyomma maculatum*. *Insect Mol. Biol.* **2016**, *25*, 283–294. [CrossRef]
83. Lee, W.J. Bacterial-modulated signaling pathways in gut homeostasis. *Sci. Signal.* **2008**, *1*, pe24. [CrossRef] [PubMed]
84. Champion, C.J.; Xu, J. The impact of metagenomic interplay on the mosquito redox homeostasis. *Free Radic. Biol. Med.* **2017**, *105*, 79–85. [CrossRef] [PubMed]
85. Romoli, O.; Gendrin, M. The tripartite interactions between the mosquito, its microbiota and *Plasmodium*. *Parasit. Vectors* **2018**, *11*, 200. [CrossRef]
86. Baron, S.; Barrero, R.A.; Black, M.; Bellgard, M.I.; van Dalen, E.M.S.; Fourie, J.; Maritz-Olivier, C. Differentially expressed genes in response to amitraz treatment suggests a proposed model of resistance to amitraz in *R. decoloratus* ticks. *Int. J. Parasitol. Drugs Drug Resist.* **2018**, *8*, 361–371. [CrossRef] [PubMed]
87. Gong, Y.; Li, T.; Zhang, L.; Gao, X.; Liu, N. Permethrin induction of multiple cytochrome P450 genes in insecticide resistant mosquitoes, *Culex quinquefasciatus*. *Int. J. Biol. Sci.* **2013**, *9*, 863–871. [CrossRef]
88. Oliver, S.V.; Brooke, B.D. The Role of Oxidative Stress in the Longevity and Insecticide Resistance Phenotype of the Major Malaria Vectors *Anopheles arabiensis* and *Anopheles funestus*. *PLoS ONE* **2016**, *11*, e0151049. [CrossRef]
89. Champion, C.J.; Xu, J. Redox state affects fecundity and insecticide susceptibility in *Anopheles gambiae*. *Sci. Rep.* **2018**, *8*, 13054. [CrossRef]
90. Koodalingam, A.; Deepalakshmi, R.; Ammu, M.; Rajalakshmi, A. Effects of NeemAzal on marker enzymes and hemocyte phagocytic activity of larvae and pupae of the vector mosquito *Aedes aegypti*. *J. Asia. Pac. Entomol.* **2014**, *17*, 175–181. [CrossRef]
91. Thanigaivel, A.; Senthil-Nathan, S.; Vasantha-Srinivasan, P.; Edwin, E.S.; Ponsankar, A.; Selin-Rani, S.; Pradeepa, V.; Chellappandian, M.; Kalaivani, K.; Abdel-Megeed, A.; et al. Chemicals isolated from *Justicia adhatoda* Linn reduce fitness of the mosquito, *Aedes aegypti* L. *Arch. Insect. Biochem. Physiol.* **2017**, *94*, e21384. [CrossRef]
92. Agwunobi, D.O.; Pei, T.; Wang, X.; Yu, Z.; Liu, J.Z. Expression profiles of glutathione *S*-transferases genes in semi-engorged *Haemaphysalis longicornis* (Acari: Ixodidae) exposed to *Cymbopogon citratus* essential oil. *Sys. Appl. Acarol.* **2020**, *25*, 918–930.

antioxidants

MDPI

Article

Immunomodulation and Antioxidant Activities as Possible Trypanocidal and Cardioprotective Mechanisms of Major Terpenes from *Lippia alba* Essential Oils in an Experimental Model of Chronic Chagas Disease

Denerieth Ximena Espinel-Mesa [1], Clara Isabel González Rugeles [2], Julio César Mantilla Hernández [3], Elena E. Stashenko [4], Carlos Andrés Villegas-Lanau [5], John Jaime Quimbaya Ramírez [1] and Liliana Torcoroma García Sánchez [1,*]

1 Infectious Diseases Postgraduate Program, Instituto de Investigación Masira, Universidad de Santander, Bucaramanga 680006, Santander, Colombia; buc19861011@mail.udes.edu.co (D.X.E.-M.); jo.quimbaya@mail.udes.edu.co (J.J.Q.R.)
2 Immunology and Molecular Epidemiology Group, School of Microbiology, Universidad Industrial de Santander, Bucaramanga 680002, Santander, Colombia; cig@uis.edu.co
3 Pathology Department, School of Medicine, Universidad Industrial de Santander, Bucaramanga 680002, Santander, Colombia; jumantil@uis.edu.co
4 National Research Center for the Agroindustrialization of Tropical Aromatic and Medicinal Plant Species—CENIVAM, Universidad Industrial de Santander, Bucaramanga 680002, Santander, Colombia; elena@tucan.uis.edu.co
5 Neurosciences Group of Antioquia, Brain Bank, Universidad de Antioquia, Medellín 050010, Antioquia, Colombia; andres.villegas@udea.edu.co
* Correspondence: l.torcoroma@udes.edu.co

Citation: Espinel-Mesa, D.X.; González Rugeles, C.I.; Mantilla Hernández, J.C.; Stashenko, E.E.; Villegas-Lanau, C.A.; Quimbaya Ramírez, J.J.; García Sánchez, L.T. Immunomodulation and Antioxidant Activities as Possible Trypanocidal and Cardioprotective Mechanisms of Major Terpenes from *Lippia alba* Essential Oils in an Experimental Model of Chronic Chagas Disease. *Antioxidants* **2021**, *10*, 1851. https://doi.org/10.3390/antiox10111851

Academic Editor: Serge Ankri

Received: 16 October 2021
Accepted: 15 November 2021
Published: 22 November 2021

Abstract: In the late phase of *Trypanosoma cruzi* infection, parasite persistence and an exaggerated immune response accompanied by oxidative stress play a crucial role in the genesis of Chronic Chagasic Cardiomyopathy (CCC). Current treatments (Benznidazole (BNZ) and Nifurtimox) can effect only the elimination of the parasite, but are ineffective for late stage treatment and for preventing heart damage and disease progression. In vivo trypanocidal and cardioprotective activity has been reported for *Lippia alba* essential oils (EOs), ascribed to their two major terpenes, limonene and caryophyllene oxide. To investigate the role of antioxidant and immunomodulatory mechanisms behind these properties, chronic-*T. cruzi*-infected rats were treated with oral synergistic mixtures of the aforementioned EOs. For this purpose, the EOs were optimized through limonene-enrichment fractioning and by the addition of exogenous caryophyllene oxide (LIMOX) and used alone or in combined therapy with subtherapeutic doses of BNZ (LIMOXBNZ). Clinical, toxicity, inflammatory, oxidative, and parasitological (qPCR) parameters were assessed in cardiac tissue. These therapies demonstrated meaningful antioxidant and immunomodulatory activity on markers involved in CCC pathogenesis (IFN-γ, TNF-α, IL-4, IL-10, and iNOS), which could explain their significant trypanocidal properties and their noteworthy role in preventing, and even reversing, the progression of cardiac damage in chronic Chagas disease.

Keywords: chronic Chagas disease; *Trypanosoma cruzi*; antioxidant; immunomodulation; immunohistochemistry; essential oils; *Lippia alba*; limonene; caryophyllene oxide

1. Introduction

Chagas disease is a parasitosis caused by *Trypanosoma cruzi* that affects approximately 10 million people worldwide, of whom around 30% will eventually develop organomegaly of the digestive tract or heart [1]. Dilated cardiomyopathy, or Chagas heart disease (CHD), is the most relevant manifestation of this infection during its chronic phase [1], making it the most prevalent cardiac infection in the world, and causing a significant public health problem in Latin American countries where it is endemic [2]. Because of the silent course of this parasitosis and the late appearance of its symptoms (usually 8–10 years after acquiring the parasite, and without any other pathognomonic signs), CHD can be confused with other etiologies, delaying timely diagnosis [3]. These factors combined have contributed to CHD being presently considered a neglected and high-cost disease, with an average cost of care for chronic cases estimated at USD $44,955 per person [4].

Progression of the disease is characterized by the persistence of the parasite in smooth muscle, and is associated with an exacerbated immune response and a permanent oxidative environment. Research indicates that these conditions together contribute to tissue, neurological, and microvasculature injury; with a consequent deterioration of contractile capacity and dilation of muscle tissue; which in turn can culminate in an eventual loss of the organ's physiological function and even result in death [5]. Current treatments available for therapeutic intervention against this infection are based on the use of two nitro-heterocyclic compounds: nitrofuran Nifurtimox (NFX) and nitroimidazole Benznidazole (BNZ). These compounds, well-established for more than 50 years as the conventional therapy against *T. cruzi* infection, have been found to exhibit limited trypanocidal activity (between 50–80% in the acute phase, and 8–20% in the chronic) [6], with high toxicity due to non-selective oxidative damage.

In general terms, patients assigned therapies based on these drugs demonstrate low treatment adherence due to the presence of multiple adverse side effects (mainly severe anorexia, digestive intolerance, and neurological disorders), as well as long treatment times; all characteristics of therapeutic regimens with high rates of treatment abandonment [7]. In response, regulatory entities such as the US Food and Drug Administration (FDA) have not approved the use of NFX in human therapy, thus limiting the available prophylactic and therapeutic options to BNZ, alone, for all clinical phases of Chagas disease.

Recent studies have reported interesting in vitro trypanocidal, antioxidant, and immunomodulatory activity for fractions derived from *Lippia alba* (mill.) N.E. Brown essential oils (EOs) rich in terpenes such as limonene, citral, and caryophyllene oxide [8,9]. In addition, trypanocidal and cardioprotective qualities have also been ascribed to these oils in animals with chronic *T. cruzi* infection [10]. The aim of the present work is to assess, in an animal model of chronic Chagas disease, potential immunomodulatory and antioxidant properties as possible mechanisms for the trypanocidal and cardioprotective activity observed for these compounds. In order to accomplish this, synergistic mixtures were generated by combining a limonene-rich fraction of *L. alba* EOs with added exogenous caryophyllene oxide (LIMOX); or by the interaction of this LIMOX with subtherapeutic doses of BNZ (LIMOXBNZ). These experimental therapies were used in a daily, oral scheme of 31 doses on a murine model (Wistar rats) infected with chronic *T. cruzi*, in which CCC had been induced.

The effect of the treatments was determined through evaluation of the clinical progression of heart disease (biochemical and morphological parameters), the trypanocidal efficacy (qPCR), and by immunohistochemical analysis of the cytokine profiles relevant to the immune response against the parasite (TNF-α, IFN-γ, IL-10, and IL-4). Likewise, a marker of oxidative stress (iNOS) was also measured. The phytotherapeutics studied showed promising data for a therapy that could be used as an adjuvant to current treatments (BNZ). Such a novel therapy would be based on standardized production technologies, with Good Agricultural Practices and environmentally sustainable EOs extracted from *L. alba*, a wild shrub from the Colombian Andean Region.

2. Materials and Methods

2.1. Plant Material

Plants belonging to the specie *L. alba*, carvone chemotype, were harvested at the National Research Center for the Agroindustrialization of Tropical Aromatic and Medicinal Plant Species (CENIVAM) (Universidad Industrial de Santander (UIS)), located in Bucaramanga, Colombia, at an altitude of 960 m above sea level. For the cultivation of plant specimens, environmental conditions and specific collection times previously standardized [8] were taken into account. A plant collection permit was obtained from the Colombian Ministry of the Environment and Sustainable Development under Resolution 1761 of November 2019. The *L. alba* identification voucher was deposited at the Colombian National Herbarium (Universidad Nacional de Colombia) under Herbarium Code COL480750.

2.2. Extraction of Essential Oils and Their Fractions

Essential oils were obtained via steam distillation in a 0.4 m^3 stainless steel column using fresh and mature *L. alba* leaves, and were separated by decantation. Subsequently, they were dried with anhydrous sodium sulfate and stored at 4 °C in amber flasks. The EOs were fractionated by reduced pressure distillation in a B/R Instruments (Easton, MD, USA) 800 high-efficiency microdistillation device, and enriched in the bioactive terpene (limonene). The F1 fraction rich in limonene was collected at a temperature of 67 °C and pressure of 12 Torr during distillation of the carvone-chemotype EOs, then stored in amber glass at 4 °C for later analysis.

2.3. GC-MS Analysis

For chromatographic and spectrometric characterization, a GC7890 Gas Chromatograph (Agilent Technologies, AT, Palo Alto, CA, USA) was used, which has an MSD 5975C mass selective detector (AT, Palo Alto, CA, USA); an electron impact ionizer (EI, 70 eV) with an aras/splitless injector (1:30 aras radius); and an MS-ChemStation G1701-DA Data System, including a WILEY, NIST, and QUADLIB 2007 spectral library. The columns used were DB-5MS fused silica capillary (J&W Scientific, Folsom, CA, USA) and DB-WAX (J&W Scientific, Folsom, CA, USA). The data obtained were reconstructed by automatic scanning, and the relative concentration of each compound was obtained by means of an AT 7890 Gas Chromatograph coupled to a flame detection system (FID, 250 °C). Identification was based upon chromatographic criteria (most importantly retention time and linear retention indices and the use of standard compounds) and spectrometric criteria (interpretation of the mass spectrum compared with standard compounds and databases) [11].

2.4. Animal Model

The animal model chosen was 30 male Wistar rats (*Rattus novergicus*) aged 38 ± 2 days old and with a mean weight of 65 ± 10 g, provided by the Universidad Industrial de Santander (UIS) Health Faculty's Bioterium in Bucaramanga, Colombia. A period of eight days of acclimatization to their new environment in the animal testing laboratory at the University of Santander (UDES) was required before their usage in research. In the lab, they were housed with two animals per cage (of dimensions 24 × 37 × 24 cm), in an individually ventilated (IVC) Easy Flow cage system, with a bed of patula pine shavings, food (commercial rodent diet, Lab-Diet®) and sterilized water ad libitum. Variables such as humidity, temperature (21 ± 22 °C), ammonia and CO_2 concentrations, and standardized dark/light cycles (lights on at 06:00 and off at 18:00) were controlled. For the experiments, animals were randomly assigned using the standard function = RAND in Microsoft Excel into six groups of six animals each. Three of these groups were controls: negative (not infected, not treated); hLIMOX-Control (not infected, treated with higher doses of LIMOX (EOs enriched in limonene and with added caryophyllene oxide)); and positive (infected, not treated). The three other groups were experimental: LIMOX (infected, treated with LIMOX); LIMOXBNZ (infected, treated with LIMOX + BNZ); and BNZ (infected, treated with BNZ). The positive control and experimental groups (LIMOX, LIMOXBNZ, and BNZ) were made up of animals infected with *T. cruzi*. The therapeutic schemes were administrated as described in Table 1 via daily, oral doses for 31 days continuously following the onset of the chronic disease phase as established through echocardiography and parasitemia absence. All experiments were carried out according to the NIH Guide for the Care and Use of Laboratory Animals to minimize animal pain and distress. The protocol used was approved by the Bioethics Committee of the Universidad Industrial de Santander and the Ethics Committee of the Universidad de Santander (Agreement Number 010-VII of 15 and 16 May 2017).

Table 1. Therapeutic schemes administrated in *Trypanosoma cruzi*-infected Wistar rats.

Group	Inoculum (Tryp)	Vehicle (Sunflower Oil) *q. s. p*	*Lippia alba* Carvone-Chemotype EO: Limonene-Enriched Fraction (mg/Kg/day)	Caryophyllene Oxide (mg/Kg/day)	Benznidazole (mg/Kg/day)
Negative	-	100 µL	-	-	-
hLIMOX-Control	-	100 µL	170	70	-
Positive	2.5×10^5	100 µL	-	-	-
LIMOX	2.5×10^5	100 µL	68.9	70	
LIMOXBNZ	2.5×10^5	100 µL	68.9	70	7.9
Benznidazole	2.5×10^5	-	-	-	100

Tryp: metacyclic trypomastigote of *Trypanosoma cruzi* clone 338Cl2 TcI; EO: Essential oil; *q. s. p*: quantity sufficient provided.

2.5. Experimental Animal Model of Chronic Chagas Disease

Animals from the positive control and experimental groups were infected using an inoculum prepared as described in the literature [10]. Briefly, 2.5×10^5 metacyclic trypomastigotes of *T. cruzi* clone 338Cl2 (isolated from a confirmed CCC patient), characterized as Discrete Typing Unit (DTU) TcI, were obtained from TAU3AAG media, resuspended in a final volume of 100 µL of phosphate buffer saline-glucose (PBS-G), and injected into the intraperitoneal cavity [10]. After infection, the rats' behavior and clinical parameters (weight, ocular perimeter, position of the vibrissae, physical condition, stool moisture, and parasitemia) were monitored every three days. Parasitemia was evaluated by observing parasites in peripheral blood obtained through puncture of the ventral coccygeal vein with a 27-gauge needle. The blood was collected in a microhematocrit tube and observed between a microscope slide and coverslip under a light optical microscope at one hundred microscopic fields (100× magnification objective) as described by Brener [12].

To determine the development and progression of CCC, the rats' heart silhouettes were evaluated via Two-Dimensional Ultrasound, measuring the Maximum Length (ML) and Maximum Diameter (MD) along the long axis of the heart with a convex transducer and a DP20 (Mindray, Madrid, Spain) ultrasound machine. This analysis was performed at three specific points during the experiment: (i) one day prior to infection to establish baseline measurements; (ii) between 60 and 70 days after infection (d.a.i) to determine the onset of the chronic phase of the infection (presence of cardiomegaly) and to define the start date for therapy; and (iii) at the end of the treatments to evaluate the effect of the therapies applied. Animals in the experimental groups (LIMOX, LIMOXBNZ, and BNZ) received doses of the assigned treatments (Table 1) in a daily, oral scheme for 31 consecutive days, and at a specific hour (7:00 a.m.).

The test subjects were euthanized one day after the end of treatment (99 d.a.i) using an anesthetic mixture of ketamine at 90 mg/kg and xylazine at 7.5 mg/kg intraperitoneally, providing hypnosis, analgesia, and muscle relaxation (the anesthetic triad). Once reflexes such as patellar, palpebral, and corneal had ceased, an additional anesthetic dose was applied for the induction of euthanasia by anesthetic overdose. Before any further procedures were effected, death was verified by establishing the absence of vital signs and reflexes (corneal, eyelash, and rhythmic breathing). Blood was then collected by cardiac puncture to measure biomarkers of liver (ALT and AST) and kidney (BUN and creatinine) function and to conduct a hemogram. Transaminases, BUN, and creatinine assays were carried out from serum samples using commercial Ccromatest kits purchased from Linear Chemicals S.L.U. (Montgat, Spain), in accordance with the manufacturer's instructions; and measurements were made using a URIT-8030 Automated Chemistry Analyzer. EDTA whole blood samples were used for hematological analysis. Total blood cell counts were performed using established URIT-3000Vet hematology equipment. Blood cell morphology and differential leukocyte counts were made on slides stained with Wright.

During the animals' necropsies, photographic and weight records were collected from solid organs (spleen, liver, heart, and kidney), which were sectioned for subsequent immunohistochemical (heart tissue) and histopathological (kidney, liver, and spleen) analysis. These sections were placed into tubes containing 10% neutral formalin stabilized for 72 h, and then embedded in paraffin blocks. A fragment of cardiac tissue from the apical region of the left ventricle was collected in RNAlater^{TM} storage reagent (Sigma-Aldrich, St. Louis, MO, USA), for parasitemia determination by real-time polymerase chain reaction (qPCR). The remainder of the heart tissue was then divided by cross sectional cuts, with each half being randomly assigned for histologic or immunohistochemical analysis. To guarantee data blindness, all procedures (control of parasitemia, echocardiography, euthanasia, and sampling) were performed by independent veterinary personal unaware of which group the animal had been assigned to. The microscopic study was also performed blind by an expert pathologist. The animal remains were properly handled in accordance with the safety and environmental responsibility protocols established by the Universidad de Santander, and disposed of by incineration in accordance with the same.

2.6. Immunohistochemistry

The hearts were completely immersed into five tissue volumes of stabilized 10% neutral formalin solution, which was refreshed after 24 h. After one week, the tissues were placed into cassettes for embedding. The dehydration, clearing, and impregnation processes were carried out in a Leica TP1020 tissue processor (Leica Biosystems, Nussloch, Germany), loading the stations with 10% formaldehyde (2 h), 70% isopropyl alcohol (2 h), 80% isopropyl alcohol (2 h), 90% isopropyl alcohol (2 h), three absolute isopropyl alcohol solutions (2 h each), three xylene solutions (1 h and 30 s for the first solution and 1 h for the others), and two of liquid paraffin (2 h each). For the paraffin embedding of the tissues, a HistoCore Arcadia modular embedding system (Leica Biosystems, Nussloch, Germany) was used. Each block was first put on ice and cut into 4 μm thick sections on a HistoCore MULTICUT microtome with high-profile blades (Leica Biosystems, Nussloch, Germany), then placed in a flotation bath with a temperature of 45 °C to deparaffinize the tissue and prepare fragments for deposition onto positively charged slides (Thermo Fisher Scientific 4951PLUS, Cleveland, OH, USA). First, antigen recovery was accomplished by heating at 58 °C for 1 h and 30 min. For tissue deparaffinization, the slides were immersed in 3 xylol solutions (at 5-min intervals), in a 50% isopropanol solution (15 immersions), and in tap water (15 immersions). Plates were immersed in a 6% H_2O_2 solution for 5 min to block endogenous peroxidases, and cleansed. Antigen recovery was carried out by immersing the slides twice into a solution of ethylenediaminetetraacetic acid (EDTA), first at 95 °C for 30 min, then at room temperature for 20 min. Each slide was then dipped in a Coplin staining jar containing TBS-T for 10 min, and gently dried with a cotton fiber towel. The tissue area was delimited with a hydrophobic marker (Macrosearch Liquid Repellent). The slides were completely covered with "Ultra V Block" reagent (Ultravision Quanto Detection System HRP DAB kit, Thermo Fisher Scientific, Cleveland, OH, USA) for 5 min, cleared with TBS-T, and dried to remove excess moisture before proceeding with the application of antibodies. For the immunohistochemical technique, 100 μL of the desired antibody was applied to each slide for 1 h at room temperature, and then washed with TBS-T. The antibodies were diluted with Antibody Diluent Ventana® (Roche) as follows: 1:1000 for Anti TNF-α (Abcam Reference ab6671, rabbit polyclonal to TNF-α); 1:250 for Anti IFN-γ (Abcam Reference ab216642, rabbit polyclonal to IFN-γ); 1:200 for Anti-Nitric Oxide Inducible Synthase iNOS (Abcam Reference ab15323, rabbit polyclonal to iNOS); 1:1000 for Anti IL-4 (Abcam Reference ab9811, rabbit polyclonal to IL-4); and 1:250 for IL-10 (Abcam Reference ab217941, and rabbit polyclonal to IL-10). To amplify the primary antibody reaction, a Primary Antibody Amplifier Quanto reagent (Ultravision

Quanto Detection System HRP DAB kit, Thermo Fisher Scientific, Cleveland, OH, USA) was applied, followed by incubation for 10 min and rinsing with TBS-T. For revealing, HRP Polymer Quanto reagent (Ultravision Quanto Detection System HRP DAB kit, Thermo Fisher Scientific, Cleveland, OH, USA) and diaminobenzidine (DAB) (provided in the kit) was applied, following the manufacturer's instructions. Finally, the slides were stained with hematoxylin-eosin. The process was completed with the adhesion of a 24 mm × 60 mm coverslip on each treated specimen. Brain, tonsils, and hearts from rats infected with *T. cruzi* were used in a standardized immunohistochemistry technique. An individual slide was used for each antibody to avoid possible contamination or cross-reaction.

For reading, the slides were scanned on a Ventana DP 200 whole-slide scanner (Roche Diagnostics, Rotkreuz, Switzerland), with the immunoreactivity calculation performed using free QuPath v. 0.2.3 software for digital pathological image analysis [13]. For this purpose, a project and a pixel classifier (px) were created with the following parameters: (i) Moderate resolution (2.00 μm/px); (ii) channel DAB; (iii) Gaussian prefilter; (iv) smoothing sigma 1; and (v) Threshold 0.25 (threshold value, positive results ≥ 0.25), defining the region of interest (ROI) as all cardiac tissue present on the slide. The results were expressed as a global percentage of immunoreactivity of the studied antibody.

2.7. DNA Extraction and Real Time PCR

Parasite loads in heart tissue were quantified post mortem by qPCR. For this procedure, parasite DNA was extracted using a commercial genomic DNA extraction kit (Invitrogen Life Technologies—Thermo Scientific, Waltham, MA, USA), in accordance with the manufacturer's instructions. The obtained elute was quantified via Nanodrop 2000 (Thermo Scientific), and the purity and integrity of the DNA was verified in a 1.2% agarose gel. For parasite quantification in tissue by qPCR, the DNA extracted from cardiac tissue was amplified following the method described by Molina-Berríos [14], for identification of *T. cruzi* satellite DNA (GenBank accession number MH884804.1), employing Cruzi I: 5′-AST CGG CTG ATC GTT TTC GA-3′ and Cruzi II: 5′AAT TCC TCC AAG CAG CGG ATA-3′, as forward and reverse primers, respectively; and Cruzi III: 6FAMCAC ACA CTG GAC ACC AAMGBNFQ, as aTaqMan probe (5′Fluorescent label 6-FAM and 3′ Quencher MGB) acquired from Applied Biosystems (Applied Biosystems, Beverly, MA, USA). Cycle reactions were carried out in a Step One Plus^{TM} 3 (Applied Biosystems, Beverly, MA, USA) under the following conditions: one initial cycle of denaturation at 2 min and 50 °C and 10 min at 95 °C, followed by 35 cycles of denaturalization (15 s at 95 °C), and annealing and extension (60 s at 50 °C).The borderline Ct (Cycle threshold) of positivity was defined by preparing a standard curve via serial dilutions of 1:10, starting from a pure culture of *T. cruzi* (strain Sylvio-X10/4 TcI), and taking DNA points at 0.1; 1.0; 10; 100; and 1000 parasites. All the experiments were carried out in duplicate and performed at least three times independently.

2.8. Statistical Analysis

The clinical, hematological, and biochemical parameter data obtained were analyzed using GraphPad Prism software (San Diego, CA, USA). In this process, a descriptive analysis of the variables was carried out, estimating the frequency and dispersion measures for the quantitative variables, and proportions for qualitative variables. The results were expressed as means ± standard error of the mean (S.E.M.). The normality of the continuous variables was evaluated using the Shapiro–Wilk test. The normality of the continuous variables was evaluated using the Shapiro–Wilk test. For comparison between infected groups (positive control and experimental groups) and negative control, the one-way ANOVA test was employed, followed by Dunnett's test; taking statistically significant differences to be $p < 0.05$. This statistical approach was applied for analysis of hematological (except polymorphonuclear (PMN) count cells), and biochemical results. For comparison between groups, the two-way ANOVA test was performed, with a post hoc analysis using Tukey, and taking significant differences to be $p < 0.05$. The results of the relative weight organ, ultrasound, and immunohistochemical assays were analyzed with a two-way ANOVA test, and a post hoc analysis using Tukey, taking significant differences to be $p < 0.05$.

3. Results

3.1. Chemical Compositions of *L. alba* Fractions

In the present study, EOs obtained by the steam distillation technique from fresh and mature carvone-chemotype *L. alba* leaves were fractionated under reduced pressure for enrichment in the bioactive terpene limonene. The relative composition of the EOs and their fractions, and the linear retention indices obtained by gas-coupled mass spectrometry, are presented in Table 2. In this work, the 90%-limonene fraction (F1) isolated from carvone-chemotype *L. alba* EOs was chosen as the base compound for the formulation of the two studied phytotherapies, LIMOX and LIMOXBNZ, as described in Table 2.

Table 2. Main compounds present in the essential oils and their fractions obtained from *Lippia alba* (carvone chemotype). Taken and adapted with permission from Quimbaya et al. [10].

Compound	Linear Retention Indices		Relative Areas GC, % (Media, $n = 3$)			
	Column		Carvone Chemotype			
	DB-5	DB-WAX	EO	F1 [106 °C]	F2 [115 °C]	F3 [120 °C]
6-Methyl-5-hepten-2-one	986	1340	0.1	0.1	-	-
p-Cymene	1024	1274	0.1	0.5	1.1	-
Limonene	1030	1202	30.6	90.5	0.7	0.1
Terpinolene	1086	1284	0.3	0.9	1.5	-
trans-Dihydrocarvone	1202	1626	0.2	0.1	1.0	2.5
Carvone	1242	1736	51.2	1.5	69.9	86.6
Geranial	1270	1728	0.1	-	-	-
Piperitone	1342	1912	1.5	0.2	2.2	2.3
α-Copaene	1376	1492	0.4	0.1	2.9	3.1
β-Elemene	1390	1592	0.5	0.3	2.6	2.9
trans-β-Caryophyllene	1420	1600	0.3	-	0.2	0.4
α-Humulene	1454	1670	0.4	-	-	0.1
Germacrene D	1482	1712	0.3	-	0.1	0.2
Bicyclogermacrene	1496	1738	7.5	0.9	6.3	7.2
β-Bisabolene	1508	1730	0.4	-	-	-
Caryophyllene oxide	1582	1990	0.2	-	0.1	-

DB-5: Gas chromatography (60 m) column; DB-WAX: Gas chromatography (60 m) column; EO: Essential oils; F: fraction; GC: Gas chromatography. Data are representative of three independent experiments and the mean of three relative GC areas was reported for each compound.

3.2. Induction of Experimental Chronic *T. cruzi* Infection in Wistar Rats

Successful induction of infection was verified by the presence of circulating parasites as observed by microscopic analysis, between 8 d.a.i. and 40 d.a.i. The levels of parasitemia between the groups of infected animals did not exhibit significant differences during the follow-up period ($p < 0.05$). Likewise, the onset and end of parasitemia demonstrated similar behavior between groups. The establishment of CCC was corroborated at 67 d.a.i. by the presence of echocardiographic characteristics of dilated cardiomyopathy (during the second ultrasound analysis) in infected animals. These cardiac imaging results were compared with the baseline established for the animals prior to infection (first ultrasound analysis). Thus, this period (67 d.a.i,) was defined as the onset of chronic disease, and therefore as the start date for administration of the oral treatment schemes.

3.3. Cardioprotective Activity

Following the completion of the treatment regimens (at 98 d.a.i.), cardiac effects were evaluated by echocardiography, recording the changes in the cardiac silhouette (maximum length or ML and maximum diameter or MD) at the third ultrasound analysis. Over this time interval, the results obtained showed no significant variations in either parameter for animals belonging to the negative control group. In contrast, animals infected and left untreated (positive control) and those infected and treated with BNZ (reference group) exhibited a significant increase in MD ($p = 0.0156$ and $p = 0.0050$, respectively). In fact, the animals assigned to these two groups were the only ones which demonstrated effects upon clinical parameters; exhibiting decay, dehydration, piloerection, and brown discharge in the conjunctiva (a stress indicator). In contrast, a reversion of both cardiac measures (ML and MD) to values similar to those found in uninfected subjects (negative control, $p > 0.05$) were evidenced in those infected animals which received the LIMOX and LIMOXBNZ therapies (with statistically significant differences when compared to the positive control, $p < 0.05$) (Figure 1).

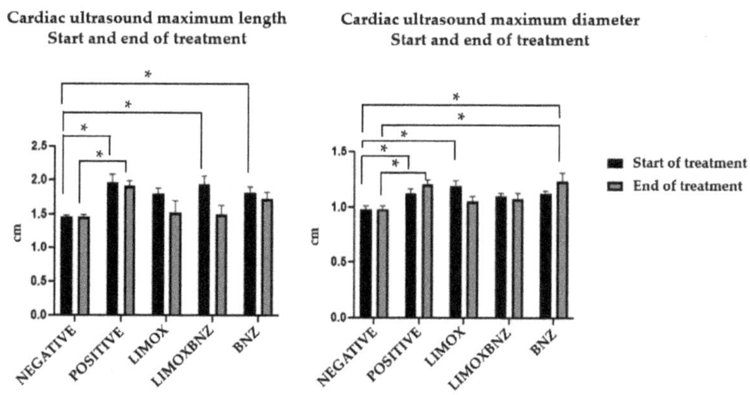

(a) (b)

	Maximum Length (cm)					Maximum Diameter (cm)				
Group	I	II	$p^{\#}$	III	$p^{\&}$	I	II	$p^{\#}$	III	$p^{\&}$
Negative	1.33 ± 0.005	1.46 ± 0.074		1.46 ± 0.069		0.80 ± 0.008	0.98 ± 0.072		0.98 ± 0.073	
Positive	1.30 ± 0.008	1.96 ± 0.296	0.003 *	1.91 ± 0.181	0.0287 *	0.82 ± 0.020	1.12 ± 0.118	0.0498 *	1.20 ± 0.103	0.0156 *
LIMOX	1.14 ± 0.005	1.80 ± 0.197	0.0548	1.53 ± 0.414	0.9854	0.70 ± 0.015	1.19 ± 0.118	0.0021*	1.06 ± 0.104	0.6778
LIMOXBNZ	1.20 ± 0.010	1.93 ± 0.301	0.0051 *	1.50 ± 0.312	0.9983	0.75 ± 0.014	1.10 ± 0.066	0.0967	1.08 ± 0.117	0.4556
BNZ	1.30 ± 0.008	1.82 ± 0.203	0.0395 *	1.72 ± 0.240	0.3091	0.78 ± 0.014	1.13 ± 0.061	0.0404 *	1.24 ± 0.183	0.0050 *

(c)

Figure 1. Echocardiography of Wistar rats infected with *Trypanosoma cruzi*. (**a**) Cardiac ultrasound maximum length start and end of treatment; (**b**) Cardiac ultrasound maximum diameter start and end of treatment; (**c**) Cardiac silhouette measurements via ultrasound (maximum length or ML and maximum diameter or MD) at the first (I), second (II), and third (III) echocardiography: (I) one day prior to infection; (II) at the start date of the therapies; and (III) at the end date of the treatments; $p^{\#}$ compared with negative control measure from ultrasound II; $p^{\&}$ compared with negative control measure from ultrasound III; * statistically significant difference at $p < 0.05$; Negative: untreated and uninfected animals; Positive: untreated and infected animals; LIMOX: infected animals treated with a mixture of an essential oil fraction of *L. alba* carvone chemotype enriched in limonene (68.9 mg/kg/day) and with added caryophyllene oxide (Sigma-Aldrich) (70 mg/kg/day); LIMOXBNZ: infected animals treated with LIMOX plus benznidazole (7.9 mg/kg/day); BNZ: infected animals treated with benznidazole (100 mg/kg/day). Data are representative of six independent experiments and values are expressed in mean ± SEM.

Post mortem macroscopic and microscopic (histopathology) heart observations are presented in Table 3 and Figure 2. The successful induction of the experimental cardiomyopathy model was confirmed by bulging and dilated forms, especially in the left ventricles, of the hearts in the positive control group (observed in 100% of the animals). Consistently, in these animals, histopathology evidenced: multiple foci of inflammatory infiltrate with cell diversity (predominantly lymphocytes, plasma cells, and histiocytes); damage to the myocardial tissue, the atria-ventricle junction sites, and the neurons of the cardiac plexus [10,15].

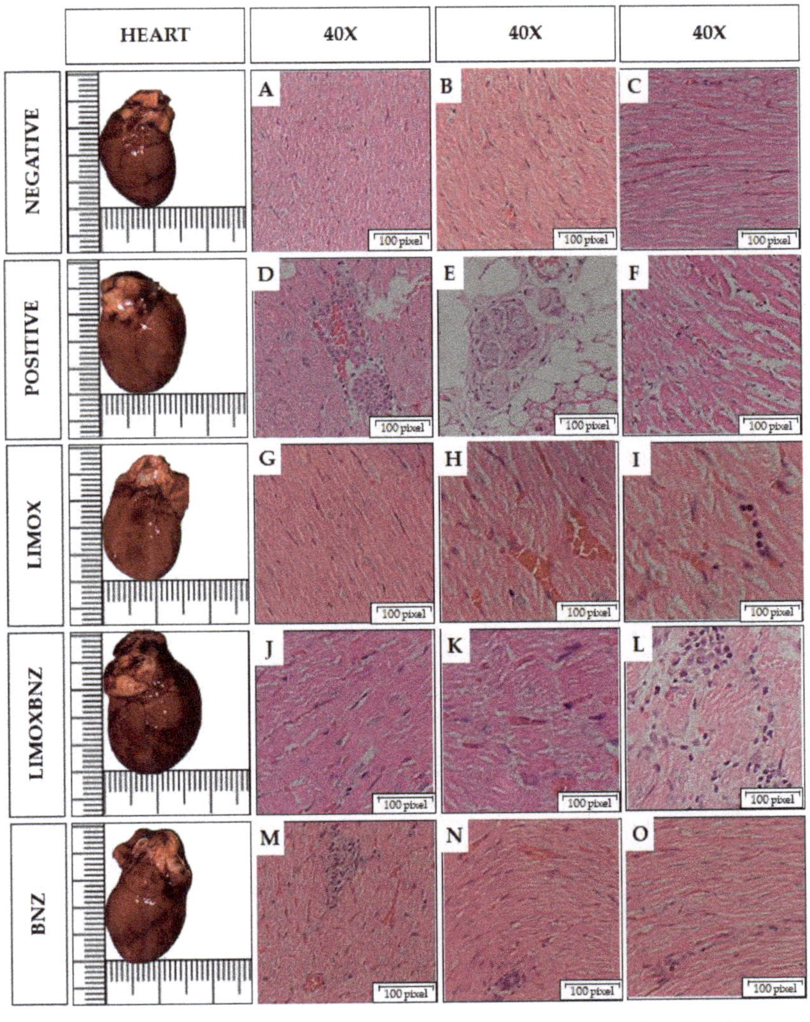

Figure 2. Macro and microscopic findings in the hearts of Wistar rats infected with *Trypanosoma cruzi*. Histological images taken using a 40× objective on Hematoxylin and Eosin stained tissue sections. Negative: untreated, uninfected animals; Positive: untreated, infected animals; LIMOX: infected animals treated with an essential oil fraction of *Lippia alba* carvone chemotype enriched in limonene (68.9 mg/kg/day) and with added caryophyllene oxide (Sigma-Aldrich) (70 mg/kg/day); LIMOXBNZ: infected animals treated with LIMOX and benznidazole (7.9 mg/kg/day); BNZ: infected animals treated with benznidazole (100 mg/kg/day). (**A–C**): Normal heart tissue. (**D**). Large focal inflammatory infiltrate in myocardium with lymphocytes, histiocytes, and plasma cells. (**E**). Aggregate

of cardiac plexus neurons, surrounded by lymphocytic infiltrate. (**F**). Multifocal infiltrate with minimal angiogenesis and loose, elongated, and sinusoidal fibers with fibrotic process. (**G**). Small histiocytic inflammatory infiltrate. (**H**). Apparent reparative process. (**I**). Lymphocytic inflammatory infiltrate with minimal fibrosis. (**J**). Loose and elongated fibers, with histiocytic lymphocyte inflammatory infiltrate. (**K**). Mild lymphohistiocytic infiltrate. (**L**). Diffuse linear inflammatory infiltrate, with minimal fibrosis. (**M**) Focal of inflammatory infiltrate in plasma lymphocytoid, loose and elongated fibers. (**N**). Foci of lymphocytic inflammatory infiltrate; loose, elongated fibers. (**O**). Focal of inflammatory infiltrate of lymphocytes and plasma cells; dilated, loose, and elongated fibers with a sinuous appearance; with fibrosis. Figures are representative of six independent experiments.

Table 3. Macroscopic and microscopic findings in the hearts of Wistar rats after receiving treatments.

Parameter	**Control Groups**			**Experimental Groups**		
	Negative	**hLIMOX-Control**	**Positive**	**LIMOX**	**LIMOXBNZ**	**BNZ**
Color	Normal	Normal	Normal	Normal	Normal	Normal
Size	Normal	Normal	Cardiomegaly	Cardiomegaly	Cardiomegaly	Cardiomegaly
RW g/Kg ($\bar{X}$ ± SEM)	400 ± 86	400 ± 30	402 ± 58	385 ± 30	410 ± 32	375 ± 34
Histopathology	Normal	Normal	Multiple foci of inflammatory infiltrate with cellular diversity and cardiac tissue damage	Heart tissue with minimal inflammatory foci and sporadic mild fibrosis	Heart tissue with mild inflammatory foci and mild fibrosis	Major cardiac damage with multiple fibrosis, cardiac fiber damage, and moderate to abundant inflammatory foci

RW: Relative weight; SEM: standard error of the mean; $\bar{X}$: mean; Negative: untreated, uninfected animals; hLIMOX-Control: not infected animals treated with higher doses of an essential oil fraction of *Lippia alba* carvone chemotype enriched in limonene (170 mg/kg/day) and with added caryophyllene oxide (Sigma-Aldrich) (70 mg/kg/day); Positive: untreated, infected animals; LIMOX: infected animals treated with an essential oil fraction of *L. alba* carvone chemotype enriched in limonene (68.9 mg/kg/day) and with added caryophyllene oxide (Sigma-Aldrich) (70 mg/kg/day); LIMOXBNZ: infected animals treated with LIMOX and benznidazole (7.9 mg/kg/day); BNZ: infected animals treated with benznidazole (100 mg/kg/day). Data are representative of six independent experiments and values are expressed in mean ± SEM.

In infected animals submitted to the array of treatments, the best performance in terms of protection against cardiac damage induced by the *T. cruzi* infection was observed in LIMOX treated animals, followed by LIMOXBNZ (Table 3 and Figure 2). In contrast, among the experimental groups, major effects upon the heart structure were apparent in the BNZ group, whose animals presented macroscopic and microscopic features very similar to infected and untreated rats (positive control).

3.4. Immunomodulation

Immunohistochemical analysis revealed a very distinct cytokine profile between uninfected (negative control) and infected animals (positive and experimental groups), as well as between untreated rats and those subjected to some types of therapy (Figure 3). In this regard, animals without *T. cruzi* infection exhibited the lowest levels of pro-inflammatory cytokines (IFN-γ and TNF-α), and the pro-oxidant marker, iNOS. On the other hand, as a result of infection, a significant increase of TNF-α was observed in infected and untreated rats with a concomitant reduction of the anti-inflammatory IL-10. With respect to treatments, LIMOX and BNZ stimulated a significant increase in IFN-γ immunoreactivity, compared to the negative control; with LIMOX as the only therapy able to reduce TNF-α levels in a significant manner (compared to positive control). Interestingly, the highest levels of the anti-inflammatory cytokine IL-4 were exhibited by the infected animals treated with LIMOX, at a significant difference with respect to the other groups (negative control, LIMOXBNZ, and BNZ). Correspondingly, this treatment was the only therapy able to reestablish the IL-10 levels to those found in uninfected animals. Finally, the iNOS analysis revealed a notable reduction of this oxidant marker (similar to the uninfected model) given by the experimental treatments (LIMOX, LIMOXBNZ, and BNZ), in comparison with the infected and untreated animals (Figure 3).

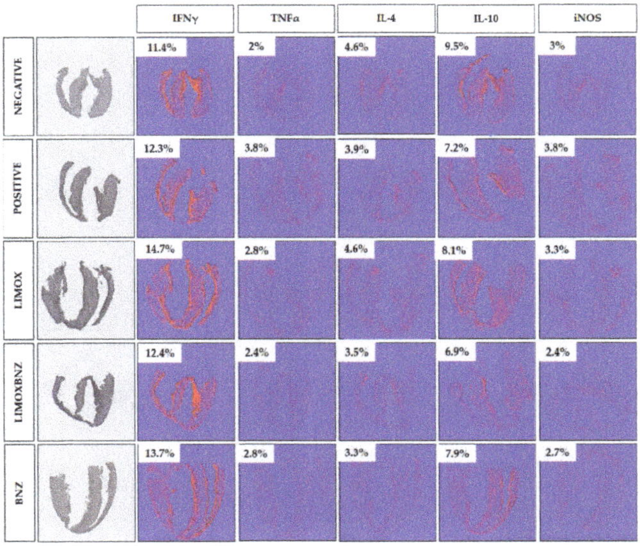

(a)

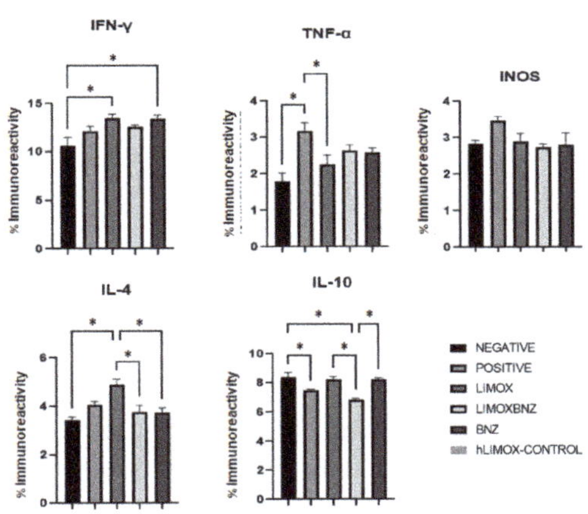

(b)

Group	Cytokine Immunoreactivity									
	IFN-γ		TNFα		iNOS		IL-4		IL-10	
	% *	SEM	% *	SEM	% *	SEM	% *	SEM	% *	SEM
Negative	10.59	0.8968	1.794	0.2255	2.826	0.09630	3.444	0.1229	8.350	0.3504
Positive	12.09	0.5301	3.161	0.2485	3.479	0.1072	4.078	0.1218	7.459	0.09430
LIMOX	13.47	0.4015	2.259	0.2489	2.895	0.2158	4.886	0.2266	8.217	0.2102
LIMOXBNZ	12.59	0.1628	2.628	0.1612	2.731	0.09743	3.758	0.2771	6.836	0.08670
BNZ	13.37	0.4215	2.575	0.1276	2.790	0.3308	3.739	0.1876	8.211	0.1207

(c)

Figure 3. Immunohistochemical analysis of cardiac tissue (**a**) Percentages of immunoreactivity obtained on cross sectional cuts of heart tissue by specific antibodies for Interferon gamma (IFN-γ), Tumor Necrosis Factor alpha (TNF-α), Interleukin (IL)-4, IL-10, and inducible Nitric Oxide Synthase (iNOS); (**b**,**c**) Comparison of immunoreactivity percentages within groups. Negative: untreated and uninfected animals; Positive: untreated and infected animals; LIMOX: infected animals treated with an essential oil fraction of *L. alba* carvone chemotype enriched in limonene (68.9 mg/kg/day) with added caryophyllene oxide (Sigma-Aldrich) (70 mg/kg/day); LIMOXBNZ: infected animals treated with LIMOX and benznidazole (7.9 mg/kg/day); BNZ: infected animals treated with benznidazole (100 mg/kg/day). All p values were calculated by comparison within groups. * $p \leq 0.05$. Data are representative of six independent experiments and values are expressed in mean ± SEM.

3.5. Trypanocidal Effect

T. cruzi DNA quantification (qPCR) analysis demonstrated that animals from the negative and positive control groups exhibited 0% (Ct ≥ 32) and 100% positivity (Ct < 32), respectively; while the applied therapies demonstrated a range of trypanocidal effectivity ranging from 66% (LIMOXBNZ) to 83% (LIMOX and BNZ) (Table 4).

Table 4. Quantification of parasite DNA by real-time PCR of heart tissue from Wistar rats infected with *Trypanosoma cruzi* and treated with 31 continuous once-daily oral doses.

	Control Groups		Experimental Groups		
Animal	**Negative (Ct)**	**Positive (Ct)**	**LIMOX (Ct)**	**LIMOXBNZ (Ct)**	**BNZ (Ct)**
1	34.833	29.855 *	38.161	33.000	38.540
2	37.341	27.391 *	29.294 *	37.937	38.598
3	38.096	26.325 *	38.545	30.839 *	32.581
4	33.399	25.425 *	37.726	38.579	31.197 *
5	35.845	27.240 *	36.843	30.323 *	33.536
6	34.347	23.875 *	35.895	33.124	38.785
Positivity (%)	0	100	17	33	17

Negative: untreated, uninfected animals;. Positive: untreated, infected animals; LIMOX: infected animals treated with an essential oil fraction of *Lippia alba* carvone chemotype (enriched in limonene) (68.9 mg/kg/day) with added caryophyllene oxide (Sigma-Aldrich) (70 mg/kg/day); LIMOXBNZ: infected animals treated with LIMOX plus benznidazole (7.9 mg/kg/day); BNZ: infected animals treated with benznidazole (100 mg/kg/day). * Positive result with Ct value > 32. Data are representative of three independent experiments performed in duplicate and values are expressed as means.

3.6. Toxicity

The toxicity of each treatment was assessed by histopathological analysis of the kidney, liver, and spleen. No abnormalities were observed in the macro and microscopic analyses of the organs obtained from negative control animals, as well as from not infected rats treated with higher doses of LIMOX (hLIMOX-Control). In contrast, large mononuclear-type inflammatory infiltrates were present in the livers of infected and untreated animals (positive control) (Figure 4). Minimal evidence of inflammation in this organ was also discovered in infected rats treated with phytotherapeutical regimens (Table 5). Correspondingly, this finding was accompanied by significantly increased AST levels (compared with the negative control), however, heightened levels of this transaminase were also observed in animals treated with BNZ (Figures 5 and 6).

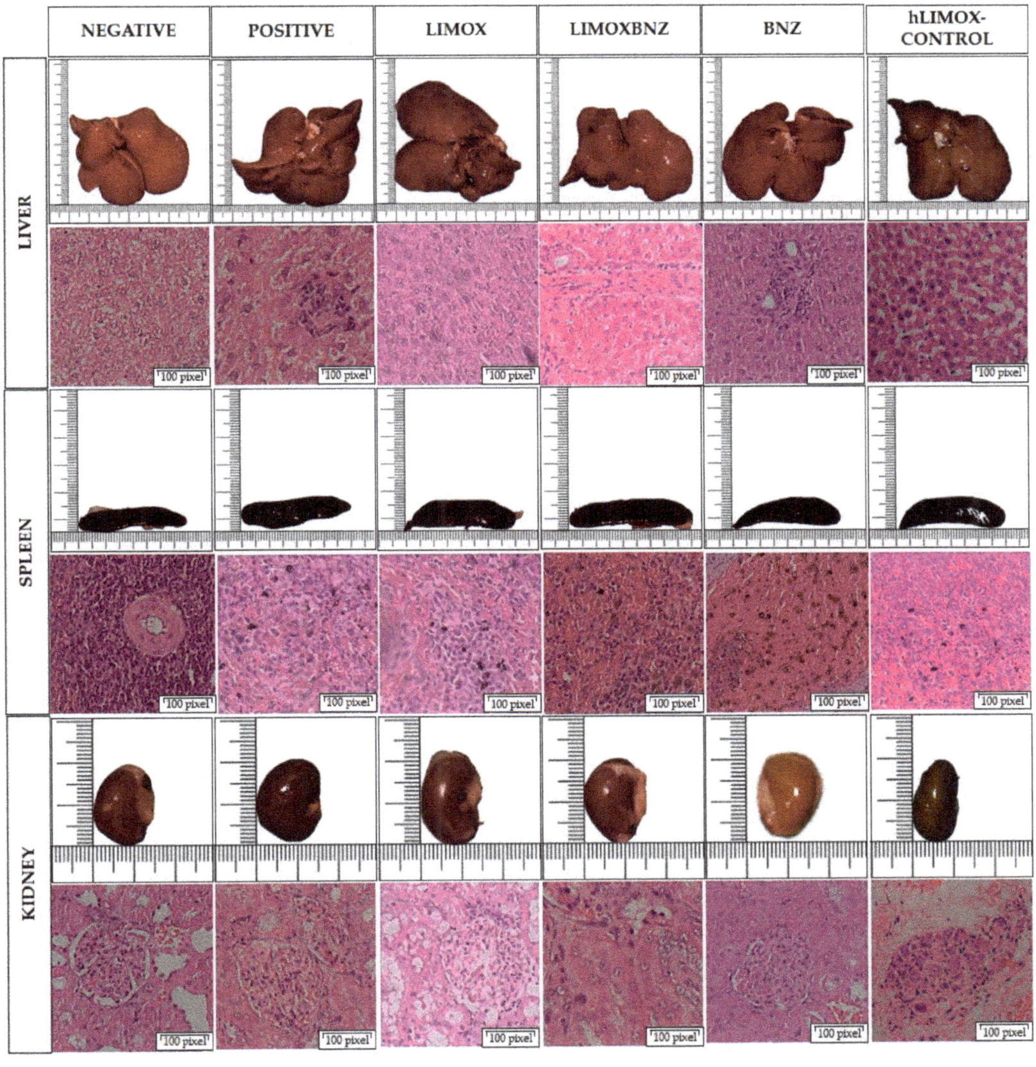

Figure 4. Organ toxicity assessment. Histological images taken using a 40× objective on Hematoxylin and Eosin stained tissue sections. Negative control: untreated, uninfected animals; hLIMOX-Control: not infected animals treated with higher doses of an essential oil fraction of *Lippia alba* carvone chemotype enriched in limonene (170 mg/kg/day) and with added caryophyllene oxide (Sigma-Aldrich) (70 mg/kg/day); Positive control: untreated, infected animals; LIMOX: infected animals treated with a mixture of an essential oil fraction of *Lippia alba* carvone chemotype (enriched in limonene) (68.9 mg/kg/day) and with added caryophyllene oxide (Sigma-Aldrich) (70 mg/kg/day); LIMOXBNZ: infected animals treated with LIMOX plus benznidazole (7.9 mg/kg/day); BNZ: infected animals treated with benznidazole (100 mg/kg/day). Figures are representative of six independent experiments.

Table 5. Macroscopic and microscopic findings in organs of Wistar rats after receiving treatments.

Organ/ Parameter	Control Groups Negative	hLIMOX-Control	Positive	Experimental Groups LIMOX	LIMOXBNZ	BNZ
			Liver			
Color	Normal	Normal	Normal	Normal	Normal	Normal
Size *	Normal	Normal	Hepatomegaly (1/6)	Hepatomegaly (2/6)	Hepatomegaly (2/6)	Normal
RW g/Kg ($\bar{X}$ ± SEM)	3.510 ± 198	3.500 ± 200	3.443 ± 210	3.327 ± 263	3.447 ± 159	3.303 ± 114
Histopathology	Normal	Normal	Lymphocytic mononuclear infiltrate (4/6)	Periportal lymphocytic mononuclear infiltrate (1/6)	Large dilation of the central vein of the lobule. Blood returns from the superior vena cava: volume and pressure overload (1/6)	Minimal lymphocytic mononuclear infiltrate around the portal (1/6)
			Spleen			
Color	Normal	Normal	Normal	Normal	Normal	Normal
Size *	Normal	Normal	Splenomegaly (3/6)	Splenomegaly (3/6)	Splenomegaly (1/6)	Hyposplenism (3/6)
RW g/Kg ($\bar{X}$ ± SEM)	266 ± 20	260 ± 30	303 ± 19	301 ± 47	300 ± 36	253 ± 25
p *	0.872	0.910	0.035	0.043	0.053	1
Histopathology	Normal	Normal	Normal	Normal	Hemosiderophages Observed (1/6)	Hemosiderophages Observed (1/6)
			Kidney			
Color	Normal	Normal	Normal	Normal	Normal	Pale brown (1/6)
Size *	Normal	Normal	Normal	Normal	Normal	Normal
RW g/Kg ($\bar{X}$ ± SEM)	681 ± 60	700 ± 100	683 ± 50	668 ± 28	717 ± 32	675 ± 71
Histopathology	Normal	Normal	Normal	Normal	Normal	Normal

RW: Relative weight; SEM: standard error of the mean; $\bar{X}$: mean; Negative: untreated, uninfected animals; hLIMOX-Control: not infected animals treated with higher doses of an essential oil fraction of *Lippia alba* carvone chemotype enriched in limonene (170 mg/kg/day) and with added caryophyllene oxide (Sigma-Aldrich) (70 mg/kg/day); Positive: untreated, infected animals; LIMOX: infected animals treated with an essential oil fraction of *Lippia alba* carvone chemotype enriched in limonene (68.9 mg/kg/day) and with added caryophyllene oxide (Sigma-Aldrich) (70 mg/kg/day); LIMOXBNZ: infected animals treated with LIMOX and benznidazole (7.9 mg/kg/day); BNZ: infected animals treated with benznidazole (100 mg/kg/day). Data are representative of six independent experiments and values are expressed in mean ± SEM. * Size defined as ratio between maximum diameter and maximum length.

Additionally, splenomegaly was evident in 16.6% of the animals in the LIMOXBNZ-treated experimental group and in 50% of the animals belonging to the LIMOX-treated and positive control groups (Figure 4). In contrast, a significant reduction in the relative weight of this organ was evident in animals subjected to BNZ treatment (Figure 4 and Table 5). In addition, therapies which included that same compound (i.e., BNZ or LIMOXBNZ) induced the presence of hemosiderophages in the spleens of some of the treated animals (16.6%) (Figure 3). In a similar manner, kidney toxicity was apparent in 16.6% of the rats treated with BNZ, presenting as a pale brown color in the macroscopic analysis (Figure 3), but without altering the histopathological or biochemical parameters in the organ. With respect to BUN, a statistically significant elevation in this marker was observed in the LIMOX group compared to the reference treatment (BNZ) (p = 0.0048), but with no difference in creatinine values (Figure 5).

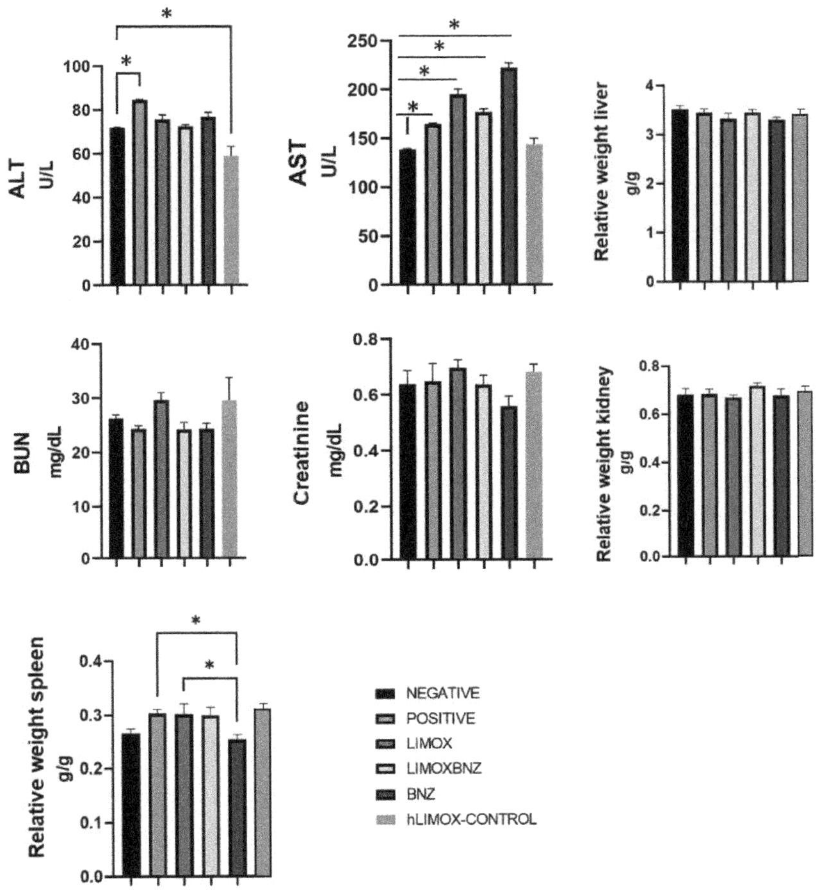

Figure 5. Toxicity of the tested therapies in the liver and spleen. ALT: Alanine aminotransferase; AST: Aspartate aminotransferase; BUN: Blood urea nitrogen. * $p \leq 0.005$ compared to the Benznidazole group. Negative control: untreated, uninfected animals; hLIMOX-Control: not infected animals treated with higher doses of an essential oil fraction of *Lippia alba* carvone chemotype enriched in limonene (170 mg/kg/day) and with added caryophyllene oxide (Sigma-Aldrich) (70 mg/kg/day); Positive control: untreated, infected animals; LIMOX: infected animals treated with a mixture of an essential oil fraction of *Lippia alba* carvone chemotype (enriched in limonene) (68.9 mg/kg/day) and with added caryophyllene oxide (Sigma-Aldrich) (70 mg/kg/day); LIMOXBNZ: infected animals treated with LIMOX plus benznidazole (7.9 mg/kg/day); BNZ: infected animals treated with benznidazole (100 mg/kg/day). * $p \leq 0.05$. Data are representative of six independent experiments and values are expressed in mean ± SEM.

With respect to hemogram parameters, thrombocytosis was associated with the LIMOX and LIMOXBNZ therapies, and was statistically significant in relation to the other treatments ($p = 0.018$ and $p = 0.017$, respectively). Regarding the leukocyte count, all infected groups presented a tendency (not statistically significant) towards higher counts than uninfected animals. In addition, the presence of atypical lymphocytes could be observed in the rats treated with BNZ, with a significant difference in relation to the other groups. The remaining hematological parameters did not exhibit significant alterations (Figure 6). An evaluation of the phytotherapies' toxicity shows that high doses of LIMOX (2.5 times the therapeutic dose) applied to not infected animals (hLIMOX-Control) caused minor signs of toxicity, particularly a change in the color of the kidney (exhibiting a pale brown shade in the macroscopic analysis), in 16.6% of the treated rats. Likewise, a statistically significant peripheral blood neutrophilic leukocytosis was also observed in this same group of animals.

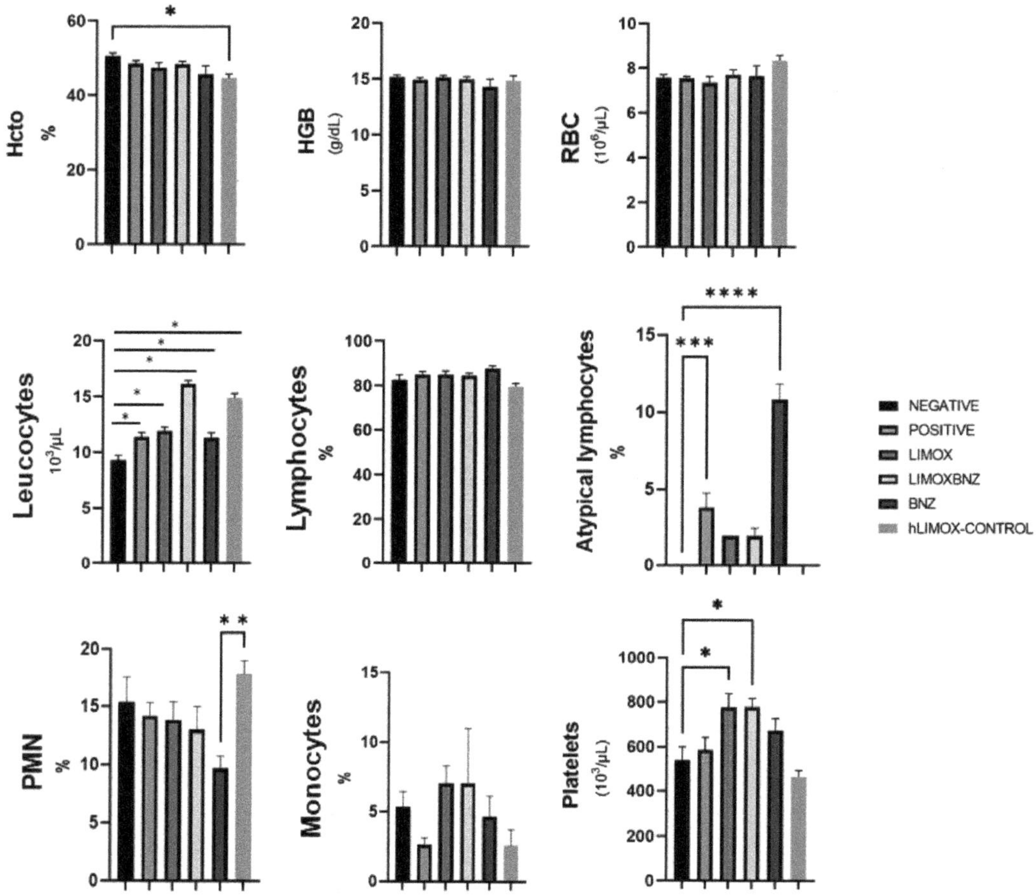

Figure 6. Hematological findings in Wistar rats infected with *Trypanosoma cruzi*. Hcto: Hematocrit; HGB: Hemoglobin; Negative: untreated, uninfected animals; hLIMOX-Control: not infected animals, treated with higher doses of an essential oil fraction of *Lippia alba* carvone chemotype enriched in limonene (170 mg/kg/day) and with added caryophyllene oxide (Sigma-Aldrich) (70 mg/kg/day); Positive: untreated, infected animals; LIMOX: infected animals treated with an essential oil fraction of *Lippia alba* carvone chemotype (enriched in limonene) (68.9 mg/kg/day) with added caryophyllene oxide (Sigma-Aldrich) (70 mg/kg/day); LIMOXBNZ: infected animals treated with LIMOX plus benznidazole (7.9 mg/kg/day); BNZ: infected animals treated with benznidazole (100 mg/kg/day). * $p < 0.05$ compared to the negative control group; ** $p < 0.05$ compared to the BNZ group; *** $p < 0.005$ compared to the negative control group **** $p < 0.0001$ compared to the negative control group. Data are representative of six independent experiments and values are expressed in mean ± SEM.

4. Discussion

Chagas heart disease (CHD) is characterized by dilated cardiomyopathy that affects the atria, ventricles, conduction system, and autonomic nervous system (ANS) [16]. In the pathogenesis of CHD, various mechanisms are implicated such as: the persistence of the parasite; denervation; both microvascular and endothelial dysfunction; a persistent and exacerbated immune response (via the imbalance between pro-inflammatory and anti-inflammatory cytokines) or even autoimmunity [17]; as well as the induction of a permanent oxidative stress caused by both reactive nitrogen species (RNS) and reactive oxygen species (ROS), which in turn directly affects the structure and function of the respiratory chain of cardiac tissue [18]. All of these factors combine to prevent the correct remodeling of heart tissue (functional tissue is replaced by non-functional fibrotic tissue), ultimately leading to heart failure, for which the only treatment option is transplantation [3].

Hence, it is urgent that new therapeutic alternatives be found to eliminate the parasite, control inflammation, and ameliorate cardiac damage, as well as minimizing the harmful side effects in human treatment posed by the current therapeutic options. In this sense, phytoderivatives obtained from aromatic plants represent a highly diverse platform for the discovery of bioactive compounds. In our group, promising in vitro trypanocidal activity was reported for the major terpenes of *L. alba* EOs [8], limonene and caryophyllene oxide; with limonene being the most effective compound against all parasite forms (IC_{50} of 9.0, 28.7, and 41.8 μg/mL for amastigotes, trypomastigotes, and epimastigotes, respectively) [8]. This activity was also exhibited in in vivo models applied to Wistar rats with chronic *T. cruzi* infection, which were treated with synergistic fractions of these EOs enriched in limonene and caryophyllene oxide [10]. Recently, in vitro immunomodulatory properties on macrophages infected with *T. cruzi* were also ascribed to these phytoderivatives [9].

In order to clarify the potential role of certain relevant anti- and pro-inflammatory cytokines and oxidative markers in the trypanocidal and cardioprotective activities observed for *L. alba* EO, the present study assessed the behavior of these markers in an in vivo model of *T. cruzi* chronic infection induced in Wistar rats. In infected and untreated animals (positive control), CHD was verified by the appearance of a statistically significant enlargement of cardiac silhouette parameters (MD and ML) at 67 d.a.i ($p < 0.05$, compared to the uninfected group). Histological findings observed in the animals belonging to the positive control group confirmed the success of CCC induction; findings such as: damage to cardiac tissue, mainly in the form of loose, elongated, and sinusoidal fibers in the myocardium; the presence of multifocal and diffuse mixed inflammatory infiltrates (predominantly of the lymphohistiocytic type); and foci of lymphohistiocytic infiltrate in cardiac plexus neurons [3,19]. At that time, continuous once-daily oral treatments with the three studied schemes (LIMOX, LIMOXBNZ, and BNZ), were administrated for 31 days.

The results showed that the treatment comprised of a mixture of limonene-enriched fraction of *L. alba* EOs and caryophyllene oxide (LIMOX) demonstrated the best performance in restoring the normal shape of the heart compared to the other therapies trialed (LIMOXBNZ and BNZ). As such, the heart dimensions of animals belonging to the LIMOX group exhibited the most reduced ML and MD measurements among infected animals, even to the point of regression to values similar to those of the uninfected control (negative). Not unexpectedly, positive control group rats (infected and untreated) presented the most notable bulging and dilated shape, followed closely those of the BNZ and LIMOXBNZ groups.

Histologically, animals treated with LIMOX evidenced very few inflammatory foci and minimal fibrosis in the heart tissue, with similar characteristics to those of the negative control, in the whole-slide scanning analysis. These results could be attributable to the trypanocidal and cardioprotective effects previously ascribed to limonene and caryophyllene oxide [8,10]. In contrast, somewhat more inflammatory foci were discovered in the tissues of the animals treated with LIMOXBNZ, in whom the signs of fibrosis were also more evident. These adverse effects could be due to the subtherapeutical BNZ doses present in the mixture, which may promote a pro-inflammatory response [10,15]. In a similar manner, animals treated with the conventional intervention (BNZ) exhibited multiple inflammatory foci in the heart, with greater cell variety (particularly lymphocytes, histiocytes, plasma cells, and monocytes), a greater area of fibrous tissue, larger cardiomyocytes, and necrosis.

In order to assess, in a histological context, the cardiac immune response modulated by chronic *T. cruzi* infection and by the therapies trialed, the present study applied an immunohistochemical technique targeting a variety of antigens relevant to CCC pathogenesis using Qupath software; an open-source whole-slide image analysis tool [13]. The results showed that the highest percentages of immunoreactivity for TNF-α and iNOS were present in the infected and untreated animals of the positive control group; as were significantly increased levels of IFN-γ. These findings were accompanied by the impairment of the anti-inflammatory IL-4. In this regard, IFN-γ regulates over a thousand genes by activating Janus tyrosine kinase (JAK) and phosphorylation of the transducer, and serving as an activator of the transcription 1 (STAT-1) pathway. This latter induces the transcription of TNF-α, interferon-inducible factor 1 (IRF1), and iNOS, among other inflammatory cytokines and chemokines [20]. Likewise, in the context of Chagas disease, IFN-γ acts synergistically with TNF-α through the activation of the nuclear transcription factor NF-kB for the positive regulation of iNOS expression; producing high levels of nitric oxide and RNS [21]. This phenomenon represents the activation of general trypanocidal mechanisms, such as the production of reactive species, through the induction of mitochondrial ROS and NADP oxidases [20,22]. These reactive species promote the production of the peroxynitrite anion, a strong oxidant mechanism that arises as a strategy for the elimination of the parasite; causing morphological disruption, severe alterations in its metabolism, calcium homeostasis, and trypanothione depletion [23,24].

Coherently, in this study, the most trypanocidal therapies LIMOX and BNZ (with 83% of parasitological cure in cardiac tissue assessed by qPCR), were accompanied by the highest IFN-γ immunoreactivity. In this regard, it was hypothesized that in *T. cruzi* infection, the elevated levels of this cytokine could be a double-edged sword; since despite its recognized role in parasite tissue clearance and as an antifibrotic agent [25], an excess of IFN-γ production could cause serious damage to cardiac tissue [25]. Nevertheless, an apparent regulatory mechanism was observed in animals subjected to LIMOX therapy (though not in the case of BNZ), represented by an increase in IL-4 production accompanied by restoration of IL-10 levels and lower percentages of TNF-α (compared to the other experimental groups, $p < 0.05$) and iNOS (whose levels did not exhibit statistically significant differences from the negative control, $p > 0.05$).

These differences in the tissue profile of pro- and anti-inflammatory mediators could explain the significant macroscopic and microscopic differences observed in the cardiac architecture between the LIMOX and BNZ groups; in which only animals treated with the terpene mixture presented regression in diameter measurements of the heart silhouette and significant reduction in inflammatory infiltrates or foci with fibrosis. In this context, it is known that TNF-α can induce collagen synthesis and fibrosis, thus contributing to the loss of cardiomyocyte contractility and its replacement by fibrotic tissue [25]. In addition, the same substance can be involved in the development of heart failure through apoptosis and induction of iNOS, with the consequent production of nitric oxide, which exerts a very strong inotropic effect [25]. These findings have been consistently documented in patients who expired from CCC [26]. Likewise, the expression of IL-10 and IL-4 has been linked to the improvement of cardiac function, as determined by the values of the left ventricular ejection fraction and the ventricular diastolic diameter [26]; factors which constitute a possible immunomodulatory tolerance mechanism which could potentially prevent cardiac damage. In this respect, IL-10 is considered a very important regulatory cytokine, and its production is associated with a better prognosis in chronic CHD, suggesting a protective effect for the Th1 response [26–29].

It is worth mentioning that these compensatory mechanisms represented by the stimulation of IL-4 and IL-10 secretion were not observed in the combined therapy of LIMOX and BNZ (LIMOXBNZ), in which both IFN-γ and TNF-α were slightly increased without a significant anti-inflammatory response. These findings suggest that BNZ, even in subtherapeutical doses, causes an antagonist effect on the Th1 response triggered by LIMOX therapy. These results align with those of Quintero et al. [9] who reported that therapies composed of BNZ, alone or in combination with *L. alba* fractions (rich in limonene and citral/caryophyllene oxide), impaired IL-4 secretion by *T. cruzi*-infected macrophages.

In the same work, Quintero et al. [9] showed that the synergistic combination of *L. alba* fractions rich in limonene and citral/caryophyllene oxide produced a significant reduction of the pro-inflammatory cytokines (IFN-γ, IL-2, and TNF-α), with a concomitant increase of the anti-inflammatory cytokines (IL-4 and IL-10), in the extracellular media of the infected macrophages [9]. Although both studies (the aforementioned as well as the current work) show evidence of an immunomodulatory effect (reduction of all pro-inflammatory cytokines with significant elevation of IL-4) by fractions derived from *L. alba* EOs enriched in limonene and caryophyllene oxide, a significant disagreement exists with respect to the behavior of IFN-γ [9]. In this context, the elevated IFN-γ levels observed only in the present in vivo study could be explained by additional compensatory cellular mechanisms converging in the global response of the innate immune system, which can be uniquely perceived in a context of cardiac tissue [25].

Regarding the toxicity of the therapies trialed, slight hepatomegaly accompanied by mild to moderate microscopic signs of periportal inflammation were reported in a percentage (16.6%) of animals belonging to the groups treated with both studied phytotherapies (LIMOX and LIMOXBNZ). These morphological features were correlated with elevated levels of AST transaminase ($p < 0.05$) in these same rats. On the other hand, kidney function assessed by biochemical, morphological, and histopathological analysis did not exhibit any significant alterations among control and experimental groups. Nevertheless, a statistically significant elevation of BUN levels was found in infected rats treated with LIMOX when compared to the reference treatment (BNZ) ($p \leq 0.005$). In addition, 16% of the animals treated with BNZ presented a macroscopic alteration in the color of the kidney (pale brown).

With reference to an effect on spleen architecture, differential responses were observed among therapies. Specifically, splenomegaly was present in the positive control animals and in those treated with both LIMOX and LIMOXBNZ, while BNZ caused a reduction in the size of this organ. This hyposplenism could be correlated with the deleterious effect on adequate balancing of the defense response, as observed herein, in animals treated with the reference therapy [30].

In hemogram analysis, a significant thrombocytosis ($p < 0.005$) was reported for both the LIMOX and LIMOXBNZ therapies. Interestingly, platelet counts in peripheral blood have been inversely associated with disease severity in patients with CCC [31]. In this regard, platelets play an important role in immune response, including protective functions via the release of chemokines that attract and activate leukocytes and, at the same time, platelet surface molecules such as P-selectin and GPIIB/IIIa (Glycoprotein IIb/IIIa) [31]. Likewise, atypical lymphocytes were observed in the peripheral blood smear analysis of all infected animals (experimental groups and positive control) with the highest percentage of these reactive cells in BNZ-treated rats (10.8% for BNZ vs. 3.8% for positive control and 2% for LIMOXBNZ and LIMOX). These cells are produced after a strong process of antigenic stimulation and stress, being classified as a nonspecific response to stimulus or as precursors of memory T and B cells [31]. However, a low percentage of these cells can normally be found in peripheral blood (2–6%) [32], thus the abnormally elevated percentages observed in BNZ therapy could be a reflex to its toxicity.

5. Conclusions

This research contributes to clarifying, in a chronic CHD model in Wistar rats, immunomodulation as a possible trypanocidal and cardioprotective mechanism of LIMOX, a therapy based on a synergistic mixture composed of caryophyllene oxide and a limonene-enriched fraction derived from *L. alba* EO. This therapy showed clear benefits in controlling parasite load, apparently through a mechanism related to the enhancement of the non-specific immune response mediated by high levels of IFN-γ. Remarkably, LIMOX also demonstrated high performance in controlling the progression of cardiac involvement in vivo, and even reversing the progression of dilated cardiomyopathy to levels similar to those found in uninfected animals. Correspondingly, a significant reduction in the severity of inflammatory foci and tissue damage was also confirmed by histopathological analysis, as well as greater tissue remodeling function. The cardioprotective effect observed via

LIMOX treatment was correlated with a protective mechanism derived from an increase in levels of the anti-inflammatory interleukin, IL-4, with a concomitant decrease of TNF-α and a reestablishment of IL-10. Thus, LIMOX becomes an interesting compound for the development of a holistic alternative therapy for the treatment of the chronic phase of Chagas disease.

Author Contributions: Conceptualization, D.X.E.-M. and L.T.G.S.; methodology, D.X.E.-M., L.T.G.S., C.I.G.R., E.E.S., C.A.V.-L., J.J.Q.R. and J.C.M.H.; formal analysis, D.X.E.-M. and E.E.S.; investigation, D.X.E.-M.; resources, L.T.G.S., C.A.V.-L., C.I.G.R. and E.E.S.; writing—original draft preparation, D.X.E.-M.; writing—review and editing, L.T.G.S.; visualization, D.X.E.-M.; supervision, L.T.G.S., E.E.S. and C.I.G.R.; project administration, D.X.E.-M.; funding acquisition, L.T.G.S. and E.E.S. All authors have read and agreed to the published version of the manuscript.

Funding: This research was funded by Ministry of National Education, the Ministry of Industry, Commerce, and Tourism, and ICETEX, call for scientific ecosystem—Scientific Colombia. Francisco José de Caldas Fund, Contract RC-FP44842-212-2018; and also supported by the Vicerrectoría de Investigaciones—Universidad de Santander UDES, under grant 001-18.

Institutional Review Board Statement: The study was approved by the the Universidad Industrial de Santander and Universidad de Santander (Agreement Number 010-VII, 15 and 16 May 2017 approved).

Informed Consent Statement: Not applicable.

Data Availability Statement: Data is contained within the article.

Acknowledgments: The authors would like to express their appreciation to Camilo Durán for his support during the collection and characterization of EOs samples; to Jorge Luis Fernández Alonso for the specimen identifications; to Martha Lucía Díaz Galvis for her assistance in the *Trypanosoma cruzi* differentiation technique; to Hector Martínez for the slide scanning; and finally, we would like to kindly thank Juan Pablo Mejía Cupajita for his support with the standardization of immunohistochemical and hematoxylin-eosin techniques.

Conflicts of Interest: The authors declare no conflict of interest.

References

1. Nunes, M.; Beaton, A.; Acquatella, H.; Bern, C.; Bolger, C.; Echeverria, L.E.; Dutra, W.O.; Gascon, I.; Morillo, C.A.; Oliveira-Filho, J.; et al. American Heart Association rheumatic fever, endocarditis and Kawasaki disease committee of the council on cardiovascular disease in the young; council on cardiovascular and stroke nursing; and stroke council, Chagas cardiomyopathy: An update of current clinical knowledge and management: A scientific statement from the American Heart Association. *Circulation* **2018**, *138*, e169–e209.
2. Bonney, K. Chagas disease in the 21st century: A public health success or an emerging threat? *Parasite* **2014**, *21*, 11. [CrossRef]
3. Vieira, J.L.; Távora, F.R.F.; Sobral, M.G.V.; Vasconcelos, G.G.; Almeida, J.R.; Fernandes, G.P.L.; da Escóssia Marinho, L.L.; de Mendonça Trompieri, D.F.; Neto, J.D.D.S.; Mejia, J.A.C. Chagas cardiomyopathy in Latin America review. *Curr. Cardiol. Rep.* **2019**, *21*, 2. [CrossRef] [PubMed]
4. Bartsch, S.M.; Avelis, C.M.; Asti, L.; Hertenstein, D.L.; Ndeffo-Mbah, M.; Galvani, A.; Lee, B.Y. The economic value of identifying and treating Chagas disease patients earlier and the impact on *Trypanosoma cruzi* transmission. *PLoS Negl. Trop. Dis.* **2018**, *12*, e0006809. [CrossRef] [PubMed]
5. Paiva, C.N.; Medei, E.; Bozza, M.T. ROS and *Trypanosoma cruzi*: Fuel to infection, poison to the heart. *PLoS Pathog.* **2018**, *14*, e1006928. [CrossRef]
6. Sales, P.A.; Molina, I.; Murta, S.M.F.; Sánchez-Montalvá, A.; Salvador, F.; Corrêa-Oliveira, R.; Carneiro, C.M. Experimental and clinical treatment of Chagas disease: A review. *Am. J. Trop. Med. Hyg.* **2017**, *97*, 1289–1303. [CrossRef]
7. Morillo, C.A.; Marin-Neto, J.A.; Avezum, A.; Sosa-Estani, S.; Rassi, A.; Rosas, F.; Villena, E.; Quiroz, R.; Bonilla, R.; Britto, C.; et al. Randomized trial of benznidazole for chronic Chagas' cardiomyopathy. *N. Eng. J. Med.* **2015**, *373*, 1295–1306. [CrossRef] [PubMed]
8. Moreno, É.M.; Leal, S.M.; Stashenko, E.E.; García, L.T. Induction of programmed cell death in *Trypanosoma cruzi* by *Lippia alba* essential oils and their major and synergistic terpenes (citral, limonene and caryophyllene oxide). *BMC Complement. Altern. Med.* **2018**, *18*, 225. [CrossRef]
9. Quintero, W.L.; Moreno, E.M.; Pinto, S.M.L.; Sanabria, S.M.; Stashenko, E.E.; García, L.T. Immunomodulatory, trypanocide, and antioxidant properties of essential oil fractions of *Lippia alba* (Verbenaceae). *BMC Complement. Med. Ther.* **2021**, *21*, 187. [CrossRef] [PubMed]

10. Quimbaya Ramírez, J.J.; González Rugeles, C.I.; Stashenko, E.E.; Mantilla Hernández, J.C.; Díaz Galvis, M.L.; García Sánchez, L.T. In vivo protection against chagasic cardiomyopathy progression using trypanocidal fractions from *Lippia alba* (Verbenaceae) essential oils. *Ind. Crop. Prod.* **2021**, *167*, 113553. [CrossRef]
11. Jennings, W.G.; Shibamoto, T. *Qualitative Analysis of Flavor and Fragrance Volatiles by Glass Capillary Gas Chromatography*; Academic Press: New York, NY, USA, 1980; p. 472.
12. Brener, Z. Therapeutic activity and criterion of cure on mice experimentally infected with *Trypanosoma cruzi*. *Rev. Inst. Med. Trop. São Paulo* **1962**, *4*, 389–396. [PubMed]
13. Bankhead, P.; Loughrey, M.B.; Fernández, J.A.; Dombrowski, Y.; McArt, D.G.; Dunne, P.D.; McQuaid, S.; Gray, R.T.; Murray, L.J.; Coleman, H.G.; et al. QuPath: Open-source software for digital pathology image analysis. *Sci. Rep.* **2017**, *7*, 16878. [CrossRef]
14. Molina-Berríos, A.; Campos-Estrada, C.; Henríquez, N.; Faúndez, M.; Torres, G.; Castillo, C.; Escanilla, S.; Kemmerling, U.; Morello, A.; López-Muñoz, R.A.; et al. Protective role of acetylsalicylic acid in experimental *Trypanosoma cruzi* infection: Evidence of a 15-epi-lipoxin a4-mediated effect. *PLoS Negl. Trop. Dis.* **2013**, *7*, e2173. [CrossRef]
15. Bonney, K.M.; Engman, D.M. Autoimmune pathogenesis of Chagas heart disease: Looking back, looking ahead. *Am. J. Clin. Pathol.* **2015**, *185*, 1537–1547. [CrossRef] [PubMed]
16. Do Nunes, M.C.P. Disfunção microvascular coronariana: ¿Isso realmente importa na doença de Chagas? *Arq. Bras. Cardiol.* **2021**, *115*, 1102–1103.
17. Rodríguez-Angulo, H.; Marques, J.; Mendoza, I.; Villegas, M.; Mijares, A.; Gironès, N.; Fresno, M. Differential cytokine profiling in Chagasic patients according to their arrhythmogenic-status. *BMC Infect. Dis.* **2017**, *17*, 221. [CrossRef]
18. Zacks, M.A.; Wen, J.; Vyatkina, G. An overview of chagasic cardiomyopathy: Pathogenic importance of oxidative stress. *An. Acad. Bras. Ciências* **2005**, *77*, 695–715. [CrossRef] [PubMed]
19. Bonney, K.M.; Luthringer, D.J.; Kim, S.A.; Garg, N.J.; Engman, D.M. Pathology and pathogenesis of Chagas heart disease. *Annu. Rev. Pathol.* **2019**, *14*, 421–447. [CrossRef]
20. Chevillard, C.; Nunes, J.P.S.; Frade, A.F.; Almeida, R.R.; Pandey, R.P.; Nascimento, M.S.; Kalil, J.; Cunha-Neto, E. Disease tolerance and pathogen resistance genes may underlie *Trypanosoma cruzi* persistence and differential progression to Chagas disease cardiomyopathy. *Front. Immunol.* **2018**, *9*, 2791. [CrossRef]
21. Lopez, M.; Tanowitz, H.B.; Garg, N.J. Pathogenesis of chronic Chagas disease: Macrophages, mitochondria, and oxidative stress. *Curr. Clin. Microbiol. Rep.* **2018**, *5*, 45–54. [CrossRef]
22. Wu, M.; Park, Y.J.; Pardon, E.; Turley, S.; Hayhurst, A.; Deng, J.; Steyaert, J.; Hol, W.G. Structures of a key interaction protein from the *Trypanosoma brucei* editosome in complex with single domain antibodies. *J. Struct. Biol.* **2011**, *174*, 124–136. [CrossRef] [PubMed]
23. Ehrt, S.; Schnappinger, D.; Bekiranov, S.; Drenkow, J.; Shi, S.; Gingeras, T.R.; Gaasterland, T.; Schoolnik, G.; Nathan, C. Reprogramming of the macrophage transcriptome in response to interferon-and *Mycobacterium tuberculosis*: Signaling roles of Nitric Oxide Synthase-2 and phagocyte oxidase. *J. Exp. Med.* **2001**, *194*, 1123–1139. [CrossRef]
24. Levick, S.P.; Goldspink, P.H. Could interferon-gamma be a therapeutic target for treating heart failure? *Heart Fail. Rev.* **2014**, *19*, 227–236. [CrossRef]
25. Rocha, I.H.; Marques, A.L.F.; Moraes, G.V.; da Silva, D.A.A.; Rodrigues, M.V.; Correia, D. Metabolic and immunological evaluation of patients with indeterminate and cardiac forms of Chagas disease. *Med. Baltim.* **2020**, *99*, e23773. [CrossRef]
26. Sousa, G.R.; Gomes, J.A.S.; Fares, R.C.G.; de Damásio, M.P.S.; Chaves, A.T.; Ferreira, K.S.; Nunes, M.C.P.; Medeiros, N.I.; Valente, V.A.A.; Correa-Oliveira, R.; et al. Plasma cytokine expression is associated with cardiac morbidity in Chagas disease. *PLoS ONE* **2014**, *9*, e87082.
27. Freyberg, Z.; Harvill, E.T. Pathogen manipulation of host metabolism: A common strategy for immune evasion. *PLoS Pathog.* **2017**, *13*, e1006669. [CrossRef]
28. Couper, K.N.; Blount, D.G.; Riley, E.M. IL-10: The master regulator of immunity to infection. *J. Immun.* **2008**, *180*, 5771–5777. [CrossRef] [PubMed]
29. Pinazo, M.J.; Thomas, M.C.; Bustamante, J.; de Almeida, I.C.; Lopez, J.; Gascon, M.C. Biomarkers of therapeutic responses in chronic Chagas disease: State of the art and future perspectives. *Mem. Inst. Oswaldo Cruz.* **2015**, *110*, 422–432. [CrossRef]
30. Golub, R.; Tan, J.; Watanabe, T.; Brendolan, A. Origin and immunological functions of spleen stromal cells. *Trends Immun.* **2018**, *39*, 503–514. [CrossRef] [PubMed]
31. Alvarez, G.; Bertocchi, G.; Pengue, C.; Cesar, G.; Eiro, M.D.C.; Lococo, B.; Viotti, R.; Natale, M.A.; Castro Eiro, M.D.; Cambiazzo, S.S.; et al. Impaired frequencies and function of platelets and tissue remodeling in chronic Chagas disease. *PLoS ONE* **2019**, *14*, e0218260.
32. Rey-Caro, L.A.; Villar-Centeno, L.A. Linfocitos atípicos en dengue: Papel en el diagnóstico y pronóstico de la enfermedad. Revisión sistemática de la literatura. *Rev. Cienc. Salud* **2012**, *10*, 323–335.

antioxidants

MDPI

Article

Insights into the Mechanisms of *Lactobacillus acidophilus* Activity against *Entamoeba histolytica* by Using Thiol Redox Proteomics

Lotem Sarid, Eva Zanditenas, Jun Ye , Meirav Trebicz-Geffen and Serge Ankri *

Department of Molecular Microbiology, Ruth and Bruce Rappaport Faculty of Medicine, Technion, Haifa 31096, Israel; lotemsarid@campus.technion.ac.il (L.S.); zanditenas@campus.technion.ac.il (E.Z.); junye@campus.technion.ac.il (J.Y.); meiravg@technion.ac.il (M.T.-G.)
* Correspondence: sankri@technion.ac.il; Tel.: +972-4829-5453

Abstract: Amebiasis is an intestinal disease transmitted by the protist parasite, *Entamoeba histolytica*. *Lactobacillus acidophilus* is a common inhabitant of healthy human gut and a probiotic that has antimicrobial properties against a number of pathogenic bacteria, fungi, and parasites. The aim of this study was to investigate the amebicide activity of *L. acidophilus* and its mechanisms. For this purpose, *E. histolytica* and *L. acidophilus* were co-incubated and the parasite's viability was determined by eosin dye exclusion. The level of ozidized proteins (OXs) in the parasite was determined by resin-assisted capture RAC (OX–RAC). Incubation with *L. acidophilus* for two hours reduced the viability of *E. histolytica* trophozoites by 50%. As a result of the interaction with catalase, an enzyme that degrades hydrogen peroxide (H_2O_2) to water and oxygen, this amebicide activity is lost, indicating that it is mediated by H_2O_2 produced by *L. acidophilus*. Redox proteomics shows that *L. acidophilus* triggers the oxidation of many essential amebic enzymes such as pyruvate: ferredoxin oxidoreductase, the lectin Gal/GalNAc, and cysteine proteases (CPs). Further, trophozoites of *E. histolytica* incubated with *L. acidophilus* show reduced binding to mammalian cells. These results support *L. acidophilus* as a prophylactic candidate against amebiasis.

Keywords: *Entamoeba histolytica*; *Lactobacillus acidophilus*; probiotic; redoxomics; cysteine proteases

Citation: Sarid, L.; Zanditenas, E.; Ye, J.; Trebicz-Geffen, M.; Ankri, S. Insights into the Mechanisms of *Lactobacillus acidophilus* Activity against *Entamoeba histolytica* by Using Thiol Redox Proteomics. *Antioxidants* **2022**, *11*, 814. https://doi.org/10.3390/antiox11050814

Academic Editor: Simone Carradori

Received: 10 March 2022
Accepted: 20 April 2022
Published: 22 April 2022

1. Introduction

Amebiasis is an enormous global medical problem because of poor sanitary conditions and unsafe hygiene practices existing in many parts of the world. According to the World Health Organization, 50 million people in India, Southeast Asia, Africa, and Latin America suffer from amebic dysentery and amebiasis causes the death of at least 100,000 individuals each year. The main mode of transmission for amebiasis is the ingestion of food or water that is contaminated with feces containing *E. histolytica* cysts. After the cyst form has been swallowed by the host, excystation occurs in the intestinal lumen, followed by colonization of the large intestine by the trophozoites where they continue to divide and encyst. Eventually, both trophozoites and cysts are excreted in stools. Only 10% of the infected individuals will develop acute intestinal and extra-intestinal diseases. One possible explanation for this observation is the difference in the gut microbiota between individuals who may significantly influence the host's immune response in amebiasis and *E. histolytica*'s virulence [1]. Over the last few decades, it has become evident that *E. histolytica*'s pathogenicity is directly linked to the parasite's interaction with the gut microbiota [2], as the parasites are reported to feed on bacteria and cellular debris found in the large intestine [1]. However, such feeding is very selective, where only those bacteria with the appropriate recognition molecules are ingested by the parasite [3]. Amebiasis is characterized by acute inflammation of the intestine with release of pro-inflammatory cytokines, reactive oxygen species (ROS), and reactive nitrogen species (RNS) from activated cells of the host's immune system. ROS and RNS are the major cytotoxic effectors for killing *E. histolytica* and cause oxidation and nitrosylation of amebic proteins, trigger stress responses, and inhibit glycolysis and the activity of some virulence factors [4–7]. Cellular means of subverting the toxicity of oxidative stress (OS) are important for the success of infectious diseases. No vaccine against amebiasis currently exists; the drug of choice for treating amebiasis is metronidazole, which may cause severe side effects such as nausea, vomiting, headaches, a metallic or bitter taste in the mouth, and more serious effects such as anorexia, ataxia, and skin rashes/itching [8,9]. Additionally, some clinical strains of *E. histolytica* are less sensitive to metronidazole, suggesting the emergence of metronidazole-resistant strains [10,11].

Probiotics are live organisms which, when administered in adequate amounts, confer a health benefit to the host [12,13]. Probiotics and commensal bacteria have been suggested to have some influence on the outcome of protozoan infections [14–16]. As an alternative bio-therapeutic for amebiasis, there are a number of studies which have been conducted, interestingly most of these studies are aimed at the efficiency of the probiotic at inhibiting adhesion of the protozoa to the intestinal mucosal surface [17,18]. Recently we have shown that *Lactobacillus acidophilus* is detrimental to *E. histolytica* [19]. This detrimental effect is associated with the transcription by the parasite of genes encoding major signaling molecules, such as kinases, regulators of small GTPases and oxidoreductases and genes encoding proteins necessary for ribosome structure. It has been suggested that the probiotic effect of certain bacteria (such as *L. acidophilus*) is mediated by the ability to produce H_2O_2 [20] via an NADH-dependent flavin reductase [21] and to maintain a normal, homeostatic microbiota [21]. In this study, we demonstrated that H_2O_2 produced by *L. acidophilus* caused the death of the parasite by oxidizing important amebic proteins. To our knowledge, this work provides the first comprehensive analysis of OXs in a protozoan parasite exposed to *L. acidophilus*.

2. Materials and Methods

2.1. *E. histolytica* and *L. acidophilus* Culture

E. histolytica trophozoites, the HM-1:IMSS strain (a gift from Samudrala Gourinath, Jawaharlal Nehru University, New Delhi, India), were grown and harvested according to a previously reported protocol [22].

L. acidophilus ATCC4356 strain was cultivated in De Man, Rogosa and Sharpe (MRS) media (Sigma-Aldrich, Jerusalem, Israel) overnight at 37 °C with agitation (200 rpm) on a New Brunswick Innova 4300 Incubator Shaker (Marshall Scientific, Hampton, NH, USA). Heat-killed *L. acidophilus* was cultivated in MRS media (Sigma-Aldrich, Jerusalem, Israel) overnight at 37 °C with agitation, followed by autoclaving at 121 °C and 1.05 kg/cm^2 for 15 min.

2.2. Reagents

Catalase from bovine liver (C9322) was purchased from Sigma-Aldrich (Jerusalem, Israel).

2.3. Ferrous Oxidation-Xylenol Orange (FOX) Assay

The amount of H_2O_2 produced by *L. acidophilus* was determined by the FOX assay according to a previously reported protocol [23].

2.4. Viability of *E. histolytica* Trophozoites

Trophozoites (~1×10^6/mL) were incubated with *L. acidophilus* (~1×10^9/mL) in serum-free Diamond's TYI S-33 medium for 120 min at 37 °C with agitation (200 rpm) in a thermoshaker (ALS-MS-100, Hangzhou Allsheng Instrument, Hangzhou, China). The viability of trophozoites was determined by the eosin dye exclusion method [6].

2.5. Detection of Oxidized Proteins (OXs) by Resin-Assisted Capture RAC (OX–RAC)

The detection of OXs by OX–RAC was performed on three biological replicates using a previously described protocol [6]. Captured proteins were eluted with 30-µL elution buffer containing 10 mM HEPES, 0.1 mM EDTA, 0.01 mM neocuproine, 0.1% sodium dodecyl sulfate (SDS), and 100 mM 2-mercaptoethanol for 20 min at room temperature. Proteins in a 10-µL aliquot of each eluent were resolved on a 12.5% SDS—polyacrylamide gel electrophoresis (PAGE) gel. Each gel was then stained with silver (Pierce Silver Stain), and each gel slice was independently analyzed by MS. A protein was considered to be oxidized when its relative amount in the DTT-treated lysates was significantly more than that in the DTT-untreated lysates ($p < 0.05$ according to the results of an unpaired t-test).

2.6. In-Gel Proteolysis and MS Analysis

The proteins in the gel were reduced with 2.8mM DTT (60 °C for 30 min), modified with 8.8 mM iodoacetamide in 100 mM ammonium bicarbonate (in the dark and at room temperature for 30 min) and digested in 10% acetonitrile and 10 mM ammonium bicarbonate with modified trypsin (Promega, Beit Haemek, Israel) overnight at 37 °C. A second trypsin digestion was carried out for another 4 h at 37 °C.

The tryptic peptides were desalted using C18 tips (Home-made, 3M) dried and re-suspended in 0.1% formic acid.

The peptides were resolved by reverse-phase chromatography on 0.075 × 180-mm fused silica capillaries (JW) packed with Reprosil reversed phase material (Dr Maisch GmbH, Ammerbuch, Germany). The peptides were eluted with linear 60 min gradient of 5 to 28% 15 min gradient of 28 to 95% and 25 min at 95% acetonitrile with 0.1% formic acid in water at flow rates of 0.15 µL/min. MS was performed by Q Exactive HF mass spectrometer (Thermo Fisher Scientific represented by BARGAL analytical instruments, Shoham, Israel) in a positive mode using a repetitively full MS scan followed by collision induces dissociation (HCD) of the 18 most dominant ions selected from the first MS scan. The mass spectrometry data were analyzed using the MaxQuant software 1.5.2.8, The Max Plank Institute of Biochemistry, Munich, Germany [24] vs. *Entamoeba histolytica* and *Lactobacillus acidophilus* proteomes from the Uniprot database with 1% FDR (false discovery rate). The data were quantified by label free analysis using the same software. Statistical analysis of the identification and quantization results was done using Perseus 1.6.7.0 software, The Max Plank Institute of Biochemistry, Munich, Germany [25]. A t-test between the groups with or without DTT was carried out, with the Benjamini–Hochberg correction for multiple testing. Proteins were considered as significantly changed if their p-value < 0.05, q-value < 0.05, and the fold change between the groups ≥ 1.

2.7. Classification of OXs According to Their Protein Class and Statistical Overrepresentation Test

The OXs were classified according to PANTHER Protein Class using the PANTHER Classification System software (http://www.pantherdb.org/, accessed on 28 July 2021) [26]. This classification of proteins derived from PANTHER/X molecular function ontology includes commonly used classes of protein families, many of which are not covered by GO molecular function.

Regarding the statistical overrepresentation test, the online system compares a list of genes of interest (in this work, genes encoding for OXs in *E. histolytica* trophozoites exposed to *L. acidophilus*) to a reference list (*E. histolytica* in database). The p-value calculation in the overrepresentation test is calculated automatically based on the number of genes expected in the test list for a particular PANTHER category, based on the reference list.

2.8. Measurement of Cysteine Proteases (CPs) Activity

CPs activity was monitored by cleavage of the synthetic substrate benzyloxycarbonyl-l-arginyl-l-arginine-p-nitroanilide (z-Arg-Arg-pNA) (Bachem, Torrance, CA, USA) using a previously described protocol [27] except that DTT was not systematically added to the reaction buffer. Briefly, z-Arg-Arg-pNA was incubated for 0–10 min at 37 °C with

E. histolytica lysate (40 µg) (prepared in phosphate buffer saline (PBS) nonidet P-40 (1%) (Sigma-Aldrich, Jerusalem, Israel) in 990 µL CP buffer (0.1 M KH2PO4, 2 mM EDTA, pH 7.0). Cleavage of Z-Arg-Arg-pNA substrate were detected at 405 nm in a Novaspec plus spectrophotometer (Sigma-Aldrich, Israel).

2.9. Adhesion Assay

The adhesion of *E. histolytica* trophozoites to HeLa cells (a kind gift from T. Kleinberger, Faculty of Medicine, Technion) was measured using a previously described protocol [28]. *E. histolytica* trophozoites (2×10^5) were incubated with live *L. acidophilus* (2×10^8), with heat-killed *L. acidophilus* (DN) (2×10^8), with paraformaldehyde-fixed *L. acidophilus* (PLA) (2×10^8) and with/without catalase (50 µg/mL) for 1 h at 37 °C and then transferred to paraformaldehyde-fixed HeLa cells monolayers for an additional hour of incubation at 37 °C. Trophozoites unattached to HeLa cell monolayers were washed once with phosphate buffer saline (PBS) buffer and the trophozoites attached to the HeLa cell monolayer were eluted with 500 µL of a solution of cold galactose (1%) in PBS and counted.

3. Results and Discussion

3.1. L. acidophilus Amebicide Activity Depends on the Formation of H_2O_2

L. acidophilus is commonly found in the gastrointestinal tract of healthy humans. It is widely used as a food preservative and as a probiotic. *L. acidophilus* antimicrobial activity is caused by the production of antimicrobial peptides, including lactacins B, organic acid production such as lactic acids and H_2O_2 (recently reviewed in [29]), and immune induction [30]. Whereas the antibacterial and antifungal activity [31,32] of *L. acidophilus* has been well illustrated, the antiparasitic properties of *L. acidophilus* have been less studied. Studies with mouse models of the diseases caused by *Giardia lamblia* [33], *Toxocara canis* [34], *Trichinella spiralis* [35], and *Cryptosporidium parvum* [36] reveal that a combination of probiotics and other probiotic strains is beneficial in the treatment and prevention of these parasites. In a recent study, we demonstrated that *L. acidophilus* is detrimental to *E. histolytica* but the amebicide mechanism was unknown [19]. In this study, we investigated whether H_2O_2 generated by *L. acidophilus* is directly responsible for the amebicide activity. We first measured the ability of *L. acidophilus* to produce H_2O_2 by the FOX assay. We found that overnight culture of *L. acidophilus* cultivated in MRS media with agitation produces 0.14 ± 0.3 mM H_2O_2. A viability assay was performed on *E. histolytica* trophozoites incubated either with *L. acidophilus* or with heat-killed *L. acidophilus*, which served as negative control. The viability of *E. histolytica* trophozoites was not affected when the parasite was incubated with *L. acidophilus* for 60 min (Figure 1). However, the viability of *E. histolytica* trophozoites was significantly decreased by 50% when the parasite was incubated with *L. acidophilus* for 120 min. In contrast, the viability of *E. histolytica* trophozoites incubated with heat-killed *L. acidophilus* for 120 min was not impaired (Figure 1). Next, we wanted to establish if the amebicidal activity of *L. acidophilus* was dependent on the formation of H_2O_2. We incubated *E. histolytica* and *L. acidophilus* in presence of catalase, an enzyme that catalyzes the decomposition of H_2O_2 to H_2O and O_2 [37]. We observed that the amebicidal activity of *L. acidophilus* was strongly reduced when catalase was added during the incubation of *L. acidophilus* with the parasite (Figure 1). Based on this finding, it strongly suggests that H_2O_2 produced by *L. acidophilus* is the primary cause of parasite death.

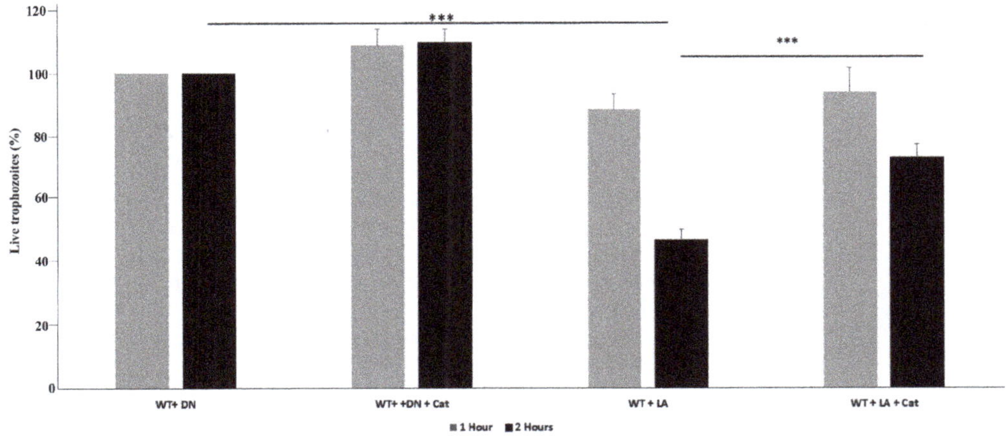

Figure 1. Viability assay of *E. histolytica* trophozoites. Note: *E. histolytica* trophozoites (WT) were incubated with live *L. acidophilus* (LA) or with heat-killed *L. acidophilus* (DN), with/without catalase (Cat) (50 μg/mL) for 60 and 120 min at 37 °C. The data represent two independent experiments performed in triplicate. *** *p* value < 0.001 by an unpaired Student's *t*-test.

3.2. Resin-Assisted Capture (RAC) of Oxidized Proteins (OX) Coupled to Mass Spectrometry (OX–RAC) Analysis of *E. histolytica* Trophozoites Exposed to *L. acidophilus*

In order to explore the amebicidal properties of *L. acidophilus*, we used OX–RAC to measure the levels of oxidized proteins (OXs) in *E. histolytica* trophozoites exposed to *L. acidophilus*. In absence of DTT treatment, OXs are not expected to bind to the thiopropyl resin [38]. We observed that the level of OXs in *E. histolytica* trophozoites exposed to heat-killed *L. acidophilus* culture is very low (Figure 2A). These results indicate that heat-killed culture of *L. acidophilus* do not trigger the formation of OXs in *E. histolytica* trophozoites. In contrast, a strong level of OXs was detected in *E. histolytica* trophozoites exposed to live *L. acidophilus* culture (Figure 2A). The addition of catalase during the interaction of *E. histolytica* trophozoites with *L. acidophilus* strongly inhibits the formation of OXs in the parasite, which confirms that the formation of OXs in the parasite is mediated by H_2O_2 produced by *L. acidophilus* (Figure 2B). These results indicate that the formation of OXs is triggered by H_2O_2 produced by *L. acidophilus*.

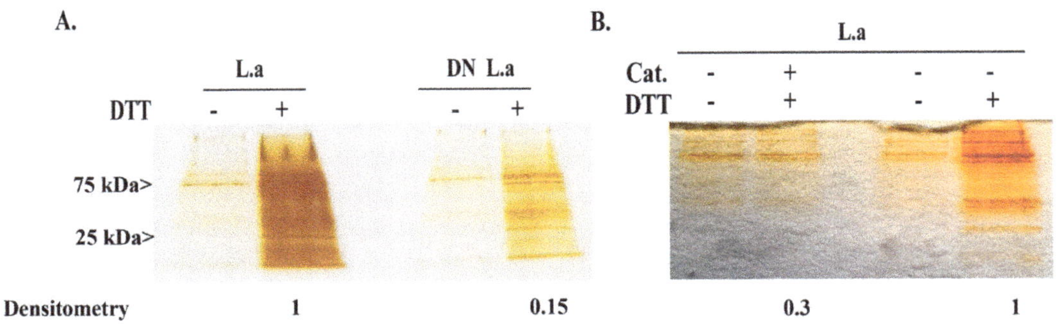

Figure 2. Detection of OXs by resin-assisted capture (OX–RAC) analysis of *E. histolytica*. Note: *E. histolytica* trophozoites were incubated with live *L. acidophilus* (L.a) or with heat-killed *L. acidophilus* (DN L.a) (**A**), with/without catalase (Cat.) (50 μg/mL) (**B**) for 2 h at 37 °C. Total protein lysate was prepared by lysing the trophozoites with 1% Igepal in PBS. The oxidized proteins in the cell lysates were subjected to RAC in the presence of 10 mM DTT (+DTT) or the absence of DTT (−DTT). The protein was resolved on a 12% SDS-PAGE and stained with silver stain.

The intensity of the protein bands were quantified by densitometry using Image J software [39]. The intensity of the OX-protein bands obtained in the presence of DTT in *E. histolytica* trophozoites incubated with live *L. acidophilus* was arbitrarily set to 1. It is important to note that the data presented in Figure 2A,B were obtained at two different times, and that the silver staining development time was different in each case.

Using MS, we identified 997 OXs in *E. histolytica* trophozoites incubated with *L. acidophilus* (Table S1), which were classified using PANTHER. The most abundant OX families belong to metabolite interconversion enzyme (PC00262), such as protein arginine N-methyltransferase (EHI_158560), the galactose-specific adhesin 170kD subunit (EHI_042370), and thioredoxin (EHI_004490) (Figure 3A). *E. histolytica* lacks glutathione, so it relies mainly on thiol for its defense against OS [40]. Thioredoxin (TRX)/thioredoxin reductase (TRXR) also contributes to redox signaling in *E. histolytica* trophozoites as well as oxidative stress responses [41]. This ubiquitous mechanism of defense is present in many parasites, including *Schistosoma mansoni*, *Plasmodium falciparum*, *Giardia lamblia*, and *Trichomonas vaginalis* [41]. TRXs are small redox proteins of around 12 kD, which act as radical scavengers. In their active site, two cysteine residues are involved in the antioxidant system. The oxidation of these cysteine residues produces disulfide bonds, which will be reduced by TRXR. The presence of TRXs as OXs in *E. histolytica* exposed to *L. acidophilus* strongly suggests that the parasite is actively responding to H_2O_2 released by the bacteria.

The other abundant OX family belongs to the protein modifying enzyme (PC00260) such as cysteine proteinase CP5 (EHI_168240), serine/threonine-protein phosphatase (EHI_031240), or E3 ubiquitin-protein ligase (EHI_050540) and the protein-binding activity modulator (PC00095) such as AIG1 family protein (EHI_176700), inhibitors of serine proteinase domain-containing protein (SERPIN) (EHI_119330), and the Rho family GTPase (EHI_070730) (Figure 3A).

SERPINs control a broad range of biological processes, including pathogen evasion of the host defense system. Cathepsin G, a pro-inflammatory enzyme released by activated neutrophils, is inhibited by serpins [42]. *E. histolytica* expresses a SERPIN that interacts with human neutrophil cathepsin G [43]. In this work, we showed that EhSERPIN is one of the OXs present in *E. histolytica* exposed to *L. acidophilus*. Studies have suggested that SERPINs are redox-regulated by oxidation of cysteine residues in the reactive site loop of these enzymes or its vicinity [44–46]. The presence of carbamidomethylated cysteine residues in the vicinity of the reactive site loop of EhSERPIN (Table S2) [43] suggests that EhSERPIN is also redox-regulated. The effect of oxidation on EhSERPIN activity has yet to be determined.

A functional motility is critical to the survival of *E. histolytica* in order to both dislodge and phagocytose host cells as well as transport virulence factors intracellularly [47]. Rho GTPases play a critical role in the regulation of motility and phagocytic activity of *E. histolytica* [48]. There are several Rho GTPases present in the parasite, and we identified six of them (EHI_126310, EHI_013260, EHI_197840, EHI_029020, EHI_129750, and EHI_070730) as OXs. EhRho1 (EHI_029020) regulates phagocytosis by regulating actin polymerization [49]. Numerous studies have shown that ROS regulate Rho GTPases activity [50]. Many Rho family GTPases contain a cysteine-containing motif (GXXXXGK[S/T]C) at their N-terminal, which is located directly adjacent to the phosphoryl-binding loop. Oxidation of the cysteine residue in this motif affects the nucleotide binding properties of these Rho GTPases [50]. According to the MS analysis of OXs (Table S2), this cysteine residue in the active site is not carbamidomethylated. Instead, we found that cysteine residues located at the C-terminal of these Rho GTPases are carbamidomethylated (Table S2). An ubiquitination region is present in the C-terminal region of many Rho GTPases that may regulate their stability [51]. In light of this information, it is tempting to speculate that the stability of these Rho GTPases is redox-dependent. An example of such regulation occurring in human endothelial cells is described here [52].

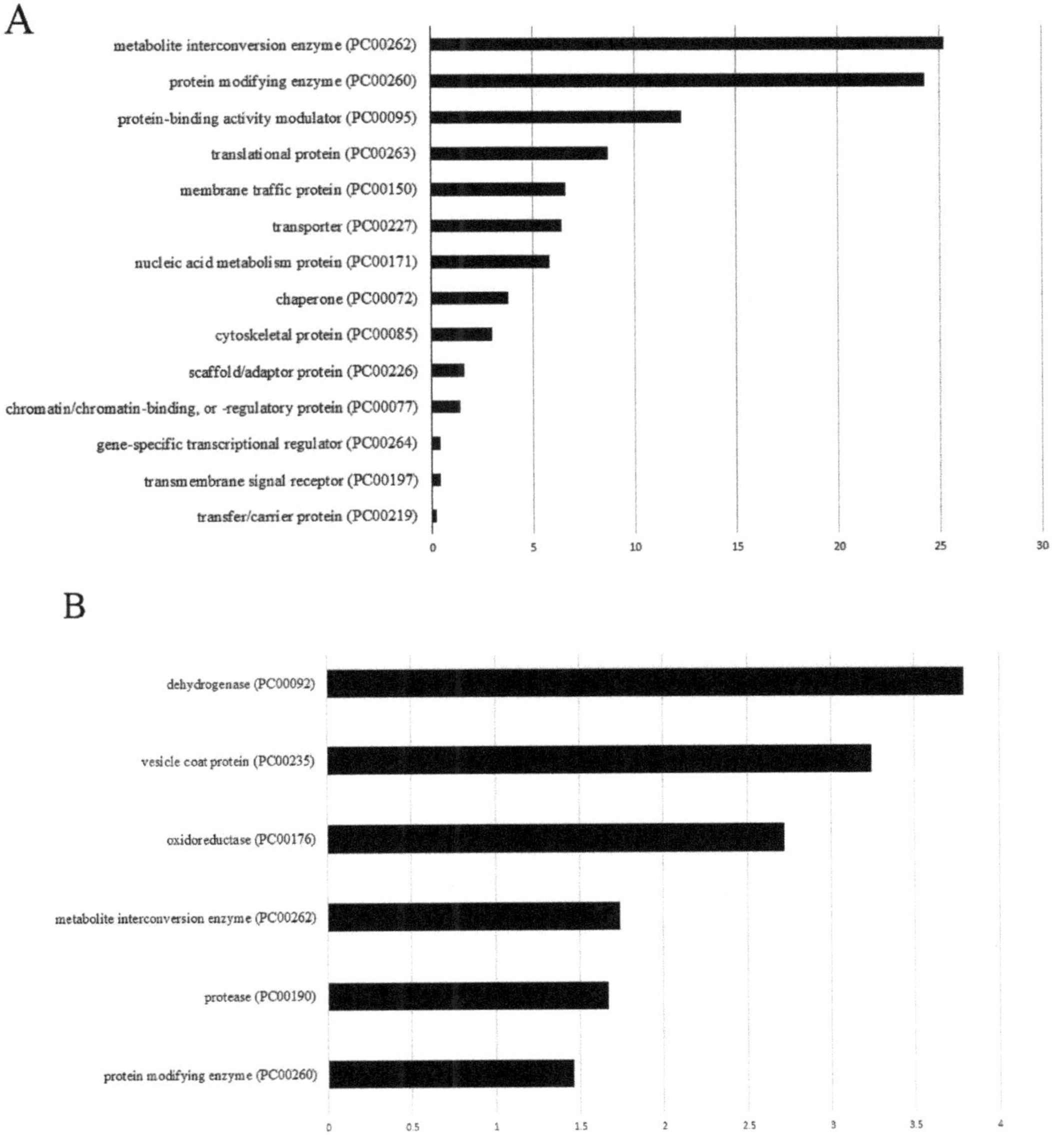

Figure 3. Protein analysis through evolutionary relationships (PANTHER) analysis of OXs in *E. histolytica* incubated with *L. acidophilus*. Note: (**A**) PANTHER sequence classification of the OXs identified in *E. histolytica* trophozoites co-incubated with *L. acidophilus*. (**B**) PANTHER statistical overrepresentation test of the OXs identified in *E. histolytica* trophozoites incubated with *L. acidophilus*.

Of the OXs in *E. histolytica* trophozoites incubated with *L. acidophilus* (Table S2), oxidoreductase (PC00176) and dehydrogenase (PC00092), such as glyceraldehyde-3-phosphate dehydrogenase (EHI_008200), NAD(FAD)-dependent dehydrogenase (EHI_099700), and pyruvate: ferredoxin oxidoreductase (EHI_051060), are significantly enriched according to the PANTHER statistical overrepresentation test (Figure 3B). Pyruvate: ferredoxin oxidoreductase (EHI_051060) is a Fe–S enzyme that catalyzes the oxidative decarboxylation of pyruvate [53]. This protein has also been identified as an OX in trophozoites exposed to

H_2O_2 [6], metronidazole, or auranofin [54]. In an oxidatively stressed parasite, pyruvate: ferredoxin oxidoreductase becomes strongly inhibited, resulting in an accumulation of pyruvate, which limits ATP production and causes parasite death [55]. Several cysteine residues present within the [4Fe–4S] clusters of close to them are carbamidomethylated suggesting that they are oxidized (Table S2). Destabilization of the Fe–S clusters integrity via oxidation of these cysteine residues in the parasite exposed to *L. acidophilus* will more certainly inactivate the enzyme and consequently contribute to the parasite death.

Other OXs, which are significantly enriched according to the PANTHER statistical overrepresentation test, include vesicle coat protein (PC00235), such as GOLD domain-containing protein (EHI_023070), beta2-COP (EHI_088220) and coatomer subunit gamma (EHI_040700) and protease (PC00190), such as EhCP-a1 (EHI_074180) and EhCP-a4 (EHI_050570) (Figure 3B).

3.3. *E. histolytica CP Activity Is Impaired by L. acidophilus*

In order to gain information on the consequence of *L. acidophilus*-mediated-oxidation on the activity of proteins that were identified in the OX–RAC analysis, we decided to focus here on the CPs. When trophozoites are incubated with live *L. acidophilus*, CPs activity is strongly inhibited (Figure 4). However, this activity is not inhibited when trophozoites are incubated with *L. acidophilus* in the presence of catalase (Figure 4). The addition of DTT in lysates of trophozoites incubated with live *L. acidophilus* partially restored CP activity. Based on these results, it could be assumed that the *L. acidophilus*-mediated-oxidation of CPs' catalytic cysteine residues inhibits CPs, while their reduction by DTT restores the activity. Indeed, the fact that adding catalase to trophozoites incubated with *L. acidophilus* prevents the inhibition of CPs confirms that H_2O_2 produced by *L. acidophilus* inhibits the CPs.

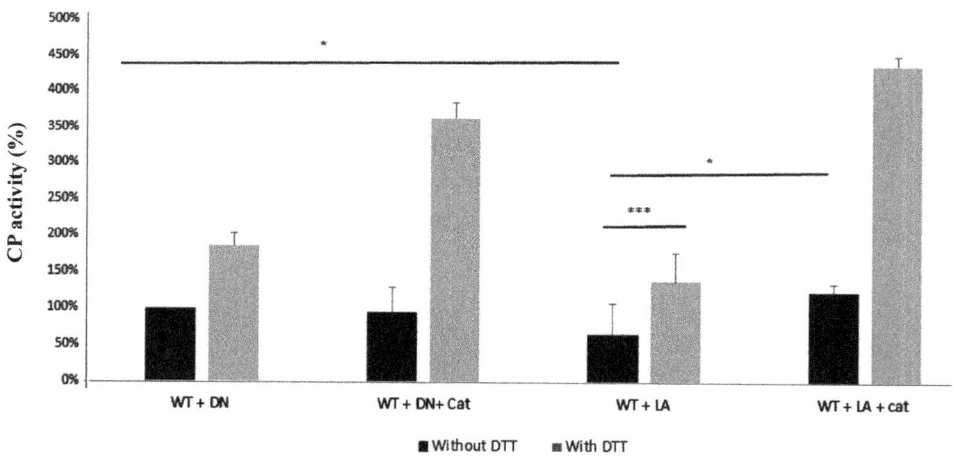

Figure 4. CPs activity of *E. histolytica* trophozoites. Note: *E. histolytica* trophozoites were incubated with heat-killed *L. acidophilus* (DN) or with live *L. acidophilus* (LA), and with/without catalase (50 μg/mL) for 2 h at 37 °C. Total protein was prepared and CPs activity was measured. One unit of CP activity was defined as the number of micromoles of substrate digested per minute per milligram of protein. CP activity performed without DTT of *E. histolytica* trophozoites incubated with heat-killed *L. acidophilus* (WT + DN) was obtained as 100% and it corresponds to 0.31 units. The data represent two independent experiments performed in triplicate. * p-value < 0.05 by an unpaired Student's t-test. *** p-value < 0.001 by an unpaired Student's t-test.

CPs are essential for the growth of *E. histolytica* trophozoites and their inhibition by inhibitors of the CPS, such as E64d, causes their death [56]. In this study, we found that many CPs, including EhCP-a1 (EHI_074180), EhCP-a4 (EHI_050570), EhCP-a5 (EHI_168240), and EhCP8 (EHI_010850), are oxidized, and that *E. histolytica* CPs activity are inhibited when the parasite is incubated with *L. acidophilus*. Some of these OXs CPs, such as EhCP-A1 and EhCP-A5, are highly expressed in *E. histolytica* [57] and are involved in rosette formation, hemolysis, and erythrocyte digestion [58]. The expression of EHI_010850 (EhCP-8) is upregulated when the parasite is incubated in the presence of hemoglobin, which suggests CP-8 is involved in iron uptake by the parasite [59]. The mechanisms that lead to oxidants inhibiting CPs have recently been examined [60]. For example, inhibition of papain by H_2O_2 results from the formation of sulfenic acid, which reacts with adjacent free thiol to form mixed disulfides. In addition, H_2O_2 inhibits cathepsin B by targeting the active site residue (Cys25) to form either sulfenic acid or sulfonic acid around 70% of the time. *E. histolytica* CPs contain four active-site residues, namely Gln, Cys, His, and Asn, the cysteine residue at the active site being present in all *E. histolytica* CPs [61]. According to the MS analysis of OXs (Table S2), this cysteine residue in the active site is carbamidomethylated, which strongly suggests that it was oxidized. By itself, this observation would explain why *E. histolytica*'s CP activity is inhibited by H_2O_2 produced by *L. acidophilus*. As opposed to *E. histolytica*, where H_2O_2 produced by *L. acidophilus* appears to inhibit CPs activity directly, in *Plasmodium* parasites, H_2O_2-mediated inhibition of CPs is dependent on the presence of free hemin, which can be released by quinoline drugs [62].

3.4. Adhesion of E. histolytica *Trophozoites to HeLa Cells Is Impaired by* L. acidophilus

E. histolytica trophozoites' ability to bind to mammalian cells is the initial step in the amebic infectious process [63]. In our experiment, trophozoites incubated with *L. acidophilus* exhibit reduced binding to HeLa cells compared to trophozoites incubated with heat-killed *L. acidophilus* or with paraformaldehyde-fixed *L. acidophilus*. However, the binding activity to HeLa cells of trophozoites incubated with *L. acidophilus* in the presence of catalase is comparable to the binding activity of heat-killed *L. acidophilus* or with paraformaldehyde-fixed *L. acidophilus* (Figure 5). These data strongly suggest that the production of H_2O_2 by *L. acidophilus* inhibits *E. histolytica*'s binding to HeLa cells rather than a competition between *L. acidophilus* and HeLa cells. The lectin Gal/GalNAc plays an essential role in parasite attachment to mammalian cells, including HeLa cells [64–66]. We previously demonstrated that oxidation of the carbohydrate-recognizing cysteine-rich domain (CRD) of Gal/GalNAc lectin renders it inactive [6]. We observed in this study that 170kDa Gal/GalNAc is one of the OXs produced in the parasite exposed to *L. acidophilus*. According to the MS analysis of OXs (Table S2), many cysteine residues are carbamidomethylated in the CRD of Gal/GalNAc lectin, which strongly suggests that they were oxidized leading to an impairment of the parasite's ability to bind mammalian cells (this work and [6]).

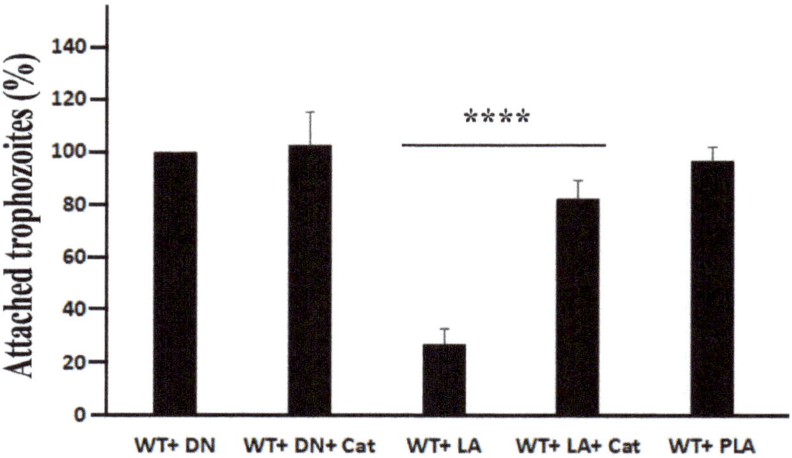

Figure 5. Binding activity assay of *E. histolytica* trophozoites. Note: *E. histolytica* trophozoites were incubated with live *L. acidophilus* (LA), with heat-killed *L. acidophilus* (DN), with paraformaldehyde-fixed *L. acidophilus* (PLA), and with/without catalase (50 μg/mL) for 1 h at 37 °C and then transferred to paraformaldehyde-fixed HeLa cell monolayers. Trophozoites attached to HeLa cells monolayers were counted. The number of trophozoites incubated with heat-killed *L. acidophilus* (WT + DN) that were bound to HeLa cells monolayer (around 75% of the original population) was obtained as 100%. The data represent two independent experiments performed in duplicate. **** p-value < 0.0001 by an unpaired Student's t-test.

4. Conclusions

The results for this study show that the production of H_2O_2 by *L. acidophilus* causes oxidation of vital proteins in *E. histolytica* and ultimately results in parasite death. The present study emphasizes *L. acidophilus'* potential as a probiotic against amebiasis. However, in vivo trials are necessary to determine whether this probiotic has health benefits on humans when it is used alone or in combination with metronidazole.

Supplementary Materials: The following are available online at https://www.mdpi.com/article/10.3390/antiox11050814/s1. Table S1: List of all OXs that were enriched by RAC in three independent experiments in *E. histolytica* trophozoites incubated with *L. acidophilus*. Table S2: List carbamidomethyl (C) sites in OXs. Table S3: Description of the parameters that are given in Tables S1 and S2.

Author Contributions: Conceptualization: L.S. and S.A.; methodology, L.S., E.Z., J.Y., M.T.-G. and S.A.; software, L.S. and S.A.; validation, L.S., E.Z., J.Y., M.T.-G. and S.A.; formal analysis, L.S., E.Z., M.T.-G. and S.A.; investigation, L.S., E.Z., J.Y., M.T.-G. and S.A.; resources, S.A.; data curation, L.S. and S.A.; writing—original draft preparation, L.S. and S.A.; writing—review and editing, L.S., E.Z., J.Y., M.T.-G. and S.A.; visualization, S.A.; supervision, S.A.; project administration, S.A.; funding acquisition, S.A. All authors have read and agreed to the published version of the manuscript.

Funding: This work was supported by the Israel Science Foundation (3208/19) and the Ministry of Science and Technology, Israel (1020546).

Institutional Review Board Statement: Not applicable.

Informed Consent Statement: Not applicable.

Data Availability Statement: Data is contained within the article and Supplementary Material.

Acknowledgments: We thank the staff of the Smoler Proteomics Center at the Technion for their technical help.

Conflicts of Interest: The authors declare no conflict of interest.

References

1. Marie, C.; Petri, W.A., Jr. Regulation of Virulence of *Entamoeba* histolytica. *Annu. Rev. Microbiol.* **2014**, *68*, 493–520. [CrossRef] [PubMed]
2. Gill, N.J.; Ganguly, N.K.; Mahajan, R.C.; Dilawari, J.B. Antibody dependent cellular cytotoxicity in experimental intestinal & hepatic amoebiasis. *Indian J. Med. Res.* **1988**, *87*, 170–175. [PubMed]
3. Schulz, T.F.; Kollaritsch, H.; Hengster, P.; Stemberger, H.; Scheiner, O.; Wiedemann, G.; Dierich, M.P. Molecular weight analysis of *Entamoeba* histolytica antigens recognized by IgG and IgM antibodies in the sera of patients with amoebiasis. *Trop. Med. Parasitol.* **1987**, *38*, 149–152. [PubMed]
4. Rastew, E.; Vicente, J.B.; Singh, U. Oxidative stress resistance genes contribute to the pathogenic potential of the anaerobic protozoan parasite, *Entamoeba* histolytica. *Int. J. Parasitol.* **2012**, *42*, 1007–1015. [CrossRef]
5. Santi-Rocca, J.; Smith, S.; Weber, C.; Pineda, E.; Hon, C.-C.; Saavedra, E.; Olivos-García, A.; Rousseau, S.; Dillies, M.-A.; Coppée, J.-Y.; et al. Endoplasmic reticulum stress-sensing mechanism is activated in *Entamoeba* histolytica upon treatment with nitric oxide. *PLoS ONE* **2012**, *7*, e31777. [CrossRef]
6. Shahi, P.; Trebicz-Geffen, M.; Nagaraja, S.; Alterzon, S.B.; Hertz, R.; Methling, K.; Lalk, M.; Ankri, S. Proteomic Identification of Oxidized Proteins in *Entamoeba* histolytica by Resin-Assisted Capture: Insights into the Role of Arginase in Resistance to Oxidative Stress. *PLoS Negl. Trop. Dis.* **2016**, *10*, e0004340. [CrossRef]
7. Hertz, R.; Ben Lulu, S.; Shahi, P.; Trebicz-Geffen, M.; Benhar, M.; Ankri, S. Proteomic Identification of S-Nitrosylated Proteins in the Parasite *Entamoeba histolytica* by Resin-Assisted Capture: Insights into the Regulation of the Gal/GalNAc Lectin by Nitric Oxide. *PLoS ONE.* **2014**, *9*, e91518. [CrossRef]
8. Roe, F.J. Metronidazole: Review of uses and toxicity. *J. Antimicrob. Chemother.* **1977**, *3*, 205–212. [CrossRef]
9. Andersson, K.E. Pharmacokinetics of Nitroimidazoles—Spectrum of Adverse Reactions. *Scand. J. Infect. Dis.* **1981**, *26*, 60–67.
10. Bansal, D.; Sehgal, R.; Chawla, Y.; Mahajan, R.C.; Malla, N. In vitro activity of antiamoebic drugs against clinical isolates of *Entamoeba histolytica* and *Entamoeba* dispar. *Ann. Clin. Microbiol. Antimicrob.* **2004**, *3*, 27. [CrossRef]
11. Iyer, L.R.B.N.; Naik, S.; Paul, J. Antioxidant enzyme profile of two clinical isolates of *Entamoeba histolytica* varying in sensitivity to antiamoebic drugs. *World J. Clin. Infect. Dis.* **2017**, *7*, 21–23. [CrossRef]
12. Hill, C.; Guarner, F.; Reid, G.; Gibson, G.R.; Merenstein, D.J.; Pot, B.; Morelli, L.; Canani, R.B.; Flint, H.J.; Salminen, S.; et al. Expert consensus document. The International Scientific Association for Probiotics and Prebiotics consensus statement on the scope and appropriate use of the term probiotic. *Nat. Rev. Gastroenterol. Hepatol.* **2014**, *11*, 506–514. [CrossRef] [PubMed]
13. Travers, M.A.; Florent, I.; Kohl, L.; Grellier, P. Probiotics for the control of parasites: An overview. *J. Parasitol. Res.* **2011**, *2011*, 610769. [CrossRef] [PubMed]
14. Bar, A.K.; Phukan, N.; Pinheiro, J.; Simoes-Barbosa, A. The Interplay of Host Microbiota and Parasitic Protozoans at Mucosal Interfaces: Implications for the Outcomes of Infections and Diseases. *PLoS Negl. Trop. Dis.* **2015**, *9*, e0004176. [CrossRef]
15. Goyal, N.; Tiwari, R.P.; Shukla, G. Lactobacillus rhamnosus GG as an Effective Probiotic for Murine Giardiasis. *Interdiscip. Perspect. Infect. Dis.* **2011**, *2011*, 795219. [CrossRef]
16. Sarjapuram, N.; Mekala, N.; Singh, M.; Tatu, U. The Potential of Lactobacillus casei and Entercoccus faecium Combination as a Preventive Probiotic against *Entamoeba*. *Probiotics Antimicrob. Proteins* **2017**, *9*, 142–149. [CrossRef]
17. Rigothier, M.C.; Maccario, J.; Gayral, P. Inhibitory activity of saccharomyces yeasts on the adhesion of *Entamoeba histolytica* trophozoites to human erythrocytes in vitro. *Parasitol. Res.* **1994**, *80*, 10–15. [CrossRef]
18. Mansour-Ghanaei, F.; Dehbashi, N.; Yazdanparast, K.; Shafaghi, A. Efficacy of saccharomyces boulardii with antibiotics in acute amoebiasis. *World J. Gastroenterol.* **2003**, *9*, 1832–1833. [CrossRef]
19. Varet, H.; Shaulov, Y.; Sismeiro, O.; Trebicz-Geffen, M.; Legendre, R.; Coppée, J.-Y.; Ankri, S.; Guillen, N. Enteric bacteria boost defences against oxidative stress in *Entamoeba histolytica*. *Sci. Rep.* **2018**, *8*, 9042. [CrossRef]
20. Collins, E.B.; Aramaki, K. Production of Hydrogen peroxide by Lactobacillus acidophilus. *J. Dairy Sci.* **1980**, *63*, 353–357. [CrossRef]
21. Hertzberger, R.; Arents, J.; Dekker, H.L.; Pridmore, R.D.; Gysler, C.; Kleerebezem, M.; de Mattos, M.J.T. H_2O_2 production in species of the Lactobacillus acidophilus group: A central role for a novel NADH-dependent flavin reductase. *Appl. Environ. Microbiol.* **2014**, *80*, 2229–2239. [CrossRef] [PubMed]
22. Diamond, L.S.; Harlow, D.R.; Cunnick, C.C. A new medium for the axenic cultivation of *Entamoeba histolytica* and other *Entamoeba*. *Trans. R. Soc. Trop. Med. Hyg.* **1978**, *72*, 431–432. [CrossRef]
23. DeLong, J.M.; Prange, R.K.; Hodges, D.M.; Forney, C.; Bishop, M.C.; Quilliam, M. Using a modified ferrous oxidation-xylenol orange (FOX) assay for detection of lipid hydroperoxides in plant tissue. *J. Agric. Food Chem.* **2002**, *50*, 248–254. [CrossRef] [PubMed]
24. Cox, J.; Hein, M.Y.; Luber, C.A.; Paron, I.; Nagaraj, N.; Mann, M. Accurate proteome-wide label-free quantification by delayed normalization and maximal peptide ratio extraction, termed MaxLFQ. *Mol. Cell. Proteom.* **2014**, *13*, 2513–2526. [CrossRef]
25. Tyanova, S.; Temu, T.; Sinitcyn, P.; Carlson, A.; Hein, M.Y.; Geiger, T.; Mann, M.; Cox, J. The Perseus computational platform for comprehensive analysis of (prote)omics data. *Nat. Methods* **2016**, *13*, 731–740. [CrossRef]
26. Mi, H.; Ebert, D.; Muruganujan, A.; Mills, C.; Albou, L.-P.; Mushayamaha, T.; Thomas, P.D. PANTHER version. 16, a revised family classification, tree-based classification tool, enhancer regions and extensive API. *Nucleic Acids Res.* **2021**, *49*, D394–D403. [CrossRef]

27. Ankri, S.; Stolarsky, T.; Mirelman, D. Antisense inhibition of expression of cysteine proteinases does not affect *Entamoeba histolytica* cytopathic or haemolytic activity but inhibits phagocytosis. *Mol. Microbiol.* **1998**, *28*, 777–785. [CrossRef]
28. Shahi, P.; Trebicz-Geffen, M.; Nagaraja, S.; Hertz, R.; Alterzon-Baumel, S.; Methling, K.; Lalk, M.; Mazumder, M.; Samudrala, G.; Ankri, S. N-acetyl ornithine deacetylase is a moonlighting protein and is involved in the adaptation of *Entamoeba histolytica* to nitrosative stress. *Sci. Rep.* **2016**, *6*, 36323. [CrossRef]
29. Dinev, T.B.G.; Denev, S.; Dermendzhieva, D.; Tzanova, M.; Valkova, E. Antimicrobial activity of Lactobacillus acidophilus against pathogenic and food spoilage microorganisms: A review. *Agric. Sci. Technol.* **2017**, *91*, 3–9. [CrossRef]
30. Weiss, G.; Rasmussen, S.; Zeuthen, L.H.; Nielsen, B.N.; Jarmer, H.; Jespersen, L.; Frøkiaer, H. Lactobacillus acidophilus induces virus immune defence genes in murine dendritic cells by a Toll-like receptor-2-dependent mechanism. *Immunology* **2010**, *131*, 268–281. [CrossRef]
31. Salari, S.; Ghasemi Nejad Almani, P. Antifungal effects of Lactobacillus acidophilus and Lactobacillus plantarum against different oral Candida species isolated from HIV/AIDS patients: An in vitro study. *J. Oral Microbiol.* **2020**, *12*, 1769386. [CrossRef]
32. Aween, M.M.; Hassan, Z.; Muhialdin, B.J.; Eljamel, Y.A.; Al-Mabrok, A.S.W.; Lani, M.N. Antibacterial activity of Lactobacillus acidophilus strains isolated from honey marketed in Malaysia against selected multiple antibiotic resistant (MAR) Gram-positive bacteria. *J. Food Sci.* **2012**, *77*, M364–M371. [CrossRef] [PubMed]
33. Al-Megrin, W.A.; Mohamed, S.H.; Saleh, M.M.; Yehia, H.M. Preventive role of probiotic bacteria against gastrointestinal diseases in mice caused by Giardia lamblia. *Biosci. Rep.* **2021**, *41*, BSR20204114. [CrossRef] [PubMed]
34. Cadore, P.S.; Walcher, D.L.; de Sousa, N.F.G.C.; Martins, L.H.R.; da Hora, V.P.; Von Groll, A.; de Moura, M.Q.; Berne, M.E.A.; Avila, L.F.D.C.D.; Scaini, C.J. Protective effect of the probiotic Lactobacillus acidophilus ATCC. 4356 in BALB/c mice infected with Toxocara canis. *Rev. Inst. Med. Trop. São Paulo* **2021**, *63*, e9. [CrossRef] [PubMed]
35. Farrag, H.M.; Huseein, E.A.; Abd El-Rady, N.M.; Mostafa, F.A.; Mohamed, S.S.; Gaber, M. The protective effect of Lactobacillus acidophilus on experimental animals challenged with Trichinella spiralis; new insights on their feasibility as prophylaxis in Trichinella spiralis endemic area. *Ann. Parasitol.* **2021**, *67*, 195–202.
36. Alak, J.I.; Wolf, B.W.; Mdurvwa, E.G.; EPimentel-Smith, G.; Kolavala, S.; Abdelrahman, H.; Suppiramaniam, V. Supplementation with Lactobacillus reuteri or *L. acidophilus* reduced intestinal shedding of cryptosporidium parvum oocysts in immunodeficient C57BL/6 mice. *Cell. Mol. Biol. (Noisy-le-grand)* **1999**, *45*, 855–863.
37. Loew, O. A New Enzyme of General Occurrence in Organismis. *Science* **1900**, *11*, 701–702. [CrossRef]
38. Shaulov, Y.; Sarid, L.; Trebicz-Geffen, M.; Ankri, S. *Entamoeba histolytica* Adaption to Auranofin: A Phenotypic and Multi-Omics Characterization. *Antioxidants* **2021**, *10*, 1240. [CrossRef]
39. Rueden, C.T.; Schindelin, J.; Hiner, M.C.; Dezonia, B.E.; Walter, A.E.; Arena, E.T.; Eliceiri, K.W. ImageJ2: ImageJ for the next generation of scientific image data. *BMC Bioinform.* **2017**, *18*, 529. [CrossRef]
40. Fahey, R.C.; Newton, G.L.; Arrick, B.; Overdank-Bogart, T.; Aley, S.B. *Entamoeba histolytica*: A eukaryote without glutathione metabolism. *Science* **1984**, *224*, 70–72. [CrossRef]
41. Jeelani, G.; Nozaki, T. *Entamoeba* thiol-based redox metabolism: A potential target for drug development. *Mol. Biochem. Parasitol.* **2016**, *206*, 39–45. [CrossRef] [PubMed]
42. Bao, J.; Pan, G.; Poncz, M.; Wei, J.; Ran, M.; Zhou, Z. Serpin functions in host-pathogen interactions. *PeerJ* **2018**, *6*, e4557. [CrossRef] [PubMed]
43. Riahi, Y.; Siman-Tov, R.; Ankri, S. Molecular cloning, expression and characterization of a serine proteinase inhibitor gene from *Entamoeba histolytica. Mol. Biochem. Parasitol.* **2004**, *133*, 153–162. [CrossRef] [PubMed]
44. Morris, E.C.; Dafforn, T.R.; Forsyth, S.L.; Missen, M.A.; Horvath, A.J.; Hampson, L.; Hampson, I.N.; Currie, G.; Carrell, R.W.; Coughlin, P.B. Murine serpin. 2A is a redox-sensitive intracellular protein. *Biochem. J.* **2003**, *371*, 165–173. [CrossRef]
45. Stief, T.W.; Aab, A.; Heimburger, N. Oxidative inactivation of purified human alpha-2-antiplasmin, antithrombin III, and C1-inhibitor. *Thromb. Res.* **1988**, *49*, 581–589. [CrossRef]
46. Mangan, M.S.; Bird, C.H.; Kaiserman, D.; Matthews, A.Y.; Hitchen, C.; Steer, D.L.; Thompson, P.E.; Bird, P.I. A Novel Serpin Regulatory Mechanism: SerpinB9 Is Reversibly Inhibited by vicinal disulfide bond formation in the reactive center loop. *J. Biol. Chem.* **2016**, *291*, 3626–3638. [CrossRef]
47. Coudrier, E.; Amblard, F.; Zimmer, C.; Roux, P.; Olivo–Marin, J.C.; Rigothier, M.C.; Guillén, N. Myosin II and the Gal-GalNAc lectin play a crucial role in tissue invasion by *Entamoeba histolytica. Cell. Microbiol.* **2005**, *7*, 19–27. [CrossRef]
48. Bosch, D.E.; Siderovski, D.P. G protein signaling in the parasite *Entamoeba histolytica. Exp. Mol. Med.* **2013**, *45*, e15. [CrossRef]
49. Bharadwaj, R.; Sharma, S.; Janhawi Arya, R.; Bhattacharya, S.; Bhattacharya, A. EhRho1 regulates phagocytosis by modulating actin dynamics through EhFormin1 and EhProfilin1 in *Entamoeba histolytica. Cell. Microbiol.* **2018**, *20*, e12851. [CrossRef]
50. Hobbs, G.A.; Zhou, B.; Cox, A.D.; Campbell, S.L. Rho GTPases, oxidation, and cell redox control. *Small GTPases* **2014**, *5*, e28579. [CrossRef]
51. Olson, M.F. Rho GTPases, their post-translational modifications, disease-associated mutations and pharmacological inhibitors. *Small GTPases* **2018**, *9*, 203–215. [CrossRef] [PubMed]
52. Kovacic, H.N.; Irani, K.; Goldschmidt-Clermont, P.J. Redox regulation of human Rac1 stability by the proteasome in human aortic endothelial cells. *J. Biol. Chem.* **2001**, *276*, 45856–45861. [CrossRef] [PubMed]
53. Reeves, R.E.; Warren, L.G.; Susskind, B.; Lo, H.S. An energy-conserving pyruvate-to-acetate pathway in *Entamoeba histolytica.* Pyruvate synthase and a new acetate thiokinase. *J. Biol. Chem.* **1977**, *252*, 726–731. [CrossRef]

54. Shaulov, Y.; Nagaraja, S.; Sarid, L.; Trebicz-Geffen, M.; Ankri, S. Formation of oxidised (OX) proteins in *Entamoeba histolytica* exposed to auranofin and consequences on the parasite virulence. *Cell. Microbiol.* **2020**, *22*, e13174. [CrossRef]
55. Pineda, E.; Encalada, R.; Rodríguez-Zavala, J.S.; Olivos-García, A.; Moreno-Sánchez, R.; Saavedra, E. Pyruvate:ferredoxin oxidoreductase and bifunctional aldehyde-alcohol dehydrogenase are essential for energy metabolism under oxidative stress in *Entamoeba histolytica*. *FEBS J.* **2010**, *277*, 3382–3395. [CrossRef]
56. Nowak, N.; Lotter, H.; Tannich, E.; Bruchhaus, I. Resistance of *Entamoeba histolytica* to the cysteine proteinase inhibitor E64 is associated with secretion of pro-enzymes and reduced pathogenicity. *J. Biol. Chem.* **2004**, *279*, 38260–38266. [CrossRef]
57. Matthiesen, J.; Bär, A.-K.; Bartels, A.-K.; Marien, D.; Ofori, S.; Biller, L.; Tannich, E.; Lotter, H.; Bruchhaus, I. Overexpression of specific cysteine peptidases confers pathogenicity to a nonpathogenic *Entamoeba histolytica* clone. *mBio* **2013**, *4*, e00072-13. [CrossRef]
58. Irmer, H.; Tillack, M.; Biller, L.; Handal, G.; Leippe, M.; Roeder, T.; Tannich, E.; Bruchhaus, I. Major cysteine peptidases of *Entamoeba histolytica* are required for aggregation and digestion of erythrocytes but are dispensable for phagocytosis and cytopathogenicity. *Mol. Microbiol.* **2009**, *72*, 658–667. [CrossRef]
59. Hernandez-Cuevas, N.A.; Weber, C.; Hon, C.C.; Guillen, N. Gene expression profiling in *Entamoeba histolytica* identifies key components in iron uptake and metabolism. *PLoS ONE* **2014**, *9*, e107102. [CrossRef]
60. Lalmanach, G.; Saidi, A.; Bigot, P.; Chazeirat, T.; Lecaille, F.; Wartenberg, M. Regulation of the Proteolytic Activity of Cysteine Cathepsins by Oxidants. *Int. J. Mol. Sci.* **2020**, *21*, 1944. [CrossRef]
61. Bruchhaus, I.; Loftus, B.J.; Hall, N.; Tannich, E. The intestinal protozoan parasite *Entamoeba histolytica* contains. 20 cysteine protease genes, of which only a small subset is expressed during in vitro cultivation. *Eukaryot. Cell* **2003**, *2*, 501–509. [CrossRef] [PubMed]
62. Herraiz, T.; Guillen, H.; Gonzalez-Pena, D.; Aran, V.J. Antimalarial Quinoline Drugs Inhibit beta-Hematin and Increase Free Hemin Catalyzing Peroxidative Reactions and Inhibition of Cysteine Proteases. *Sci. Rep.* **2019**, *9*, 15398. [CrossRef] [PubMed]
63. Mirelman, D.; Kobiler, D. Adhesion properties of *Entamoeba histolytica*. In *Adhesion and Microorganism Pathogenicity*; Pittman Med London: London, UK, 1981; Volume 80, pp. 17–35.
64. Aguirre Garcia, M.; Gutierrez-Kobeh, L.; Lopez Vancell, R. *Entamoeba histolytica*: Adhesins and lectins in the trophozoite surface. *Molecules* **2015**, *20*, 2802–2815. [CrossRef] [PubMed]
65. Petri, W.A., Jr.; Haque, R.; Mann, B.J. The bittersweet interface of parasite and host: Lectin-carbohydrate interactions during human invasion by the parasite *Entamoeba histolytica*. *Annu. Rev. Microbiol.* **2002**, *56*, 39–64. [CrossRef] [PubMed]
66. Vines, R.R.; Ramakrishnan, G.; Rogers, J.B.; Lockhart, L.A.; Mann, B.J.; Petri, W.A., Jr. Regulation of adherence and virulence by the *Entamoeba histolytica* lectin cytoplasmic domain, which contains a beta2 integrin motif. *Mol. Biol. Cell* **1998**, *9*, 2069–2079. [CrossRef] [PubMed]

antioxidants

MDPI

Article

Two Distinct Superoxidase Dismutases (SOD) Secreted by the Helminth Parasite *Fasciola hepatica* Play Roles in Defence against Metabolic and Host Immune Cell-Derived Reactive Oxygen Species (ROS) during Growth and Development

Nichola Eliza Davies Calvani [1,*], Carolina De Marco Verissimo [1], Heather Louise Jewhurst [1], Krystyna Cwiklinski [1,2], Andrew Flaus [3] and John Pius Dalton [1]

1 Molecular Parasitology Laboratory (MPL), Centre for One Health and Ryan Institute, School of Natural Sciences, University of Galway, H91 DK59 Galway, Ireland
2 Institute of Infection, Veterinary and Ecological Sciences, University of Liverpool, Liverpool L69 3BX, UK
3 Centre for Chromosome Biology, School of Natural Science, University of Galway, H91 TK33 Galway, Ireland
* Correspondence: nichola.calvani@universityofgalway.ie

Abstract: The antioxidant superoxide dismutase (SOD) catalyses the dismutation of superoxide, a dangerous oxygen free radical, into hydrogen peroxide and molecular oxygen. Superoxide generation during the oxidative burst of the innate immune system is considered a key component of the host defence against invading pathogens. We demonstrate the presence and differential expression of two SODs in *Fasciola hepatica*, a leaderless cytosolic (FhSOD1) and an extracellular (FhSOD3) form containing a secretory signal peptide, suggesting that the parasites exploit these enzymes in distinct ways to counteract reactive oxygen species (ROS) produced by cellular metabolism and immune defences. Both enzymes are highly expressed by the infective newly excysted juvenile (NEJ) stages and are found in abundance in their excretory–secretory products (ES), but only FhSOD1 is present in adult ES, suggesting that the antioxidants have different functions and pathways of secretion, and are under separate temporal expression control during the migration, growth, and development of the parasite. Functionally, the recombinant FhSOD1 and FhSOD3 exhibit similar activity against superoxide to their mammalian counterparts. Confocal immuno-localisation studies demonstrated the presence of FhSOD1 and FhSOD3 on the NEJ tegument and parenchyma, supporting our suggestion that these enzymes are secreted during host invasion to protect the parasites from the harmful oxidative bursts produced by the activated innate immune response. By producing superoxide enzymatically in vitro, we were able to demonstrate robust killing of *F. hepatica* NEJ within 24 h post-excystment, and that the lethal effect of ROS was nullified with the addition of SOD and catalase (the antioxidant enzyme responsible for the dismutation of hydrogen peroxide, a by-product of the SOD reaction). This study further elucidates the mechanism by which *F. hepatica* protects against ROS derived from cellular metabolism and how the parasite could mitigate damage caused by the host's immune response to benefit its survival.

Keywords: antioxidants; excretory–secretory products; helminth; immune defence; oxidative burst; ruminants; trematode; parasite; worm

Citation: Calvani, N.E.D.; De Marco Verissimo, C.; Jewhurst, H.L.; Cwiklinski, K.; Flaus, A.; Dalton, J.P. Two Distinct Superoxidase Dismutases (SOD) Secreted by the Helminth Parasite *Fasciola hepatica* Play Roles in Defence against Metabolic and Host Immune Cell-Derived Reactive Oxygen Species (ROS) during Growth and Development. *Antioxidants* **2022**, *11*, 1968. https://doi.org/10.3390/antiox11101968

Academic Editor: Serge Ankri

Received: 31 August 2022
Accepted: 23 September 2022
Published: 30 September 2022

1. Introduction

Fasciolosis, a zoonotic disease of humans and livestock, is caused by infection with the digenean trematodes, *Fasciola hepatica* and *Fasciola gigantica*. The economic impact of fasciolosis on livestock production is expected to exceed USD 3 billion/year [1]. The parasite also infects an estimated 17 million people globally, and 180 million people live in endemic regions where they are at risk of infection [2].

Infection occurs when the mammalian host ingests encysted parasites, metacercariae, carried on plant material or water. The metacercariae excyst in the host's intestine and the newly excysted juveniles (NEJ) penetrate the intestinal wall and migrate through the abdominal cavity to the liver. Upon reaching the liver, the immature parasites spend 8–12 weeks burrowing through the parenchymal tissue, where they rapidly increase in size and mature. After taking up residence in the bile ducts and gall bladder, the parasites complete their development into egg-laying adults [3]. During this period of migration and growth, the infected host mounts an acute proliferative cellular and humoral immune response to block the parasites and mitigate damage and haemorrhaging caused by their migration. This acute response is characterised by marked eosinophilia, along with the infiltration of macrophages and lymphocytes to the parasite tracks, resulting in the local production of damaging reactive oxygen species (ROS) [4].

Genomic, transcriptomic, and proteomic analyses of several life stages of *F. hepatica* (infective metacercariae, NEJ at 1, 3, and 24 h post-excystment, immature juvenile flukes from 21 days post-infection, and mature adults) have identified many developmentally regulated proteins, thereby shedding light on the complex interactions these parasites have with their hosts throughout the infection process [5–7]. The major proteins secreted during these early infection stages include cathepsin-like proteases, protease inhibitors, and a slew of antioxidant enzymes, including superoxide dismutase (FhSOD), peroxiredoxin (FhPrx), thioredoxin (FhTrx), glutathione peroxidase (FhGPx), and glutathione-S-transferase (FhGST) [8]. Together, these proteins are assumed to help tissue invasion, macromolecule digestion, and defend against the onslaught of host-generated ROS [7].

SODs are a class of metalloenzyme antioxidants that, in pathogens, play a role in defence against exogenous ROS produced by the host by catalysing the two-step disproportionation of superoxide anions ($O_2^{\bullet-}$) into hydrogen peroxide (H_2O_2) [9]. Three isoforms of SOD have been described in helminths (worms) and their mammalian hosts based on their localisation and metal co-factors: (1) a cytosolic Cu/Zn SOD, (2) a mitochondrial Mn-SOD, and (3) an extracellular Cu/Zn SOD characterised by the presence of a hydrophobic N-terminal signal peptide [10–14]. To date, only a cytosolic form of Cu/Zn SOD has been described in *F. hepatica* and *F. gigantica* [15–17]. However, our interrogation of the existing *Fasciola* spp. genomes revealed the presence of multiple SOD sequences in these parasites, including the cytosolic form, a mitochondrial form, and a novel extracellular SOD possessing a characteristic N-terminal signal peptide sequence. This extracellular SOD is observed at greater protein abundance in the secretome of *F. hepatica* NEJ relative to that of the mature adults, suggesting that it acts as a specialised enzyme protecting the invading parasites against the oxygen-mediated killing mechanisms of their hosts.

Here, we characterised the role of SODs in the defence of *F. hepatica* against ROS via the production of functional recombinant cytosolic (rFhSOD1) and extracellular (rFhSOD3) forms excreted/secreted by the parasite during early invasion. We show that these antioxidants are highly homologous to their mammalian host counterparts, which accounts for their lack of immunogenicity in infected sheep. Specific anti-rFhSOD1 and anti-rFhSOD3 antibodies were produced and used in immuno-localisation experiments to identify the site of enzyme production and secretion in *F. hepatica* NEJ and adults. Finally, we developed an in vitro biological assay that enzymatically produces ROS and demonstrates the susceptibility of NEJ to both superoxide and hydrogen peroxide. This killing of NEJ was obviated via the addition of SOD and catalase, revealing the power of this cascade in defence against ROS during invasion.

2. Materials and Methods

2.1. Identification and Phylogenetic Analysis of F. hepatica Superoxide Dismutases

The *F. hepatica* SOD gene sequences were identified by BLAST analysis using the previously reported *F. hepatica* SOD sequence (Kim et al., 2000 [16]; AF071229) against the *F. hepatica genome* (WormBase ParaSite Version WBPS16 (WS279): PRJEB6687 and PRJEB25283; Cwiklinski et al., 2015 [5]; FhSOD1: BN1106_s3189B000243; maker-scaffold10x_61_pilon-snap-gene-0.36, FhSOD2: snap_masked-scaffold10x_1664_pilon-processed-gene-0.2, FhSOD3: BN1106_s4478B000037; maker-scaffold10x_713_pilon-snap-gene-0.105). Annotation of the resulting sequences was confirmed using in silico tools (Uniprot, Gene Ontology (GO), and InterProScan).

Homologous trematode SOD DNA sequences were identified and retrieved using BLAST analysis of publicly available genome databases at WormBase Parasite (http://parasite.wormbase.org/index.html Version WBPS16 (WS280), accessed on 7 April 2021). Genomic DNA sequences were imported, manually inspected, translated, and aligned in CLC Main Workbench 21.0.3 (Table S1). Identification of N-terminal signal peptide sequences was carried out using the SignalP and TMHMM plugin (Version 21.0) in CLC Main Workbench (Version 21.0.3). Previously characterised mammalian SOD sequences (*Ovis aries, Bos taurus, Bos indicus*, and *Homo sapiens*) were downloaded from GenBank (National Center for Biotechnology Information, NCBI) (Table S1). Ambiguous regions (i.e., containing gaps and/or poorly aligned) were removed from the resultant amino acid alignment of all mammalian and trematode sequences with Gblocks (v0.91b) using the following parameters: minimum length of a block after gap cleaning: 10; positions with a gap in less than 50% of the sequences were selected in the final alignment if they were within an appropriate block; all segments with contiguous non-conserved positions longer than 8 were rejected; minimum number of sequences for a flank position: 85% [18,19]. The resultant sequence alignment spanned 107 amino acids (Gly 67-Gly173, relative to FhSOD1), containing 43 amino acid sequences, which was submitted for Smart Model Selection (SMS) using PhyML 3.0 (Figure S1) [20]. Evolutionary history was inferred using the maximum likelihood method and the Whelan and Goldman (WAG + G+I) model, with 1000 bootstrap support in MEGA11 [21,22]. Initial tree(s) for the heuristic search were obtained automatically by applying Neighbor-Join and BioNJ algorithms to a matrix of pairwise distances estimated using the JTT model, and then selecting the topology with superior log likelihood value. A discrete Gamma distribution was used to model evolutionary rate differences among sites (5 categories (+G, parameter = 1.3434)). The rate variation model allowed for some sites to be evolutionarily invariable ([+I], 11.21% sites). Amino acid sequence similarity and identity was determined using UniProt ClustalO [23].

2.2. *Transcriptomic and Proteomic Expression Analysis of FhSOD1 and FhSOD3*

Stage-specific *F. hepatica* transcriptome datasets, previously described by Cwiklinski et al. (2015; PRJEB6904) [5] were interrogated to determine the differential transcription of the *FhSOD1* and *FhSOD3* genes, represented as the number of transcripts per million (TPM). *F. hepatica* proteomic datasets were interrogated to determine the FhSOD1 and FhSOD3 protein abundance, represented by the exponentially modified protein abundance index (emPAI) within the somatic proteome and the secreted protein fraction (secretome/ES proteins), using Scaffold (version Scaffold_5.1.2, Proteome Software, Portland, OR, USA). Analysis of the somatic proteome was carried out for the metacercariae, NEJ (3, 24, and 48 h post-excystment) and immature life cycle stages using the data reported by Cwiklinski et al. (2018, 2021) [6,7]. Secretome analysis was carried out on the 24 h NEJ, immature, and adult liver fluke secreted proteins using the data reported by Cwiklinski et al. (2018, 2021) [6,7] and Murphy et al. (2020).

2.3. *Preparation of Excretory/Secretory (E/S) Products and Somatic Extracts from Adult F. hepatica*

Adult liver flukes were collected from the livers of sheep during post-mortem abattoir surveillance in Roscommon, Ireland, as previously described [24]. The parasites were washed with sterile 1× PBS before culturing in RPMI medium (containing 0.1% glucose, 100 U penicillin and 100 mg/mL streptomycin) at a ratio of 1 worm/2 mL at 37 °C and 5% CO_2. After 2 h, culture media (containing the E/S) was collected and centrifuged at 300× *g* for 10 min and then 700× *g* for 30 min to eliminate large debris. The supernatant was subsequently concentrated using a spin column protein concentrator with a molecular weight cut-off of 3 kDa (Amicon ultra, Merck Millipore, Burlington, MA, USA), aliquoted and frozen at −80 °C prior to use.

Adult somatic proteins were extracted from a frozen adult fluke via homogenisation in 100 μL DPBS followed by centrifugation at 300× *g* for 15 min and then 800× *g* for a further 15 min. The protein concentration in the resultant supernatant of the somatic extract and E/S was measured using the Bradford Protein Assay (Bio-Rad).

2.4. Expression and Purification of Functional Recombinant FhSOD1 and FhSOD3 in Escherichia coli

The cytoplasmic (*FhSOD1*) and extracellular (*FhSOD3*, with the signal peptide sequence removed) sequences were codon optimised for expression in *Escherichia coli* and individually cloned into pET-28a(+) vectors with a C-terminal His-tag (GenScript, Piscataway, NJ, USA). The vectors were electro-transformed into kanamycin-resistant ClearColi BL21 (DE3) (ThermoFisher Scientific, Waltham, MA, USA) cells, with transformants selected on LB-Miller + kanamycin (50 μg/mL) agar plates after growth overnight at 37 °C.

Recombinant cells harbouring *FhSOD1* and *FhSOD3* were grown in LB-Miller broth at 37 °C and 180 rpm with 50 μg/mL kanamycin until the OD600 was between 0.7 and 0.8. Isopropyl-β-d-thiogalactopyranoside (IPTG; ThermoFisher Scientific) was added to the culture medium at 0.5 mM to induce protein expression, and cultures were incubated at 21 °C for a further 21 h. A 1 mL aliquot of each culture was removed at T_0 and T_{21} to monitor protein production. Following centrifugation at 10,000× *g* for 10 min at 4 °C, the recovered bacteria were resuspended in sterile ST buffer (10 mM Tris, 150 mM NaCl, pH 8.0) and stored at −20 °C overnight. Cell pellets were thawed on ice then treated with 10 mg/mL lysozyme for 30 min. After incubation, 750 μL 10% Sodium lauroyl sarcosinate (sarcosyl) was added to the pellets prior to sonication at 70% amplitude with six cycles of 10 s burst/10 s rest on ice.

To recover soluble protein, the supernatant was collected after centrifugation at 15,000× *g* for 30 min at 4 °C, diluted to a final volume of 50 mL in lysis buffer (sodium phosphate buffer, pH 8, 10 mM imidazole) and purified using the Profinia Affinity Chromatography Protein Purification System (Bio-Rad, Hercules, CA, USA) with the corresponding mini profinity IMAC and mini Bio-Gel P—6 desalting cartridges (Bio-Rad). Proteins were eluted into 4 mL of 1× PBS, aliquoted, and stored at –70 °C until use. Protein concentration and purity were verified immediately after purification using the Bradford Protein Assay (Bio-Rad) and by 4–20% SDS-PAGE gels (Bio-Rad) stained with Biosafe Coomassie (Bio-Rad), respectively. To further confirm the expression and purification of the recombinant proteins, Western blots were performed using a monoclonal mouse anti-polyhistidine antibody (1:10,000) (Sigma-Aldrich, St. Louis, MO, USA) as a primary antibody, followed by incubation with a secondary antibody alkaline phosphatase conjugated goat to mouse-anti-IgG diluted 1:5000 (Sigma-Aldrich). The gels and Western blots were visualised using a G:BOX Chemi XRQ imager (Syngene, Bengaluru, India).

The tertiary states of the recombinant FhSOD1 and FhSOD3 proteins were resolved by size-exclusion chromatography (gel filtration) performed on a high performance Superdex 75 10/300 GL (Tricorn) column, with a flow rate of 400 μL/min and eluted into 1× PBS. Three proteins of different molecular sizes were resolved in the column as standards, namely conalbumin (76 kDa), carbonic anhydrase (29 kDa), and aprotinin (6.5 kDa) (GE Healthcare). Upon determination of the retention parameters, rFhSOD1 and rFhSOD3 were added to the column, with 200 μL aliquots of each purification fraction collected and stored at 4 °C for enzyme activity (see below). The 3D structure of the corresponding native FhSOD1 (D915_003308) and FhSOD3 (D915_009739) amino acid sequences was predicted by the AlphaFold Protein Structure Database [25,26].

2.5. Analysis of rFhSOD1 and rFhSOD3 Enzyme Activity

The enzymatic activity of the recombinant FhSODs was measured via the adaptation of an existing xanthine oxidase (XOD)-based SOD assay, whereby $O_2^{\bullet -}$ is enzymatically produced because of the conversion of xanthine to H_2O_2 and uric acid, which in turn transforms nitroblue tetrazolium (NBT) to NBT-diformazan dye [27]. To determine the activity of our recombinant enzymes under physiological conditions compatible with the parasite, our assay was conducted at 37 °C in 200 µL assay buffer (1 × PBS; Sigma-Aldrich, 0.1 mM Hypoxanthine; Sigma-Aldrich, 0.1 mM DTPA; Sigma-Aldrich, 2.5 µL/mL tetrazolium salt; Cayman Chemical, Ann Arbor, MI, USA). All other enzymes (native XOD from bovine milk; Roche, used at a final concentration of 8.0×10^{-3} U/mL), including the standard curve (native bovine erythrocyte SOD, BS; Sigma-Aldrich), were diluted in PBS prior to use. Recombinant proteins and the standard curve (BS) were serially diluted 1:1 in PBS (rFhSOD1 and rFhSOD3: 10–0.3125 µg/mL; BS: 100–3.125 µg/mL, corresponding to 4–0.125 U/mL) and run in duplicate at a final assay volume of 250 µL. The assay was read continuously for 30 min at 450 nm using a microplate reader (PolarStar Omega Spectrophotometer; BMG LabTech, Ortenburg, Germany) immediately after the addition of XOD.

2.6. Immune Recognition of FhSOD1 and FhSOD3 in Sera from F. hepatica Experimentally Infected Sheep

Sera were collected from *F. hepatica* experimentally infected sheep and assayed as previously described by Lopez Corrales et al. (2020). The infections were carried out by Agri-Food and Biosciences Institute (AFBI; Belfast, UK) under license from the Department of Health, Social Services and Public by the *Animal (Scientific Procedures) Act* 1986 (License No. PPL 2771; PPL 2801). FhSOD1 and FhSOD3 total IgG antibodies were analysed by ELISA and Western blot according to standard methods. Briefly, for the ELISA, flat-bottom 96-well microtitre plates (Nunc MaxiSorp, Biolegend, San Diego, CA, USA) were coated with rFhSOD1, rFhSOD3, or recombinant *F. hepatica* cathepsin L1 (rFhCL1) (5 µg/mL in 0.05 M carbonate buffer, pH 9.6) and incubated overnight at 4 °C [28]. After incubation in blocking buffer (2% bovine serum albumin in PBS-0.05% Tween-20 (*v*/*v*), PBST, pH 7.4) and washing three times in PBST, sheep sera collected from 10 animals at 0, 3, 7, 11, 15, and 23 weeks post-infection (WPI) was diluted 1:100 in serum dilution buffer (PBS, 0.5% Tween 80, 0.5 M NaCl), added to the antigen-coated wells in triplicate, and allowed to incubate for 1 h at 37 °C. After washing five times, 100 µL/well of HRP-conjugated donkey anti-sheep IgG (ThermoFisher Scientific) diluted 1:50,000 in blocking buffer was added, and the plates were incubated for 1 h at 37 °C. Following five washes, 100 µL/well of 3,3′,5,5′-Tetramethylbenzidine (TMB; Sigma-Aldrich) was added, and the plates were incubated at room temperature (RT) for 4.5 min. The reaction was stopped by the addition of 100 µL/well of 1 M sulphuric acid. The optical density was determined at a wavelength of 450 nm (OD450) in a PolarStar Omega spectrophotometer (BMG LabTech, Ortenburg, Germany).

Western blot analysis was performed using standard methods [29]. Briefly, rFhSOD1 and rFhSOD3 (1 µg/lane) was resolved by electrophoresis in 4–20% precast SDS-PAGE gels and electro-transferred onto nitrocellulose membranes prior to incubation in blocking buffer (5% milk in PBST, pH 7.4) for 1 h at RT. rFhCL1 was used as a positive control at 7 WPI. After washing five times with PBST, the membranes were probed with pooled sera from experimentally infected sheep at 0, 7, and 20 WPI diluted 1:1000 in blocking buffer (2.5% milk in PBST, pH 7.4) for 1 h at RT. After washing five times in PBST, the membranes were incubated with the secondary antibody alkaline phosphatase conjugated donkey-anti-sheep IgG at a 1:10,000 dilution for 1 h at RT. Following a final wash, the immune-reactive bands were visualised using the substrate SigmaFast BCIP/NBT (Sigma-Aldrich).

2.7. Production of Specific Antibodies against rFhSOD1 and rFhSOD3

Non-homologous sequences at the N-terminal of FhSOD1 (VMSGSSGVQGTVKFVQE-SET) and FhSOD3 (NASYSGQIFVNADGNLLTVR) were identified and protein-specific peptides (pFhSOD1 and pFhSOD3) were synthetically produced coupled with ovalbumin and used to immunise rabbits to generate FhSOD1 and FhSOD3-specific antibodies (Anti-pFhSOD1 and Anti-pFhSOD3) (Eurogentec, Seraing, Belgium). In addition, polyclonal antibodies against the purified recombinant rFhSOD1 and rFhSOD3 proteins (AntirFhSOD1 and Anti-rFhSOD3) were produced in rabbits (Eurogentec). Anti-rFhSOD1 and anti-rFhSOD3 antibodies were adsorbed against recombinant rFhSOD3 and rFhSOD1, respectively, prior to use to ensure specificity. The reactivity of the anti-pFhSOD1 and pFhSOD3 antibodies and anti-rFhSOD1 and rFhSOD3 polyclonal antibodies was determined by Western blot analysis against 0.05 μg/lane rFhSOD1 and rFhSOD3. Native bovine erythrocyte SOD (BS; 0.05 μg/lane) was used as a negative control. Immuno-detection was conducted as described above, with the following exceptions: membranes were probed with rabbit pre-immune sera, anti-pFhSOD1, anti-pFhSOD3, anti-rFhSOD1 or anti-rFhSOD3 raised in rabbits at a 1:10,000 dilution for 1 h at RT. After washing, the membranes were further probed with the secondary antibody, alkaline phosphatase conjugated goat-anti-rabbit IgG, at a 1:5000 dilution for 1 h at RT. Following final washes, the immune-reactive bands were visualised using the substrate SigmaFast BCIP/NBT (Sigma-Aldrich).

2.8. Immuno-Detection of Native FhSOD1 and FhSOD3 in NEJ, Adult Parasites and Their Extracts

Localisation of FhSOD1 and FhSOD3 in NEJ was conducted as follows: *F. hepatica* metacercariae (Italian isolate; Ridgeway Research, St. Briavels, UK) were excysted and cultured in RPMI 1640 medium containing 2 mM L-glutamine, 30 mM HEPES, 0.1% (*w*/*v*) glucose, 2.5 μg/mL gentamycin, and 10% foetal calf serum (ThermoFisher Scientific) for 24 h, as previously described [29]. NEJ were then fixed with 4% paraformaldehyde (PFA) in 0.1 M PBS (Sigma-Aldrich) pH 7.4, for 1 h at RT. After three washes in antibody diluent (PBS containing 0.1% (*v*/*v*) Triton X-100, 0.1% (*w*/*v*) bovine serum albumin and 0.1% (*w*/*v*) sodium azide; AbD buffer), the NEJ were incubated in rabbit pre-immune sera, anti-rFhSOD1, anti-rFhSOD3, or anti-rFhCL3 polyclonal antibodies (used as a non-related positive control) diluted 1:500 in AbD buffer overnight at 4 °C. After three washes in AbD, the NEJ were incubated in a 1:200 dilution of the secondary antibody, fluorescein isothiocyanate (FITC)-labelled goat-anti-rabbit IgG (Sigma-Aldrich) overnight at 4 °C in the dark. To counter-stain muscle tissue, the samples were incubated in AbD containing phalloidin-tetramethylrhodamine isothiocyanate (TRITC) (200 μg/mL) overnight in the dark at 4 °C. The NEJ were then whole-mounted onto slides using 10% glycerol solution containing 0.1 M propyl gallate and visualised under an Olympus Fluoview 3000 laser scanning confocal microscope using a PL APO CS 6 × 0 oil objective lens. Olympus type F immersion oil was used in viewing and all images were taken at room temperature.

Localisation of FhSOD1 and FhSOD3 in adult *F. hepatica* was conducted on parasites collected from the livers of infected sheep during post-mortem surveillance in Roscommon, Ireland. After collection, the parasites were washed in 1 × PBS and fixed in 4% PFA for 4 h at RT. After 4 h, the PFA was removed, replaced with 1 × PBS followed by an incubation for 1 h at RT. This washing process was repeated twice before dehydration in ascending ethanol and subsequent infiltration with JB-4 resin (Sigma-Aldrich EM0100). Serial 0.5 μM sections were mounted onto slides and probed with rabbit pre-immune sera, anti-rFhSOD1, anti-rFhSOD3, or anti-rFhCL1 polyclonal antibodies (used as a non-related positive control) diluted 1:1000 in PBST for 5 h at RT in a humid container. The sections were washed three times in PBST before addition of a 1:1000 dilution of the secondary antibody, FITC-labelled goat-anti-rabbit IgG (Sigma-Aldrich) in PBST, after which they were placed in a humid container and allowed to incubate overnight in the dark at 4 °C. The slides were once again washed three times in PBST and dried, and cover slips were mounted using 10% glycerol solution containing 0.1 M propyl gallate. The sections were visualised under a Leica DM2500 LED optical fluorescent microscope (Leica Microsystems, Wetzlar, Germany).

Adult *F. hepatica* E/S and somatic extract (10 μg/lane) were resolved by gel electrophoresis in 4–20% SDS-PAGE gels and electro-transferred onto nitrocellulose membranes prior to incubation in blocking buffer for 1 h at RT. The membranes were probed overnight at 4 °C with rabbit pre-immune sera, anti-rFhSOD1, or anti-rFhSOD3 at a 1:1000 dilution. After washing, the membranes were further probed with the secondary antibody alkaline phosphatase conjugated goat-anti-rabbit IgG at a 1:10,000 dilution for 1 h at RT. Following final washes, the immune-reactive bands were visualised using the substrate SigmaFast BCIP/NBT (Sigma-Aldrich).

2.9. Killing of F. hepatica NEJ with Superoxide and Protection with SOD and Catalase

To determine the susceptibility of *F. hepatica* NEJ to superoxide in vitro, the SOD activity assay described in Section 2.5 was repurposed to examine the effect of enzymatically generated superoxide on live NEJ. In this application, assay conditions and enzyme concentrations, except for catalase (CAT, recombinant from Serratia; Abcam, Cambridge, UK) and BS, were as described above in Section 2.5. Tetrazolium salt was omitted from the assay buffer (AB) to avoid any potential toxicity to the NEJ. *F. hepatica* metacercariae were excysted as described above, washed in 1 × PBS and incubated as described below.

At total of 10 NEJ per replicate were incubated in (i) PBS; (ii) AB with XOD; (iii) AB with XOD and BS (0.16 U/mL); (iv) AB with XOD and CAT (43 U/mL); (v) AB with XOD, BS and CAT (43 U/mL); (vi) AB with XOD and CAT (4.3 U/mL); and (vii) AB with XOD, BS and CAT (4.3 U/mL). NEJ were incubated at 37 °C and 5% CO_2 immediately after addition of XOD. After 24 h incubation, the number of dead NEJ (characterised by a lack of movement, including the absence of gut activity or a complete breakdown of the tegument and internal structure, after a one-minute observation) were counted. All treatments were conducted in duplicate over several days, giving a total of six biological replicates per treatment. To test the capacity of our recombinant enzymes to counteract the impacts of ROS on NEJ, we repeated the assay with the inclusion of rFhSOD1 and rFhSOD3 at concentrations of equal enzyme activity to BS, with and without the addition of catalase.

2.10. Statistical Analysis

Data were collected, stored, and analysed in Microsoft Excel version 16, Microsoft, Corporation, Redmond, WA, USA and GraphPad Prism version 5, GraphPad, San Diego, CA, USA. Differences between treatment groups were assessed by one-way ANOVA and Tukey's post hoc test with 95% confidence intervals.

3. Results

3.1. The F. hepatica Genome Contains Developmentally Regulated Cytosolic and Extracellular SODs

Analysis of the *F. hepatica* genome identified three superoxide dismutase genes, corresponding to the three types of SOD molecules identified within helminth parasites. Here, we report the two SODs that are known to be secreted by *F. hepatica* and, therefore, act at the host-parasite interface; a cytosolic SOD referred to herein as FhSOD1; and an extracellular SOD identified according to the presence of an N-terminal signal peptide sequence referred to herein as FhSOD3 (Figure 1). FhSOD1 corresponds to the previously reported cytosolic SOD [15,16].

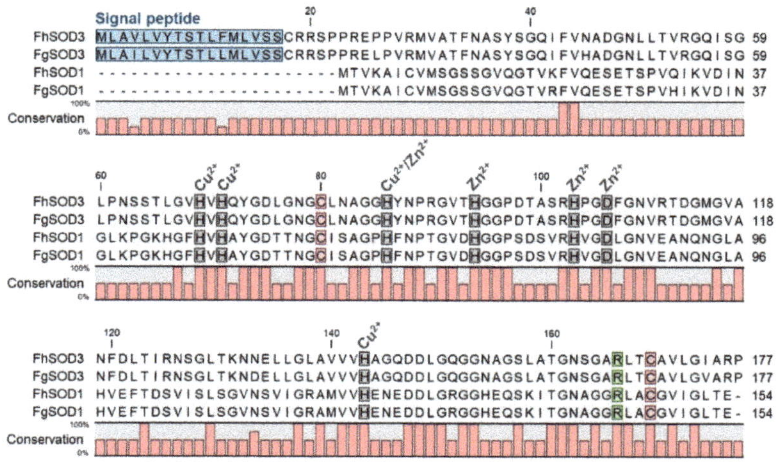

Figure 1. Comparative analysis of *F. hepatica* and *F. gigantica* SOD1 and SOD3 amino acid sequences. Metal binding sites are highlighted in grey and predicted signal peptides in blue. The arginine residue responsible for guiding the superoxide anion into the active site is indicated in green and the cysteine residues involved in disulphide bond formation are shown in red. Gaps are indicated by dashes. Amino acid conservation is shown below the alignment.

The two SOD genes display different transcriptional profiles across the developmental parasitic stages found within the mammalian host (Figure 2A). FhSOD1 is constitutively expressed at low levels across all the life cycle stages analysed. In contrast, FhSOD3 displays a markedly higher level of transcription by the infective stage metacercariae and the NEJ, which then drops to comparable levels to FhSOD1 expression within the immature and adult flukes.

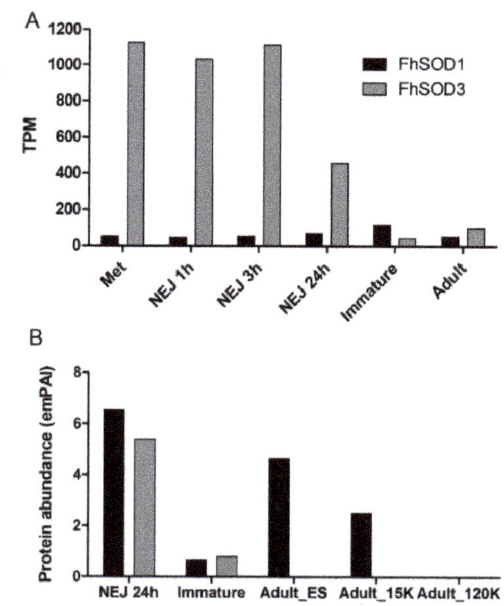

Figure 2. *Cont.*

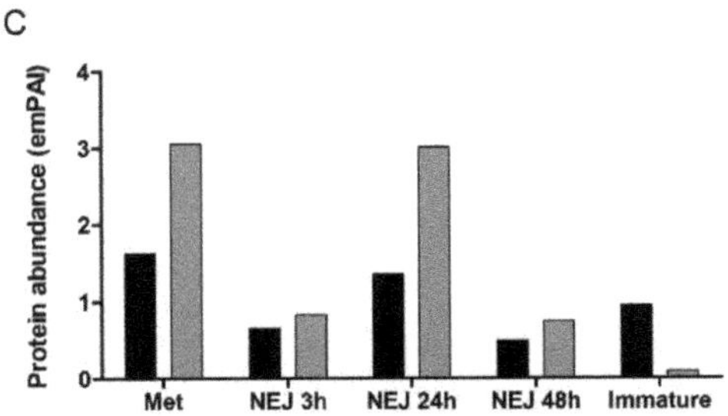

Figure 2. Expression profile of FhSOD1 and FhSOD3 throughout the *F. hepatica* life cycle within the mammalian host. (**A**) Graphical representation of the stage-specific transcription of the FhSOD1 and FhSOD3 genes displayed as transcripts per million (TPM) extrapolated from the transcriptome study of the metacercariae (Met), newly excysted juveniles at 1, 3, and 24 h post-excystment (NEJ 1 h, NEJ 3 h, NEJ 24 h), immature and adult liver flukes by Cwiklinski et al. [5]. (**B**) Graphical representation of the FhSOD1 and FhSOD3 proteins secreted within the ES protein fraction by the NEJ, immature, and adult parasites, represented by the exponentially modified protein abundance index (emPAI). The adult secretome data are displayed as the EV-depleted fraction (Adult_ES), and the microvesicle (Adult_15K) and exosome (Adult_120K) sub-fractions of the EV component recovered following centrifugation at 15,000× *g* and 120,000× *g*, respectively. (**C**) Graphical representation of the FhSOD1 and FhSOD3 protein abundance within the somatic proteome of the metacercariae (Met); newly excysted juvenile (NEJ) 3, 24, and 48 h post-excystment (NEJ 3 h; NEJ 24 h; NEJ 48 h); and immature liver flukes. The proteomic data for the NEJ, immature, and adult parasites are extrapolated from Cwiklinski et al. [6], Cwiklinski et al. [7], and Murphy et al. [30], respectively.

Despite the relatively lower level of gene transcription, proteomic analysis revealed that FhSOD1 is more abundantly secreted than its extracellular counterpart, FhSOD3, predominantly by the NEJ and adult parasite stages (Figure 2B). Analysis of the adult secretome revealed that, in addition to being secreted within the protein soluble fraction, FhSOD1 is also found within extracellular vesicles (EVs), specifically the microvesicles that are known to be released by the gastrodermal cells of the gut. This is consistent with other *F. hepatica* proteins that lack a signal peptide sequence for classical secretion, highlighting that the parasite uses non-classical routes of secretion to release multiple molecules that interact with its host [31]. FhSOD3 is abundantly secreted by the NEJ parasites consistent with its gene transcription profile, with low levels secreted by immature flukes and no protein detected within the adult stages (Figure 2B). The somatic proteome profile for both FhSOD proteins is also consistent with that observed at the gene level, with comparable levels of FhSOD1 being detected across all the life cycle stages analysed, in comparison to FhSOD3, which displays a higher protein abundance within the metacercariae and NEJ 24 h post-excystment (Figure 2C).

3.2. Cytosolic and Extracellular SODs Are Distinct Yet Highly Conserved in Trematodes

Interrogation of available Platyhelminth genomes revealed a number of homologous SOD1 and SOD3 genes (SOD1: n = 14; SOD3: n = 19) from parasitic flatworms of class Trematoda and free-living flatworms of class Turbellaria. Phylogenetic analysis separates the SOD1 and SOD3 sequences into two distinct clades (Figure S2). The *F. hepatica* SOD1 and SOD3 sequences cluster with homologous sequences from F. gigantica and Echinostoma caproni, as is expected in these closely related parasite species. The majority of the SOD genes are present within their respective genomes as single copy genes, with the exception of Paragonimus westermani, which appears to contain two SOD1 sequences, while several *Schistosoma* spp. appear to contain two SOD3 sequences that cluster into distinct sub-clades (Figure S2). The SOD1 sequence from the free-living flatworm *Schmidtea mediterranea* does not cluster with its parasitic counterparts within the SOD1 clade.

All the SOD1 sequences analysed were comparable with that observed for the *F. hepatica* and *F. gigantica* SOD1 sequences, in that they all lacked an N-terminal signal peptide for classical secretion in line with the predicted cytosolic function of these proteins (data not shown). The majority of the Platyhelminth SOD3 sequences contain a signal peptide implying a common extracellular role for these proteins. Three Platyhelminth SOD3 sequences were missing the signal peptide sequence in the current respective genome assemblies (*E. caproni* SOD3, *T. regenti* SOD3, *S. rodhaini* SOD3; Table S1), with further investigation required to confirm their annotation.

Alignment of FhSOD1 with FhSOD3 revealed 34.64% sequence identity and 63.28% sequence similarity (ClustalO/uniprot) (Table S2). Comparison with the predicted *F. gigantica* SOD1 and SOD3 amino acid sequences shows a high level of conservation between the two species, with 98.70 and 100.00% identity and similarity, respectively, between the SOD1 sequences, and 96.61 and 99.44% identity and similarity, respectively, between the SOD3 sequences (Table S2). FhSOD1 has an average sequence identity of 69.81% to other Platyhelminth SOD1 sequences, and an average sequence identity of 59.42% with the four selected mammalian (*O. aries*, *B. taurus*, *B. indicus*, and *H. sapiens*) SOD1 sequences (Table S2). The trematode SOD1 sequences showed an average similarity of 90.20%, whereas the SOD3 sequences were more diverse, with an average similarity of 68.40% (Table S2). Collectively, this information suggests that FhSOD1 is more closely related to the human cytosolic enzyme than its own extracellular enzyme. Alignment against the respective *O. aries*, *B. taurus*, *B. indicus*, and *H. sapiens* SOD1 and SOD3 sequences shows 100% conservation of predicted copper and zinc metal binding sites (His69, His71, His86, His94, His103, Asp106, His143) (Figure S1; residue numbering relative to FhSOD3) [11,14,16].

3.3. Recombinant FhSOD1 and FhSOD3 Are Highly Active

To investigate the role of *F. hepatica* SODs in parasite metabolism and host defence, we exploited E. coli to recombinantly express both FhSOD1 and FhSOD3 antioxidant proteins. SDS-PAGE and Western blot analysis using monoclonal anti-histidine antibodies revealed soluble rFhSOD1 and rFhSOD3 purified at ~16 and ~17 kDa, respectively, which are predicted to have a conserved secondary structure consisting of a β-sheet made up of eight antiparallel β-strands typical of SODs (Figures S3 and S4). rFhSOD1 and rFhSOD3 had similar activity against superoxide (~400 U/mg each, as defined by the standard curve), where one unit of enzyme activity is defined as the amount of protein required to exhibit 50% dismutation of superoxide under physiological conditions (150 mM salt, pH 7.4, 37 °C), albeit they were ~10 times less active than the positive control (Figure S3). Size exclusion (gel filtration) chromatography revealed distinct peaks corresponding to molecular weights of ~36 kDa (rFhSOD1), ~31 kDa (rFhSOD3), and ~76 kDa (rFhSOD3), indicative of homodimers and a mix of homodimers and homotetramers for rFhSOD1 and rFhSOD3, respectively (Figure 3). Analysis of the individual fractions collected from each peak demonstrate that both the dimeric (rFhSOD1 and rFhSOD3) and tetrameric (rFhSOD3) forms of the recombinant proteins are enzymatically active (Figure 3).

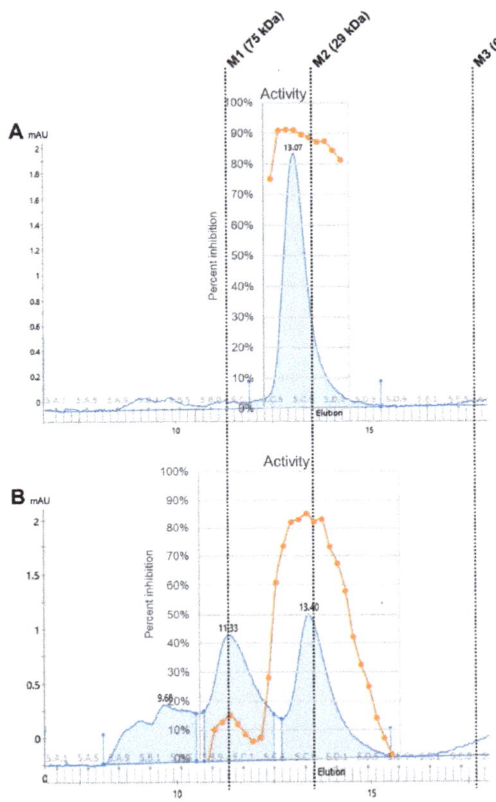

C

	Large peak		Small peak	
	Predicted size (kDa)	**Divided by monomer**	**Predicted size (kDa)**	**Divided by monomer**
FhSOD1	-	-	36.43	2.2
FhSOD3	76.05	4.4	30.54	1.8

Figure 3. Structural organisation and enzymatic activity of rFhSOD1 and rFhSOD3. Size exclusion chromatography of (**A**) FhSOD1 and (**B**) FhSOD3. The corresponding enzyme activity of each elution fraction is expressed as the percentage inhibition of formazan dye formation. (**C**) The predicted molecular sizes of each recombinant protein calculated against the molecular weight markers. Markers are indicated by dotted lines—M1; conalbumin, M2; carbonic anhydrase, M3; aprotinin.

3.4. Detection of Native *F. hepatica* SOD

Western blot analysis of adult parasite E/S and somatic extract using polyclonal antibodies raised in rabbits to the recombinant FhSODs revealed detectable FhSOD1 in both extracts at the expected size (Figure 4). In agreement with transcriptome and proteome data from adult parasites, native FhSOD3 was not detected in either fraction by polyclonal antibodies against rFhSOD3. These results are consistent with the immuno-detection of the proteins in adult parasite sections, where FhSOD1 was distributed throughout the muscle, tegument, and parenchyma, while limited FhSOD3 was visible in the tegument (Figure 5). The anti-pFhSOD1 and pFhSOD3 antibodies demonstrated protein-specific binding against rFhSOD1 and rFhSOD3 by Western blot analysis but did not reveal detectable fluorescence when applied to fixed parasite specimens, suggesting recognition of a linear epitope in NEJ (Figure S5).

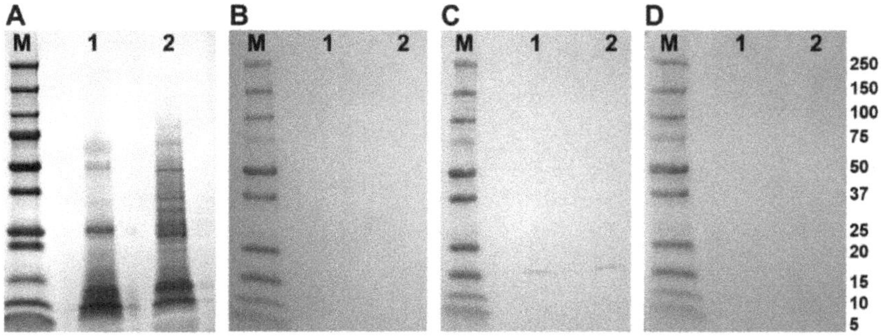

Figure 4. Immune detection of native *F. hepatica* SODs in adult parasite E/S and somatic extracts. (**A**) Native (10 ug/well) *F. hepatica* adult worm E/S and somatic extracts were resolved in a 4–20% SDS-PAGE gel and stained with Biosafe Coomassie. Lane 1: adult E/S; lane 2: adult somatic extract. (**B–D**) Western blot analysis of native FhSOD1 and FhSOD3. Lane 1: adult E/S; lane 2: adult somatic extract. Immuno blots were probed with (**B**) rabbit pre-immune sera (negative control), (**C**) anti-rFhSOD1 polyclonal antibodies raised in rabbit, and (**D**) anti-rFhSOD3 polyclonal antibodies raised in rabbit. M molecular weight in kDa (Precision Plus Protein Dual, Bio-Rad).

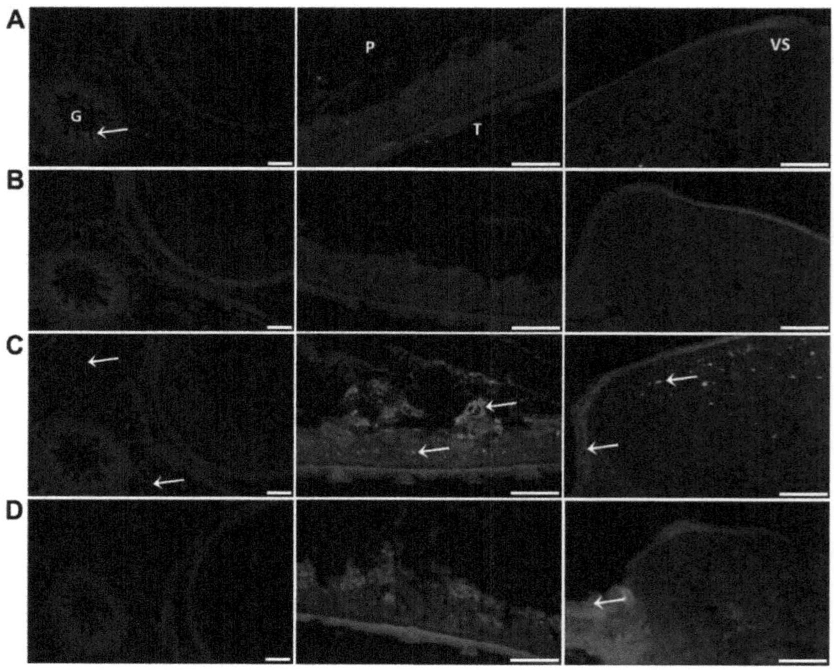

Figure 5. Immuno-localisation of native FhSOD1 and FhSOD3 in *F. hepatica* adult sections. Sections were probed with (**A**) anti-rFhCL1 polyclonal antibodies (non-related positive control), (**B**) rabbit pre-immune sera (negative control), (**C**) anti-rFhSOD1 polyclonal antibodies, and (**D**) anti-rFhSOD3 polyclonal antibodies raised in rabbit. Immuno-localisation of native *F. hepatica* proteins is represented by green fluorescence (FITC staining) and indicated with white arrows. G; gut, P; parenchyma, T; tegument, VS; ventral sucker, scale bars 50 μM.

Whole mount immuno-localisation of NEJ 3 h post-excystment using anti-rFhSOD1 and anti-rFhSOD3 antibodies revealed the presence of FhSOD1 and FhSOD3 on the outer surface, gut and tegument as indicated by diffuse fluorescence (Figure 6). This is in contrast to NEJ stained with anti-rFhCL3 polyclonal antibodies, which locates the FhCL3 cysteine peptidase solely within the bifurcated gut (Figure 6). Despite a potential cross-reaction between the two FhSOD polyclonal antibodies, these results are consistent with the transcriptome and proteome analyses of the juvenile parasites, which show abundant expression by the metacercariae and 1, 3, and 24 h post-excystment NEJ (Figure 2A,C).

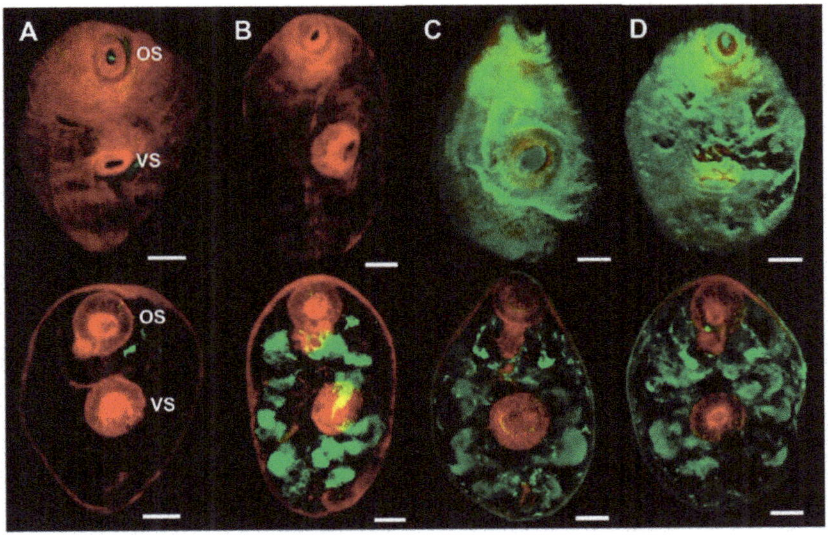

Figure 6. Immuno-localisation of native FhSOD1 and FhSOD3 in newly excysted juveniles (NEJ). Whole-mount *F. hepatica* NEJ 3 h post-excystment were probed with (**A**) rabbit pre-immune sera (negative control), (**B**) anti-rFhCL3 polyclonal antibodies (non-related positive control), (**C**) anti-rFhSOD1 polyclonal antibodies, and (**D**) anti-rFhSOD3 polyclonal antibodies. Immuno-localisation of native *F. hepatica* proteins is represented by green fluorescence (FITC staining). All samples were counter-stained with phalloidin-TRITC to stain muscle tissue (red fluorescence). OS; oral sucker, VS; ventral sucker, scale bars; 25 μM.

3.5. Homology with Host SOD Facilitates Immune Evasion

Western blot analysis of rFhSOD1 and rFhSOD3 probed with pooled sera from experimentally infected sheep demonstrated that these proteins are not immunogenic during *F. hepatica* infection (Figure 7A–D). These results were confirmed by analysis of individual sheep sera collected at 0, 3, 7, 11, 15, and 23 WPI by ELISA against rFhSOD1 and rFhSOD3, using the immunogenic cathepsin peptidase, rFhCL1, as a positive control (Figure 7E).

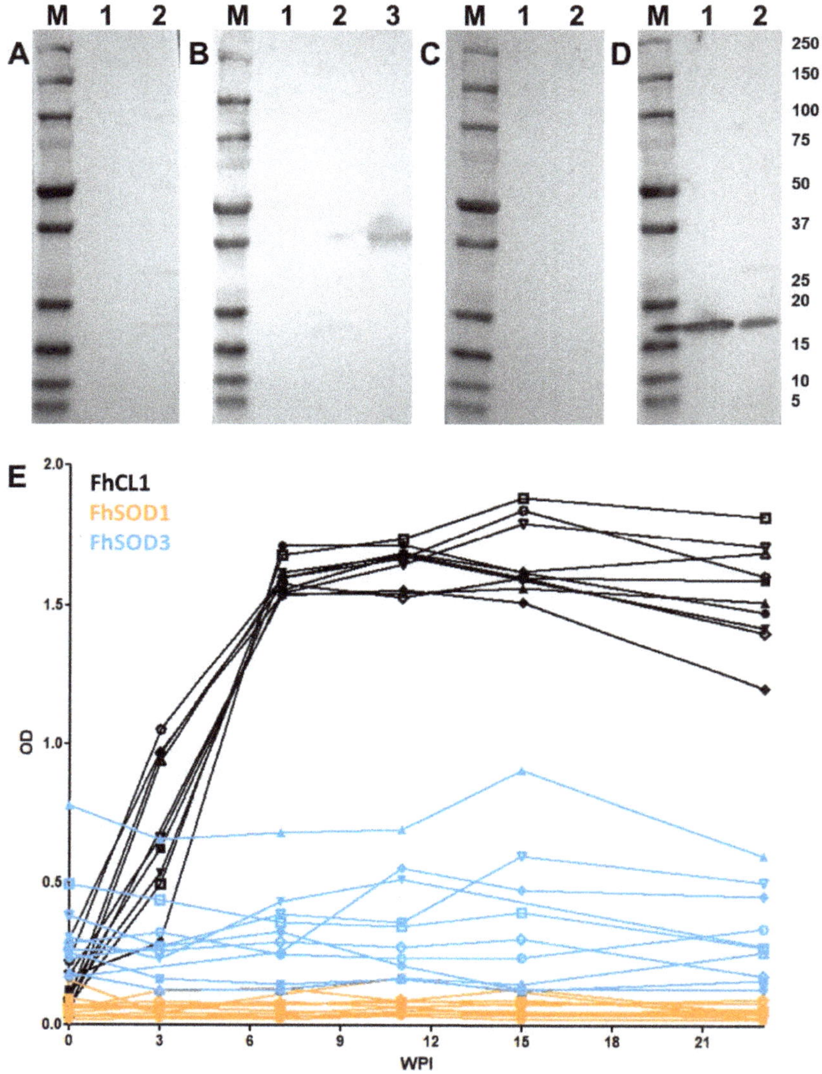

Figure 7. Antibody responses to recombinant FhSOD1 and FhSOD3 in experimentally infected sheep. (**A–C**) Recombinant (1 ug/well) FhSOD1 and FhSOD3 were transferred to a nitrocellulose membrane and probed with pooled sera from experimentally infected sheep (**A**) pre-infection (negative control), (**B**) 7 weeks post-infection (WPI), and (**C**) 20 WPI. (**D**) Recombinant FhSOD1 and FhSOD3 were probed with anti-SOD1 and anti-SOD3 polyclonal antibodies raised in rabbits (positive control). Lane 1; recombinant FhSOD1, lane 2; recombinant FhSOD3, lane 3; recombinant FhCL1 zymogen mutant (positive control), M; molecular weight in kDa (Precision Plus Protein Dual, Bio-Rad. (**E**) The optical density of IgG antibodies to rFhSOD1, rFhSOD3 and rFhCL1 from experimentally infected sheep 0, 3, 7, 11, 15 and 25 WPI.

3.6. The Addition of SOD and Catalase Negate the Lethal Effects of Reactive Oxygen Species against F. hepatica NEJ In Vitro

Co-incubation of *F. hepatica* NEJ with enzymatically generated superoxide killed 98% of exposed NEJ within 24 h. There was no significant difference in NEJ survival after the addition of BS or rFhSOD1 ($p > 0.99$); however, rFhSOD3 prevented death in 30% of

NEJ ($p = 0.0041$) (Figure 8; Table S3). The addition of catalase protected NEJ in a dose-dependent manner, with 73 and 43% of NEJ alive when incubated with 43 and 4.3 U/mL catalase, respectively, at 24 h (Figure S6), the effect of which was enhanced when used in the conjunction with SODs (Table S3).

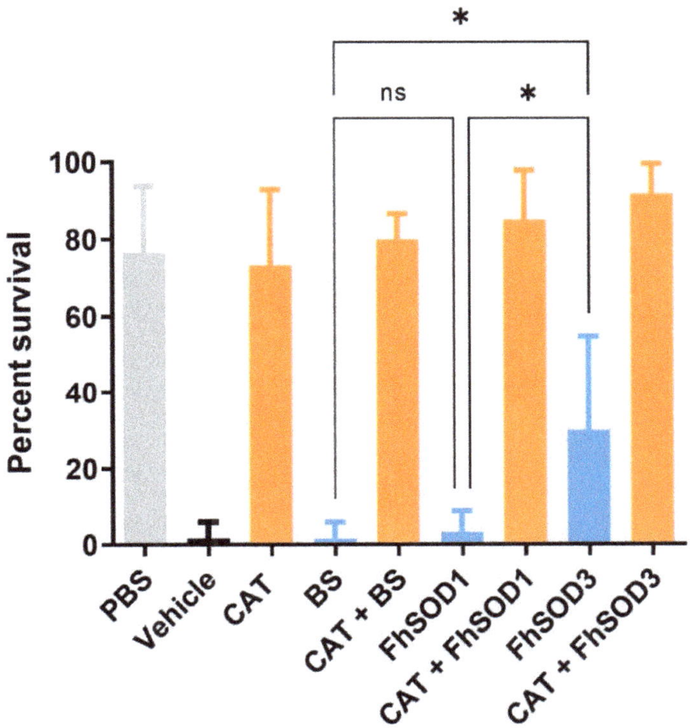

Figure 8. Susceptibility of *F. hepatica* NEJ to superoxide and their subsequent protection via the addition of SOD and catalase. Six replicates of 10 NEJ/well were incubated in the presence of enzymatically generated superoxide for 24 h at 37 °C and 5% CO_2. PBS; phosphate buffered saline (negative control), Vehicle; assay buffer containing enzymatically generated superoxide via the step-wise reduction of hypoxanthine into xanthine and superoxide by xanthine oxidase (positive control), CAT; recombinant catalase, BS; native bovine erythrocyte SOD, ns; not significant (0.1234), * $p < 0.0332$.

4. Discussion

Superoxide dismutases (SODs) are amongst a collection of antioxidants that play essential roles in parasite defence against oxygen free radicals generated physiologically by cellular metabolism and externally by innate immune cells, such as macrophages and neutrophils, during invasion and infection [4,7,32]. Similar to many trematodes, the liver fluke *F. hepatica* possesses a complex system involved in host immuno-modulation and immuno-evasion wherein a repertoire of proteins is excreted/secreted to simultaneously defend against and distract the host's immune response [30,31]. By interrogating existing genomic, transcriptomic, and proteomic analysis of various intra-mammalian life stages of *F. hepatica,* we identified several antioxidants in the protein cocktail deployed by the parasite during the early invasive and migratory processes, including both a cytosolic (FhSOD1) and an extracellular (FhSOD3) SOD [5–7]. Identified and characterised in *F. hepatica* for the first time herein, we demonstrate that FhSOD3 is NEJ-associated and, thus, likely to play a key role in invasion of the mammalian host; for example, the penetration of the intestine and liver tissue. FhSOD1 on the other hand, which is released into host tissues through alternative secretion pathways, may function simultaneously in parasite metabolism and defence against exogenous ROS. Through a series of immuno-localisation, serological, and functional in vitro experiments, we propose that FhSOD1 and FhSOD3 play distinct roles in the development of *F. hepatica* and its interaction with mammalian definitive hosts.

As is observed with other proteins excreted/secreted by *F. hepatica* throughout its life cycle, such as the cathepsin L and B proteases, the transcription and expression of FhSOD1, and FhSOD3 are tightly regulated [6,7]. FhSOD1 is constitutively transcribed across all *F. hepatica* life stages at relatively low levels. FhSOD3, however, is highly expressed in the infective metacercariae and early NEJ, before declining to transcription levels, similar to FhSOD1 in immature and adult worms. The lower level of FhSOD1 transcription in metacercariae and NEJ compared to FhSOD3 is in contrast to the relative protein abundance in these early stages and suggests that the parasite produces and stores FhSOD1 prior to excystment in the small intestine. Both FhSOD1 and FhSOD3 are at their most abundant in *F. hepatica* E/S and somatic extract early in the life cycle, which implies that they are vital during early invasion of the mammalian host. Macrophages and neutrophils play an essential role in the early innate immune response against invading pathogens via the production of ROS, and thus it is reasonable to suggest that *F. hepatica* NEJ, which are vulnerable and living on limited glycogen stores, would have a pre-prepared store of antioxidants to defend against this onslaught of ROS [33]. The abundance of both FhSODs in metacercariae and NEJ may also reflect a defence against ROS arising from normal metabolic processes occurring in the most environmentally robust stage of the parasite.

Of the two proteins, only FhSOD3 possesses an N-terminal signal peptide and is, thus, transported to the extracellular environment via the classical secretory pathway. FhSOD1, on the other hand, is found in both the microvesicular fraction and the EV-depleted supernatant of adult *F. hepatica* E/S. This is an interesting finding given that cytosolic SOD enzymes are thought to act solely on endogenous ROS produced during cellular metabolism, and not interact with host-derived ROS [32]. Although the mechanism(s) by which this compartmentalisation of FhSOD1 occurs remains unknown, the presence of FhSOD1 in different vesicular and non-vesicular fractions of the E/S suggests distinct methods of delivery to host immune cells and, thus, different defensive or immuno-modulatory roles. Given the observed homology between the cytosolic SOD of *F. hepatica* and other trematode parasites, it is worth investigating if these findings are unique to this species or if they are indicative of an as-of-yet unexplored generalised helminth defence mechanism. Contrary to the cytosolic SOD, the diversity observed between the signal-peptide containing extracellular SOD, including the apparent duplication of this isoform within *Schistosoma* spp., could indicate unique host-specific adaptations in these worms.

In the current study, neither rFhSOD1 nor rFhSOD3 were immunogenic in experimentally infected sheep. Our findings reflect those previously observed when both experimentally and naturally infected buffalo sera were probed for antibodies to FgSOD1 [17]. Similarly, there has been a lack of detectable antibodies when sera from experimentally infected rats were probed with recombinant serine protease inhibitors (serpins; rFhSrp1/2), and when sera from experimentally infected sheep was probed with recombinant cathepsin B1 (rFhCB1) [28,29]. FhSOD1 is highly homologous to its mammalian counterpart, with an average similarity of 87.34% compared to SOD1 sequences from *O. aries, B. taurus, B. indicus*, and *H. sapiens*. It is possible that the parasite exploits these similarities as yet another strategy of immune evasion, wherein their high host homology and, thus, low immunogenicity allows them to operate undetected and unimpeded. Indeed, this host mimicry has been observed in *Schistosoma mansoni*, another trematode parasite of humans, leading to concerns that this high homology with host molecules may induce autoimmune responses during vaccine trials [34].

Unlike *F. hepatica*, the cytosolic SOD of *S. mansoni* (SmCT-SOD) is expressed at the highest levels in adult worms, leaving the larval stages susceptible to immune elimination via host-generated superoxide [34]. In these parasites, SmCT-SOD was localized to the tegument and gut epithelium of adults. This contrasts with the findings of the current study, which showed that FhSOD1 and FhSOD3 localise to the tegument and gut of NEJ and show a marked decline in FhSOD-specific immuno-localisation in adult parasites. The differential life-stage expression of cytosolic SOD between *F. hepatica* and *S. mansoni* may reflect different host migration routes and tissue tropism between these parasites. *S. mansoni* infects mammalian hosts via penetration of the skin before eventually residing in the mesenteric venules, where they would be more available to host-generated ROS than *F. hepatica*, which is located within the bile ducts and gallbladder upon maturation [35]. Nevertheless, in both parasites, immuno-localisation occurs at the host:parasite interface and supports the theory that these proteins are employed in defence of exogenous superoxide.

Previous studies have shown that *F. hepatica* NEJ are resistant to killing by superoxide produced in vitro chemically or by peritoneal lavage cells isolated from Indonesian Thin-Tailed (ITT) sheep [36,37]. ITT sheep are resistant to infection by *F. gigantica* but have been shown to be susceptible to infection with *F. hepatica* [38]. It was suggested that this susceptibility is related to the higher gene expression of SOD1 mRNA in *F. hepatica* NEJ compared to *F. gigantica* [36]. In the current study, we demonstrated robust in vitro killing of *F. hepatica* NEJ by enzymatically generated ROS. Interestingly, ROS-associated killing of NEJ was partially prevented via the addition of rFhSOD3 only, but completely inhibited via the introduction of catalase. Given that all three SODs were utilized at concentrations of equal enzyme activity, these results suggest that rFhSOD1 and rFhSOD3 do not behave similarly under physiological conditions. Further, the production of hydrogen peroxide in our assay results from the (a) two-step reduction of hypoxanthine into xanthine and superoxide and (b) the reduction of superoxide by SOD. It is well known, however, that the production of superoxide by macrophages during the oxidative burst is part of a cascade of highly damaging reactive oxygen and nitrogen species; thus, our data imply that *F. hepatica* needs to employ an array of antioxidant proteins to defend against cell-mediated immune responses [9,32]. Helminths do not express catalase, but rather exploit a thiol-independent antioxidant system to detoxify hydrogen peroxide wherein thioredoxin/glutathione reductase (FhTGR) reduces thioredoxin (FhTrx), which then reduces and activates peroxiredoxin (FhPrx), all of which are up-regulated in *F. hepatica* NEJ [7,8,39]. In our in vitro assay, we utilized catalase instead of FhPrx, and thus circumvented this complex cascade and provided proof-of-principal evidence that *F. hepatica* NEJ exploit a series of antioxidants to defend against host ROS during early invasion. Future work will focus on the complex interactions between each of these antioxidant proteins and their collective role in combatting damaging ROS.

5. Conclusions

We propose that *F. hepatica* exploits two distinct SODs to defend against host-generated ROS during early invasion and infection. We have shown that these proteins have unique expression profiles and secretory pathways, and are, thus, likely to play divergent roles in the development and growth of the parasite in its mammalian host. Going forwards, it will be imperative to define how these proteins interact with each other and with the slew of other antioxidant proteins secreted by the parasite during early invasion, and thus work in concert to detoxify ROS intracellularly and in the extracorporeal environment.

Supplementary Materials: The following supporting information can be downloaded at: https://www.mdpi.com/article/10.3390/antiox11101968/s1, Table S1. Trematode and mammalian amino acid sequence details for phylogenetic analysis. Table S2. Similarity and identity of trematode and mammalian SOD1 and SOD3 amino acid sequences. Table S3. Significance of *F. hepatica* NEJ survival rates when co-cultured with enzymatically generated superoxide dismutase in vitro. Figure S1. Alignment of mammalian and trematode SOD1 and SOD3 amino acid sequences for phylogenetic analysis. Selection of conserved blocks from the complete SOD alignment was determined by Gblocks 0.91b [18,19], with 41% of the positions conserved. Metal binding sites are indicated by grey bars. Gaps are shown as dashes. Variable residues are highlighted in red. *Fasciola hepatica* and *Fasciola gigantica* sequences are highlighted in bold. The first amino acid is numbered 67 (relative to FhSOD3—

see Figure 1 in main text). Figure S2. Phylogenetic analysis of trematode and mammalian SOD amino acid sequences. Phylogenetic analysis was carried out using the maximum likelihood method and the Whelan and Goldman model [21]. The tree with the highest log likelihood (−3031.18) is shown. The percentage of trees in which the associated taxa clustered together (out of 1000 iterations) is shown below the branches (Bootstrap support values less than 50% are omitted). The tree is drawn to scale, with branch lengths measured in the number of substitutions per site. Figure S3. Expression and enzyme activity of rFhSOD1 and rFhSOD3. (A) Recombinant proteins expressed in *E. coli* ClearColi cells and purified by affinity column, resolved in a 4–20% SDS-PAGE gel and stained with Biosafe Coomassie. (B) Recombinant proteins were electro-transferred onto a nitrocellulose membrane and probed with mouse monoclonal antibodies to poly-histidine (1:10,000). P; *E. coli* pellet, S; purified supernatant. M; molecular weight in kDa (Precision Plus Protein Dual, BioRad). (C) Activity of recombinant FhSOD1 and FhSOD3 compared to native bovine erythrocyte SOD (BS). Both rFhSOD1 and rFhSOD3 exhibited a specific activity of ~400 U/mg relative to the standard curve, where one unit is the amount of enzyme required to exhibit 50% dismutation of the superoxide radical. Figure S4. The predicted 3D structure of (A) FhSOD1 and (B) FhSOD3. Structure predicted by the AlphaFold Protein Structure Database [25,26]. The colour of each amino acid indicates the per-residue confidence score (pLDDT) where dark blue resembles very high confidence (> 90 pLDDT), light blue resembles confident (90 > pLDDT > 70), yellow indicates low confidence (70 > pLDDT > 50). Figure S5. Specificity of polyclonal and anti-peptide antibodies against recombinant *F. hepatica* SODs. (A) Purified recombinant (0.05 ug/well) FhSOD1 and FhSOD3 were resolved in a 4–20% SDS-PAGE gel and stained with Biosafe Coomassie. Lane 1: BS; lane 2: rFhSOD1; lane 3: rFhSOD3. (B–F) Western blot analysis of rFhSOD1 and rFhSOD3. Lane 1: BS; lane 2: rFhSOD1; lane 3: rFhSOD3. Immuno blots were probed with (B) rabbit pre-immune sera (negative control), (C) anti-FhSOD1 peptide antibodies, (D) anti-FhSOD3 peptide antibodies, (E) anti-rFhSOD1 polyclonal antibodies, and (F) anti-rFhSOD3 polyclonal antibodies raised in rabbit. Figure S6. SOD and catalase inhibit the lethal effects of superoxide and hydrogen peroxide on *F. hepatica* NEJ in vitro. (A) Principal of the assay demonstrating the step-wise reduction of hypoxanthine into xanthine and superoxide by xanthine oxidase and the subsequent production of hydrogen peroxide and uric acid as a by-product of the reaction. (B) Enzyme activity of rFhSOD1 and rFhSOD3 vs. BS prior to their inclusion in the NEJ assay at 0.01 mg/mL (rFhSODs) and 0.001 mg/mL (BS) (C) Killing of *F. hepatica* NEJ using enzymatically generated superoxide is prevented via the addition of catalase (CAT) and BS in a dose-dependent manner. PBS; negative control, Vehicle; positive control (superoxide generated in vitro). (D) The impact of superoxide and hydrogen peroxide on *F. hepatica* NEJ after 24 h of co-culture is shown. Black scale bars; 100 μM, white scale bar; 50 μM, ns; not significant.

Author Contributions: Conceptualization, N.E.D.C. and J.P.D.; methodology, N.E.D.C., C.D.M.V., H.L.J., K.C., A.F. and J.P.D.; formal analysis, N.E.D.C. and K.C.; investigation, N.E.D.C.; resources, J.P.D.; data curation, N.E.D.C. and K.C.; writing—original draft preparation, N.E.D.C.; writing—review and editing, N.E.D.C., C.D.M.V., K.C. and J.P.D.; visualization, N.E.D.C., H.L.J. and K.C.; supervision, J.P.D.; project administration, N.E.D.C.; funding acquisition, J.P.D. All authors have read and agreed to the published version of the manuscript.

Funding: This research was funded by Science Foundation Ireland (SFI, Ireland) Research Professorship grant 17/RP/5368.

Institutional Review Board Statement: Not applicable.

Informed Consent Statement: Not applicable.

Data Availability Statement: The transcriptome data sets used to extrapolate the FhSOD gene transcription were previously reported by Cwiklinski et al. [5] and are available in the European Nucleotide Archive repository, PRJEB6904; http://www.ebi.ac.uk/ena/data/view/PRJEB6904 (30 August 2022). The mass spectrometry proteomics data analysed as part of this study have been deposited to the ProteomeXchange Consortium via the PRIDE partner repository with the following data set identifiers (a) NEJ specific datasets Cwiklinski et al. [6]: PXD007255, PXD016561; (b) immature fluke Cwiklinski et al. [7]: PXD021221; (c) adult ES and EV datasets Murphy et al. [30]: PXD002570 and PXD016561.

Acknowledgments: The authors would like to thank Siobhán Gaughan for her invaluable administrative assistance in her role as Project Manager of the Molecular Parasitology Laboratory at the University of Galway, Ireland.

Conflicts of Interest: The authors declare no conflict of interest. The funders had no role in the design of the study; in the collection, analyses, or interpretation of data; in the writing of the manuscript; nor in the decision to publish the results.

References

1. Torgerson, P.R.; Macpherson, C.N. The socioeconomic burden of parasitic zoonoses: Global trends. *Vet. Parasitol.* **2011**, *182*, 79–95. [CrossRef] [PubMed]
2. Keiser, J.; Utzinger, J. Emerging foodborne trematodiasis. *Emerg. Infect. Dis.* **2005**, *11*, 1507–1514. [CrossRef] [PubMed]
3. Thomas, A.P. The life history of the liver-fluke (*Fasciola hepatica*). *Q. J. Microsc. Sci.* **1883**, *s2–s23*, 99. [CrossRef]
4. Bottari, N.B.; Mendes, R.E.; Lucca, N.J.; Schwertz, C.I.; Henker, L.C.; Olsson, D.C.; Piva, M.M.; Sangoi, M.; Campos, L.P.; Moresco, R.N.; et al. Oxidative stress associated with pathological lesions in the liver of rats experimentally infected by *Fasciola hepatica*. *Exp. Parasitol.* **2015**, *159*, 24–28. [CrossRef]
5. Cwiklinski, K.; Dalton, J.P.; Dufresne, P.J.; La Course, J.; Williams, D.J.; Hodgkinson, J.; Paterson, S. The *Fasciola hepatica* genome: Gene duplication and polymorphism reveals adaptation to the host environment and the capacity for rapid evolution. *Genome Biol.* **2015**, *16*, 71. [CrossRef]
6. Cwiklinski, K.; Jewhurst, H.; McVeigh, P.; Barbour, T.; Maule, A.G.; Tort, J.; O'Neill, S.M.; Robinson, M.W.; Donnelly, S.; Dalton, J.P. Infection by the helminth parasite *Fasciola hepatica* requires rapid regulation of metabolic, virulence, and invasive factors to adjust to its mammalian host. *Mol. Cell Proteom.* **2018**, *17*, 792–809. [CrossRef]
7. Cwiklinski, K.; Robinson, M.W.; Donnelly, S.; Dalton, J.P. Complementary transcriptomic and proteomic analyses reveal the cellular and molecular processes that drive growth and development of *Fasciola hepatica* in the host liver. *BMC Genom.* **2021**, *22*, 46. [CrossRef]
8. Dorey, A.; Cwiklinski, K.; Rooney, J.; De Marco Verissimo, C.; López Corrales, J.; Jewhurst, H.; Fazekas, B.; Calvani, N.E.D.; Hamon, S.; Gaughan, S.; et al. Autonomous non antioxidant roles for *Fasciola hepatica* secreted Thioredoxin-1 and Peroxiredoxin-1. *Front. Cell Infect Microbiol.* **2021**, *11*, 667272. [CrossRef]
9. Radi, R. Oxygen radicals, nitric oxide, and peroxynitrite: Redox pathways in molecular medicine. *Proc. Natl. Acad. Sci. USA* **2018**, *115*, 5839–5848. [CrossRef]
10. Sanchez-Moreno, M.; Leon, P.; Salas-Peregrin, J.M.; Garcia-Ruiz, M.A.; Monteoliva, M. Superoxide dismutase in trematodes. Isoenzymatic characterization and studies of inhibition by a series of benzimidazoles and by pyrimidines of recent syntheses. *Arzneim. Forsch.* **1987**, *37*, 903–905.
11. Cardoso, R.M.; Silva, C.H.; Ulian de Araújo, A.P.; Tanaka, T.; Tanaka, M.; Garratt, R.C. Structure of the cytosolic Cu,Zn superoxide dismutase from *Schistosoma mansoni*. *Acta Crystallogr. D Biol. Crystallogr.* **2004**, *60*, 1569–1578. [CrossRef] [PubMed]
12. Hong, Z.; Kosman, D.J.; Thakur, A.; Rekosh, D.; LoVerde, P.T. Identification and purification of a second form of Cu/Zn superoxide dismutase from *Schistosoma mansoni*. *Infect Immun.* **1992**, *60*, 3641–3651. [CrossRef] [PubMed]
13. Henkle-Dührsen, K.; Tuan, R.S.; Wildenburg, G.; Eschbach, M.L.; Tawe, W.; Zipfel, P.; Walter, R.D. Localization and functional analysis of the cytosolic and extracellular CuZn superoxide dismutases in the human parasitic nematode *Onchocerca volvulus*. *Mol. Biochem. Parasitol.* **1997**, *88*, 187–202. [CrossRef]
14. Nguyen, N.H.; Tran, G.B.; Nguyen, C.T. Anti-oxidative effects of superoxide dismutase 3 on inflammatory diseases. *J. Mol. Med.* **2020**, *98*, 59–69. [CrossRef] [PubMed]
15. Piacenza, L.; Radi, R.; Goñi, F.; Carmona, C. CuZn superoxide dismutase activities from *Fasciola hepatica*. *Parasitology* **1998**, *117 Pt 6*, 555–562. [CrossRef]
16. Kim, T.S.; Jung, Y.; Na, B.K.; Kim, K.S.; Chung, P.R. Molecular cloning and expression of Cu/Zn-containing superoxide dismutase from *Fasciola hepatica*. *Infect Immun.* **2000**, *68*, 3941–3948. [CrossRef]
17. Lalrinkima, H.; Raina, O.K.; Chandra, D.; Jacob, S.S.; Bauri, R.K.; Chandra, S.; Yadav, H.S.; Singh, M.N.; Rialch, A.; Varghese, A.; et al. Isolation and characterization of Cu/Zn-superoxide dismutase in *Fasciola gigantica*. *Exp. Parasitol.* **2015**, *151–152*, 1–7. [CrossRef]
18. Dereeper, A.; Guignon, V.; Blanc, G.; Audic, S.; Buffet, S.; Chevenet, F.; Dufayard, J.F.; Guindon, S.; Lefort, V.; Lescot, M.; et al. Phylogeny.fr: Robust phylogenetic analysis for the non-specialist. *Nucleic Acids Res.* **2008**, *36*, W465–W469. [CrossRef]
19. Castresana, J. Selection of conserved blocks from multiple alignments for their use in phylogenetic analysis. *Mol. Biol. Evol.* **2000**, *17*, 540–552. [CrossRef]
20. Guindon, S.; Dufayard, J.F.; Lefort, V.; Anisimova, M.; Hordijk, W.; Gascuel, O. New algorithms and methods to estimate maximum-likelihood phylogenies: Assessing the performance of PhyML 3.0. *Syst. Biol.* **2010**, *59*, 307–321. [CrossRef]
21. Whelan, S.; Goldman, N. A general empirical model of protein evolution derived from multiple protein families using a maximum-likelihood approach. *Mol. Biol. Evol.* **2001**, *18*, 691–699. [CrossRef] [PubMed]
22. Tamura, K.; Stecher, G.; Kumar, S. MEGA11: Molecular Evolutionary Genetics Analysis Version 11. *Mol. Biol. Evol.* **2021**, *38*, 3022–3027. [CrossRef] [PubMed]

23. The UniProt Consortium. UniProt: The universal protein knowledgebase in 2021. *Nucleic Acids Res.* **2021**, *49*, D480–D489. [CrossRef] [PubMed]
24. Cwiklinski, K.; Drysdale, O.; López Corrales, J.; Corripio-Miyar, Y.; De Marco Verissimo, C.; Jewhurst, H.; Smith, D.; Lalor, R.; McNeilly, T.N.; Dalton, J.P. Targeting secreted protease/anti-protease balance as a vaccine strategy against the helminth *Fasciola hepatica*. *Vaccines* **2022**, *10*, 155. [CrossRef] [PubMed]
25. Jumper, J.; Evans, R.; Pritzel, A.; Green, T.; Figurnov, M.; Ronneberger, O.; Tunyasuvunakool, K.; Bates, R.; Žídek, A.; Potapenko, A.; et al. Highly accurate protein structure prediction with AlphaFold. *Nature* **2021**, *596*, 583–589. [CrossRef] [PubMed]
26. Varadi, M.; Anyango, S.; Deshpande, M.; Nair, S.; Natassia, C.; Yordanova, G.; Yuan, D.; Stroe, O.; Wood, G.; Laydon, A.; et al. AlphaFold Protein Structure Database: Massively expanding the structural coverage of protein-sequence space with high-accuracy models. *Nucleic Acids Res.* **2022**, *50*, D439–D444. [CrossRef]
27. McCord, J.M.; Fridovich, I. Superoxide dismutase. An enzymic function for erythrocuprein (hemocuprein). *J. Biol. Chem.* **1969**, *244*, 6049–6055. [CrossRef]
28. López Corrales, J.; Cwiklinski, K.; De Marco Verissimo, C.; Dorey, A.; Lalor, R.; Jewhurst, H.; McEvoy, A.; Diskin, M.; Duffy, C.; Cosby, S.L.; et al. Diagnosis of sheep fasciolosis caused by *Fasciola hepatica* using cathepsin L enzyme-linked immunosorbent assays (ELISA). *Vet. Parasitol.* **2021**, *298*, 109517. [CrossRef]
29. De Marco Verissimo, C.; Jewhurst, H.L.; Tikhonova, I.G.; Urbanus, R.T.; Maule, A.G.; Dalton, J.P.; Cwiklinski, K. *Fasciola hepatica* serine protease inhibitor family (serpins): Purposely crafted for regulating host proteases. *PLoS Negl. Trop. Dis.* **2020**, *14*, e0008510. [CrossRef]
30. Murphy, A.; Cwiklinski, K.; Lalor, R.; O'Connell, B.; Robinson, M.W.; Gerlach, J.; Joshi, L.; Kilcoyne, M.; Dalton, J.P.; O'Neill, S.M. *Fasciola hepatica* Extracellular Vesicles isolated from excretory-secretory products using a gravity flow method modulate dendritic cell phenotype and activity. *PLoS Negl. Trop. Dis.* **2020**, *14*, e0008626. [CrossRef]
31. Cwiklinski, K.; de la Torre-Escudero, E.; Trelis, M.; Bernal, D.; Dufresne, P.J.; Brennan, G.P.; O'Neill, S.; Tort, J.; Paterson, S.; Marcilla, A.; et al. The Extracellular Vesicles of the helminth pathogen, *Fasciola hepatica*: Biogenesis pathways and cargo molecules involved in parasite pathogenesis. *Mol. Cell Proteom.* **2015**, *14*, 3258–3273. [CrossRef] [PubMed]
32. Schatzman, S.S.; Culotta, V.C. Chemical warfare at the microorganismal level: A closer look at the superoxide dismutase enzymes of pathogens. *ACS Infect Dis.* **2018**, *4*, 893–903. [CrossRef] [PubMed]
33. Ruiz-Campillo, M.T.; Molina Hernandez, V.; Escamilla, A.; Stevenson, M.; Perez, J.; Martinez-Moreno, A.; Donnelly, S.; Dalton, J.P.; Cwiklinski, K. Immune signatures of pathogenesis in the peritoneal compartment during early infection of sheep with *Fasciola hepatica*. *Sci. Rep.* **2017**, *7*, 2782. [CrossRef]
34. Carvalho-Queiroz, C.; Cook, R.; Wang, C.C.; Correa-Oliveira, R.; Bailey, N.A.; Egilmez, N.K.; Mathiowitz, E.; LoVerde, P.T. Cross-reactivity of *Schistosoma mansoni* cytosolic superoxide dismutase, a protective vaccine candidate, with host superoxide dismutase and identification of parasite-specific B epitopes. *Infect Immun.* **2004**, *72*, 2635–2647. [CrossRef]
35. Anderson, T.J.C.; Enabulele, E.E. *Schistosoma mansoni*. *Trends Parasitol.* **2021**, *37*, 176–177. [CrossRef]
36. Piedrafita, D.; Estuningsih, E.; Pleasance, J.; Prowse, R.; Raadsma, H.W.; Meeusen, E.N.; Spithill, T.W. Peritoneal lavage cells of Indonesian thin-tail sheep mediate antibody-dependent superoxide radical cytotoxicity in vitro against newly excysted juvenile *Fasciola gigantica* but not juvenile *Fasciola hepatica*. *Infect Immun.* **2007**, *75*, 1954–1963. [CrossRef] [PubMed]
37. Piedrafita, D.; Spithill, T.W.; Dalton, J.P.; Brindley, P.J.; Sandeman, M.R.; Wood, P.R.; Parsons, J.C. Juvenile *Fasciola hepatica* are resistant to killing in vitro by free radicals compared with larvae of *Schistosoma mansoni*. *Parasite Immunol.* **2000**, *22*, 287–295. [CrossRef]
38. Pleasance, J.; Raadsma, H.W.; Estuningsih, S.E.; Widjajanti, S.; Meeusen, E.; Piedrafita, D. Innate and adaptive resistance of Indonesian Thin Tail sheep to liver fluke: A comparative analysis of *Fasciola gigantica* and *Fasciola hepatica* infection. *Vet. Parasitol.* **2011**, *178*, 264–272. [CrossRef]
39. McGonigle, S.; Curley, G.P.; Dalton, J.P. Cloning of peroxiredoxin, a novel antioxidant enzyme, from the helminth parasite *Fasciola hepatica*. *Parasitology* **1997**, *115 Pt 1*, 101–104. [CrossRef]

antioxidants

MDPI

Article

New Insights on NETosis Induced by *Entamoeba histolytica*: Dependence on ROS from Amoebas and Extracellular MPO Activity

César Díaz-Godínez [1], Joshue Fabián Jorge-Rosas [1], Mario Néquiz [2], Santiago Martínez-Calvillo [3], Juan P. Laclette [1], Carlos Rosales [1,*] and Julio C. Carrero [1,*]

[1] Departamento de Inmunología, Instituto de Investigaciones Biomédicas, Universidad Nacional Autónoma de México (UNAM), Ciudad de México 04510, Mexico; cesar_rha32@hotmail.com (C.D.-G.); a.fabian.sfrayer@gmail.com (J.F.J.-R.); laclette@biomedicas.unam.mx (J.P.L.)

[2] Unidad de Investigación en Medicina Experimental, Facultad de Medicina, Universidad Nacional Autónoma de México, Ciudad de México 06720, Mexico; manequiz@yahoo.com.mx

[3] Unidad de Biomedicina, Facultad de Estudios Superiores Iztacala, Universidad Nacional Autónoma de México, Av. de los Barrios 1, Col. Los Reyes Iztacala, Tlalnepantla, Estado de México 54090, Mexico; scalv@unam.mx

* Correspondence: carosal@iibiomedicas.unam.mx (C.R.); carrero@unam.mx (J.C.C.); Tel.: +525-56228945 (C.R.); +525-56229220 (J.C.C.)

Abstract: NETosis is a neutrophil process involving sequential steps from pathogen detection to the release of DNA harboring antimicrobial proteins, including the central generation of NADPH oxidase dependent or independent ROS. Previously, we reported that NETosis triggered by *Entamoeba histolytica* trophozoites is independent of NADPH oxidase activity in neutrophils, but dependent on the viability of the parasites and no ROS source was identified. Here, we explored the possibility that *E. histolytica* trophozoites serve as the ROS source for NETosis. NET quantitation was performed using SYTOX® Green assay in the presence of selective inhibitors and scavengers. We observed that respiratory burst in neutrophils was inhibited by trophozoites in a dose dependent manner. Mitochondrial ROS was not also necessary, as the mitochondrial scavenger mitoTEMPO did not affect the process. Surprisingly, ROS-deficient amoebas obtained by pre-treatment with pyrocatechol were less likely to induce NETs. Additionally, we detected the presence of MPO on the cell surface of trophozoites after the interaction with neutrophils and found that luminol and isoluminol, intracellular and extracellular scavengers for MPO derived ROS reduced the amount of NET triggered by amoebas. These data suggest that ROS generated by trophozoites and processed by the extracellular MPO during the contact with neutrophils are required for *E. histolytica* induced NETosis.

Keywords: *Entamoeba histolytica*; neutrophil extracellular traps (NETs); neutrophils; reactive oxygen species (ROS); myeloperoxidase

Citation: Díaz-Godínez, C.; Jorge-Rosas, J.F.; Néquiz, M.; Martínez-Calvillo, S.; Laclette, J.P.; Rosales, C.; Carrero, J.C. New Insights on NETosis Induced by *Entamoeba histolytica*: Dependence on ROS from Amoebas and Extracellular MPO Activity. *Antioxidants* **2021**, *10*, 974. https://doi.org/10.3390/antiox10060974

Academic Editor: Serge Ankri

Received: 15 May 2021
Accepted: 6 June 2021
Published: 18 June 2021

1. Introduction

Neutrophil extracellular traps (NETs) are DNA fibers associated with histones and antimicrobial proteins that are released by neutrophils into the extracellular space in a process denominated NETosis [1]. Initially, NETs were described as a strategy used for neutrophils to entrap and kill microorganisms [2–5]; nevertheless, these structures have been also related to other processes such as coagulation or complement activation, and even with autoimmune pathological processes such as erythematous systemic lupus [6–8]. Since its discovery, one of the central issues in the study of NETs has been the understanding of the mechanism associated with its release. Death of neutrophils through NETosis is not a random event; on the contrary, it is a highly regulated process involving a series of sequential steps [9–11]. Initially, neutrophils detect pathogens through receptors on cell surface such as TLRs, dectins, integrins or antibody receptors [12–15]. Then, diverse signaling pathways are activated to drive events to lead NETotic commitment, resulting in reactive oxygen species (ROS) generation, histone processing, disassembling of a nuclear envelope and dissolution of internal membranes [16–19]. Decondensed DNA is then associated with antimicrobial proteins from cytoplasmic granules such as neutrophil elastase (NE), myeloperoxidase (MPO), cathepsin G (CG), proteinase 3 (PR3) and others [1]. Finally, cytoplasmic membrane ruptures and NETs are released [20].

Multiple mechanisms of NETosis have been identified depending on the stimulus used to trigger the process, including nuclear, mitochondrial or blebbing NETs [1,21,22]. The diverse mechanisms seem to differ from each other in the receptors involved, the signaling pathways activated or the ROS source required.Phorbol 12-myristate 13-acetate (PMA) and calcium ionophores (A23187 and ionomycin) are widely used stimuli to study NETosis; nevertheless, they have the disadvantage that they are stimuli with little biological relevance [23,24]. Despite this, the findings obtained with PMA and calcium ionophores have allowed us to decipher part of the process, mainly the critical need for ROS generation [25,26]. While ROS derived from NADPH oxidase was linked to PMA-induced NETosis [27], NETosis triggered by calcium ionophores does not require NADPH oxidase activity, although that depends partially on mitochondrial ROS (mitROS) [28]. Therefore, NETosis mechanisms have been classified into two groups: NADPH oxidase dependent or independent [29]. NAPDH-independent NETosis is not limited to the generation of mitROS as reactive nitrogen species (RNS), as well as exogenous ROS sources could also serve to trigger NET release [30,31]. In this context, it has been demonstrated that microorganisms such as *Escherichia coli* and *Mycobacterium smegmatis* produce ROS in stress conditions [32]; moreover, ROS from *Candida albicans* are responsible for triggering NET release in neutrophils from patients with chronic granulomatous disease [33].

Entamoeba histolytica is the protozoan responsible for causing intestinal (amoebic colitis or dysentery) and extraintestinal (amoebic liver abscess) amoebiasis in humans [34]. This parasite represents a public health problem, especially in developing countries where a prevalence of 1% to 20% in the population and up to 50% of diarrhea due to amoebiasis in infants have been reported [35–37]. Immune response activation against amoeba implicates a rapid recruitment of neutrophils to the infection site [38]; however, the exact role of these cells in amoebiasis remains unknown. Thus, some experiments suggest that neutrophils are required for clearance of the infection [39], whereas other evidence suggests that these cells could play a pathological role [40]. We previously showed that *E. histolytica* trophozoites induce a rapid NETosis in human neutrophils that is dependent on the viability of the parasite but independent of NADPH oxidase and PAD4 activities [41]. Nevertheless, a ROS source has not been identified. In this report, we investigated the source of ROS that leads to *E. histolytica* trophozoites-induced NETosis.

2. Materials and Methods

2.1. *E. histolytica Trophozoites*

E. histolytica trophozoites (HM1:IMSS strain) were axenically cultured in TYI-S-33 medium supplemented with Diamond vitamin tween solution (Merck) and 15% heat-inactivated adult bovine serum (Microlab). Trophozoites were grown for 72 h at 37 °C until they reached the log phase and harvested by chilling on ice for 5 min and centrifugation at 1400 rpm during 5 min at 10 °C. The pellet was resuspended in PBS pH 7.4 and the trophozoites counted in hemocytometer and preserved at room temperature until use. For selected experiments, amoebas (5×10^5) were resuspended in 500 µL of PBS and fixed with formaldehyde (3.7%) or heat-inactivated (56 °C) for 30 min in both cases.

2.2. *Neutrophil Isolation*

Neutrophils were obtained from peripheral blood of healthy volunteers in line with the approach of García–García et al. [42] using Ficoll-PaqueR gradient (GE Healthcare) and hypertonic shock to lyse erythrocytes. Cells were resuspended in PBS pH 7.4, counted in hemocytometer and reserved at 4 °C until use. This study was carried out in accordance with the recommendations and approval of the Ethical Committee for Studies on Humans of the Instituto de Investigaciones Biomédicas, UNAM (Ethical approved number: FMED/CI/RGG/ 013/01/2008). All subjects signed a written informed consent.

2.3. NET Quantitation Assay

NET quantitation was performed as described before [43]. In brief, neutrophils (5×10^5) were centrifuged at 4000 rpm for 2 min and resuspended in 500 µL of RPMI-1640 medium (Biological Industries) supplemented with 5% fetal bovine serum (FBS, Gibco) and 500 nM SYTOX® Green (Invitrogen). A volume of 100 µL of cell suspension (1×10^5 neutrophils) was added to a 96 well plate, allowed to sediment for 20 min at 37 °C and then, stimulated with 1×10^3, 2×10^3, 5×10^3 or 1×10^4 *E. histolytica* viable trophozoites (trophozoite:neutrophil ratios 1:100, 1:50, 1:20 and 1:10, respectively). In other experiments, neutrophils were stimulated with 5×10^3 formaldehyde-fixed or heat-inactivated trophozoites. Co-cultures were incubated at 37 °C and fluorescence was measured during 4 h from the well bottom using a spectrofluorometer Synergy HTX (BioTek) with 485 nm excitation and 528 nm emission filters. NETosis induced by PMA (50 nM, Merck) and A23187 (10 µM, Merck) were used as positive controls.

To determine the role of NADPH-oxidase in amoeba-induced NETosis, neutrophils (5×10^5) were resuspended in 500 µL of PBS and pretreated with the inhibitor apocynin (400 µM) or vehicle DMSO (0.1%) for 30 min at 4 °C (all reagents were supplied by Merck). After pretreatment, neutrophils were induced to NETosis with trophozoites as described above. To determine the role of ROS from neutrophils, these cells (5×10^5) were resuspended in 500 µL of PBS and pretreated separately for 30 min at 4 °C with the ROS scavengers pyrocatechol (200 µM), catalase (200 UI/mL), luminol (50, 100 and 200 µM), isoluminol (50, 100 and 200 µM) or mitoTEMPO (400 µM). As negative controls, neutrophils were pretreated with the corresponding vehicles (all reagents were supplied by Merck). After pretreatments, neutrophils were immediately tested for amoeba-induced NETosis as described above in culture media added with the inhibitors or scavengers at concentrations indicated previously (except for mitoTEMPO). All experiments were performed three times in triplicates.

2.4. NET Visualization

NET immunofluorescence was performed as described previously [41] with some modifications. In brief, neutrophils (2×10^5) were resuspended in 100 µL of RPMI-1640 medium supplemented with 5% FBS and seeded on coverslips pretreated with poly-L-lysine solution (Merck). After sedimentation for 20 min at room temperature, the neutrophils were stimulated with 1×10^4 viable trophozoites or positive controls PMA (50 nM) and A23187 (10 µM). Co-cultures were incubated for 4 h at 37 °C and then fixed with 3.7% formaldehyde during 10 min. Fixed cells were permeabilized using 0.2% Triton X-100 (BioRad) in PBS for 5 min. Detergent was washed out two times with cold PBS and blocking was carried out with a 1% BSA, 0.3 M glycine and 0.1% Tween 20 in PBS, for 30 min at 37 °C. Samples were then incubated with primary antibodies against NETs constituents: anti-NE (Santa Cruz Biotechnology, sc-365950), anti-MPO (Abcam, ab16886) or anti-acetylated histone H4 (Abcam, ab61238) antibodies diluted 1:100 in 1% BSA, 0.1% Tween 20 in PBS for 1 h at room temperature. Samples were washed two times with cold PBS and then incubated with secondary anti-mouse IgG-FITC (Merck, F5387) or anti-rabbit IgG-TRITC (Zymax) antibodies diluted 1:50 in the same solution as primary antibodies for 1 h at room temperature in the dark. After two washes with cold PBS, samples were stained with 5 µg/mL DAPI (Merck) and the coverslips were mounted on slides using Fluoroshield (Merk) before observation in a fluorescence microscope (Olympus BX51). Images were processed using ImageJ software.

For detection of MPO on the surface of amoebic trophozoites, immunofluorescence was performed as above but cell cultures were fixed for 5 and 10 min after the addition of parasites.

2.5. Intracellular ROS Quantitation in Neutrophils

Neutrophils (5×10^5) were resuspended in 500 μL of PBS added with 10 μM 2′,7′-dichlorofluorescein diacetate (H_2DCFDA) and incubated for 30 min at 37 °C in the dark. Cells were centrifuged at 4000 rpm for 2 min and resuspended in 500 μL of RPMI-1640 supplemented with 5% FBS. Subsequently, each 100 μL of suspension (1×10^5 neutrophils) was transferred to 96 well plate, allowed to sediment for 20 min at 37 °C and then stimulated with 1×10^3, 2×10^3 or 5×10^3 viable *E. histolytica* trophozoites (trophozoite:neutrophil ratios 1:100, 1:50 and 1:20). Fluorescence intensity was measured after incubation during 1 h at 37 °C from the well bottom in the spectrofluorometer Synergy HTX using 485 nm excitation and 528 nm emission filters. PMA (50 nM) and A23187 (10 μM) were used as positive controls for ROS production.

2.6. Mitochondrial ROS Quantitation in Neutrophils

Isolated neutrophils (5×10^5) were resuspended in 500 μL of PBS added with MitoSOX™ Red (10 μM) and incubated for 30 min at 4 °C. Cells were centrifuged at 4000 rpm for 2 min and resuspended in 500 μL of RPMI-1640 medium supplemented with 5% FBS. Each 100 μL of MitoSOX pretreated neutrophils (1×10^4) were transferred to 96 well plate and stimulated with 1×10^3, 2×10^3, 5×10^3 or 1×10^4 viable amoebas (trophozoite:neutrophil ratios 1:100, 1:50, 1:20 and 1:10). Fluorescence was read from the well's bottom after 2 h using a spectrofluorometer Synergy HTX (BioTek) with 485 nm excitation and 580 nm emission filters. PMA (50 nM) and A23187 (10 μM) were used as negative and positive controls of mitochondrial ROS, respectively.

2.7. ROS Quantitation in E. histolytica Trophozoites

Viable, formaldehyde-fixed or heat-inactivated trophozoites (5×10^5) were resuspended in 500 μL of PBS added with H_2DCFDA (100 μM) and incubated for 1 h at 37 °C. After incubation, amoebas were centrifuged at 4000 rpm for 2 min and resuspended in 500 μL of RPMI-1640 medium supplemented with 5% FBS. Each 100 μL of the cell suspension (1×10^5 H_2DCFDA-pretreated trophozoites) was added to 96 well plate and allowed us to sediment for 10 min at 37 °C. Fluorescence was read from well bottom using a spectrofluorometer Synergy HTX with 485 nm excitation and 528 nm emission filters.

2.8. ROS Scavenging from Amoebic Trophozoites

Trophozoites (5×10^5) were resuspended in 500 μL of PBS added with the ROS scavengers pyrocatechol (50, 100 or 200 μM) or luminol (50, 100 or 200 μM), as well as with the vehicle DMSO (0.1%), and incubated during 30 min at 37 °C. Then, the cell suspension was added with H_2DCFDA (100 μM) and ROS quantitation determined as mentioned above. On the other hand, to determine the role of amoebas-derived ROS in the NETosis process, trophozoites (5×10^5) were resuspended in 500 μL of PBS added with the ROS scavengers pyrocatechol (50, 100 and 200 μM) or luminol (50, 100 and 200 μM). Trophozoites were incubated for 1.5 h at 37 °C, centrifuged at 4000 rpm for 2 min and then resuspended in 500 μL of PBS to be used immediately for NET induction.

2.9. Visualization of Amoebas-Derived ROS

Trophozoites treated for ROS quantitation as well as trophozoites treated for ROS depletion with the scavenger pyrocatechol as mentioned above, were fixed with formaldehyde (3.5%) and counterstained with DAPI (5 μg/mL). An aliquot of 20 μL was observed under fluorescence microscope Olympus BX51. Images were processed using ImageJ software.

2.10. Detection of MPO Activity

Neutrophils (5×10^5) were centrifuged at 4000 rpm for 2 min and resuspended in 500 μL of RPMI-1640 medium supplemented with 5% FBS (Gibco) and luminol (200 μM) or isoluminol (200 μM). Each 100 μL of cell suspension (1×10^5 neutrophils) was placed in a 96 well plate and cells were incubated for 20 min at 37 °C for sedimentation. Posteriorly, neutrophils were stimulated with 1×10^3, 2×10^3, 5×10^3 or 1×10^4 *E. histolytica* viable trophozoites. Co-cultures were incubated at 37 °C for 4 h and luminescence was measured every 5 min from the well bottom using a spectrofluorometer Synergy HTX. PMA (50 nM, Merck) and A23187 (10 μM, Merck) were used as controls.

2.11. Statistical Analysis

Statistical significance was tested with paired two-tailed Student's *t*-test. Data are reported as mean ± SD. A p value ≤ 0.05 was considered statistically significant.

3. Results

3.1. E. histolytica Trophozoites Induce NETosis in a Dose-Dependent Manner

Herein, we evaluate amoeba-induced NETosis under different amoeba:neutrophil ratios. As shown in Figure 1A, amoebas trigger NET release in a dose-dependent manner at ratios between 1:100 to 1:20. The maximum level of DNA release was obtained at ratio 1:20, as no significant differences were observed with respect to the 1:10 ratio and the positive controls PMA and A23187.

NET release was confirmed by detecting MPO in the DNA by immunofluorescence. As shown in Figure 1B, untreated control neutrophils have condensed, multilobed nucleus and MPO located in the cytoplasmic compartment as expected. The positive controls PMA and A23187 induced nuclear decondensation and formation of DNA fibers that were colocalized with MPO. On the other hand, *E. histolytica* trophozoites induced cloudy NETs with heterogenous distribution of MPO. To ensure that these structures correspond to NET, we observed that acetylated histone H4 and NE also colocalized with released DNA in response to trophozoites (Figure 1C).

3.2. NETosis Induced by E. histolytica Trophozoites Is Independent of Neutrophil's ROS

As shown in Figure 2A, NADPH oxidase inhibitor apocynin significantly reduced NETosis induced by PMA, whereas NETosis induced by A23187 was not affected, as expected. Amoebas-induced NETosis was not affected by apocynin at any trophozoite:neutrophil ratio tested, suggesting that this process is certainly independent of ROS from neutrophils NADPH oxidase (Figure 2A). Then, we evaluated general production of ROS by neutrophils at the different ratios and found that although the respiratory burst was completely suppressed at a 1:20 ratio, as we previously reported, the neutrophils produce ROS at 1:100 and 1:50 ratios (Figure 2B). Taken together, the data strongly suggest that amoeba-induced NETosis is independent of ROS from neutrophils.

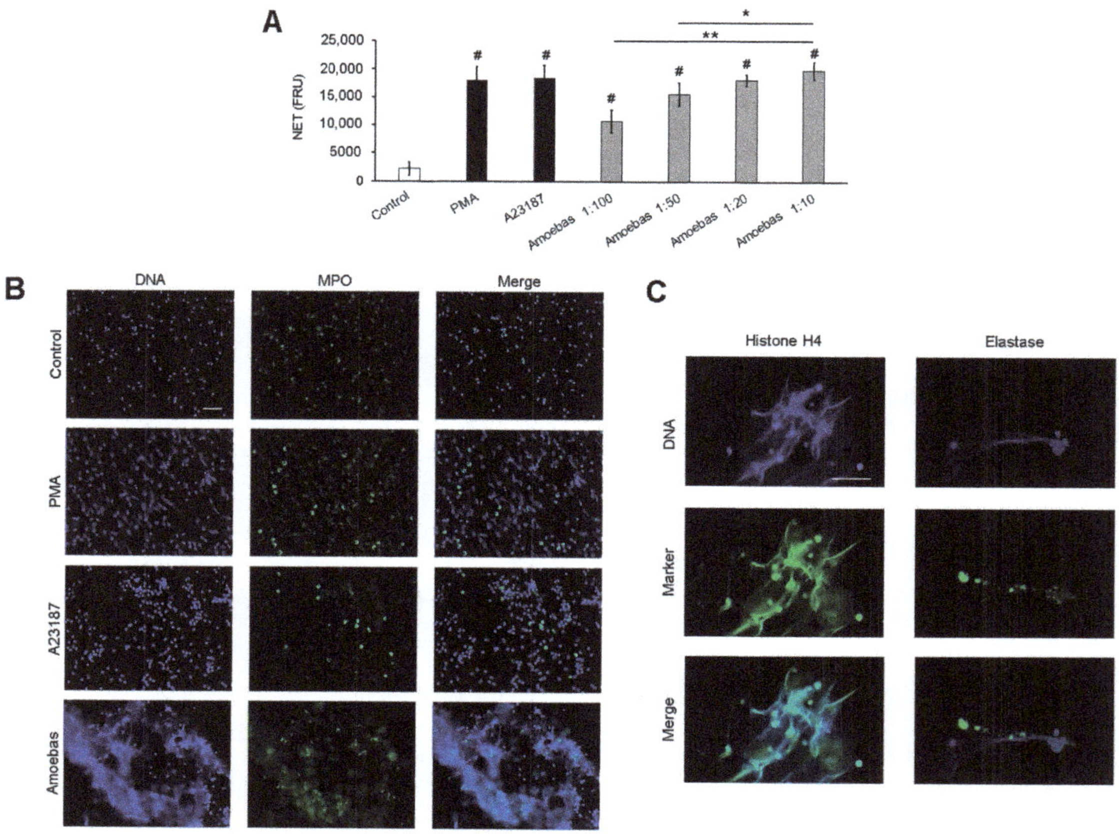

Figure 1. *Entamoeba histolytica* induces NETosis on human neutrophils in a dose-dependent manner. (**A**) Human neutrophils (1×10^5) were cultured with *E. histolytica* trophozoites at ratios amoeba:neutrophil of 1:100, 1:50, 1:20 and 1:10 in RPMI-1640 medium added with 5% FBS and 500 nM SYTOX® Green. PMA (50 nM) and A23187 (10 μM) were used as positive controls of NETosis. Finally, fluorescence was read at 4 h. NETs amount is expressed in fluorescence relative units (FRU). Values are means ± SD of three independent experiments. * $p < 0.05$, ** $p < 0.001$, # $p < 0.001$ respect to control. (**B**) Neutrophils (2×10^5) were stimulated with PMA (50 nM), A23187 (10 μM) or *E. histolytica* trophozoites (1×10^4) during 4 h. After fixation, cells were marked using anti-MPO primary antibody followed by anti-mouse IgG-FITC secondary antibody. DNA was counterstained with DAPI. Images were taken at 40× magnification. Scale bar 100 μm. (**C**) Neutrophils (2×10^5) were co-cultured with 1×10^4 *E. histolytica* trophozoites for 4 h. Cells were fixed and immunofluorescence was performed using anti-NE and or anti-acetylated histone H4 primary antibodies followed by anti-mouse IgG-FITC (for NE) or anti-rabbit-TRITC (for histone) secondary antibodies. DNA was counterstained with DAPI. Images were taken at 100× magnification. Scale bar: 50 μm.

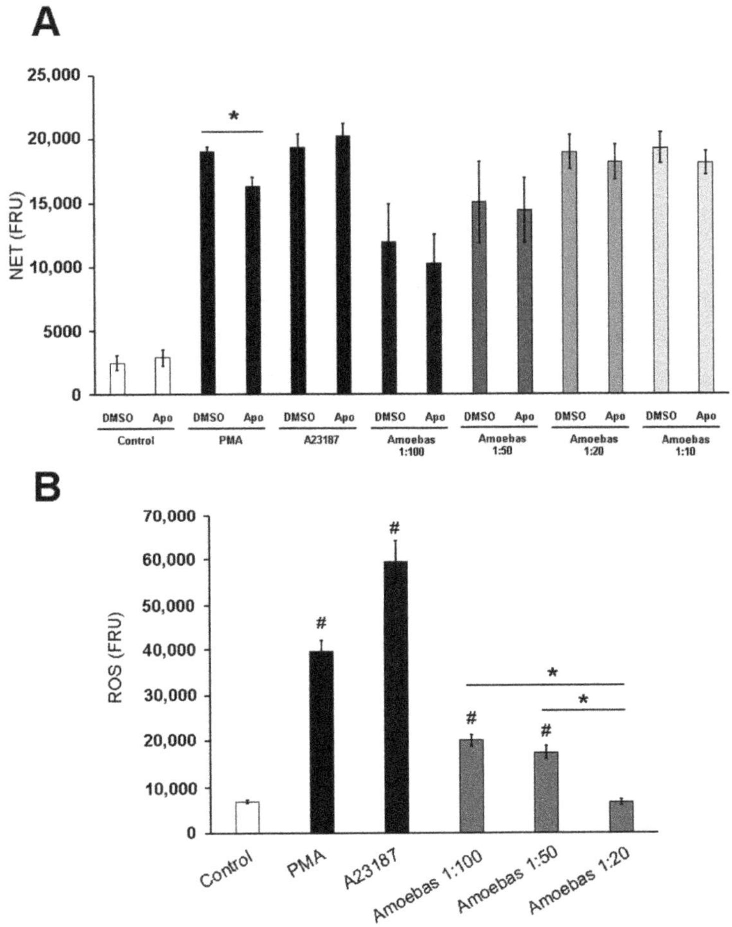

Figure 2. NETosis induced by *E. histolytica* occurs independently of NADPH oxidase. (**A**) Neutrophils (1×10^5) were pretreated with 400 μM apocynin (Apo) or DMSO for 30 min. Posteriorly, cells were transferred to RPMI-1640 medium added with 5% FBS and 500 nM SYTOX® Green and then stimulated with PMA (50 nM), A23187 (10 μM) or *E. histolytica* trophozoites at ratios of 1:100, 1:50, 1:20 and 1:10. Fluorescence was read at 4 h. (**B**) H_2DCFDA-pretreated neutrophils (1×10^5) were culture in RPMI-1640 medium supplemented with 5% FBS and then stimulated with PMA (50 nM), A23187 (10 μM) or *E. histolytica* trophozoites at ratios 1:100, 1:50 or 1:20. Fluorescence was read at 1 h. NET and ROS amounts are expressed in fluorescence relative units (FRU). Values are means ± SD of three independent experiments. * $p< 0.0001$, # $p < 0.001$ with respect to control.

3.3. Dead Trophozoites Do Not Induce NET Release and Contain Scarce ROS

As shown in Figure 3A, heat-killed or fixed trophozoites did not induce NETosis on human neutrophils, in contrast to viable amoebas that induced NET levels like PMA and A23187. This result suggests that metabolically active trophozoites are required for triggering NETosis. Since most studies suggest that NETosis requires a source of ROS, which is not related to neutrophils in this case, we explored the possibility that ROS from trophozoites was involved. So, we decided to quantify ROS in living and dead amoebas and found, as expected, that formaldehyde-fixed and heat-killed trophozoites exhibited far fewer ROS than live amoebas (Figure 3B).

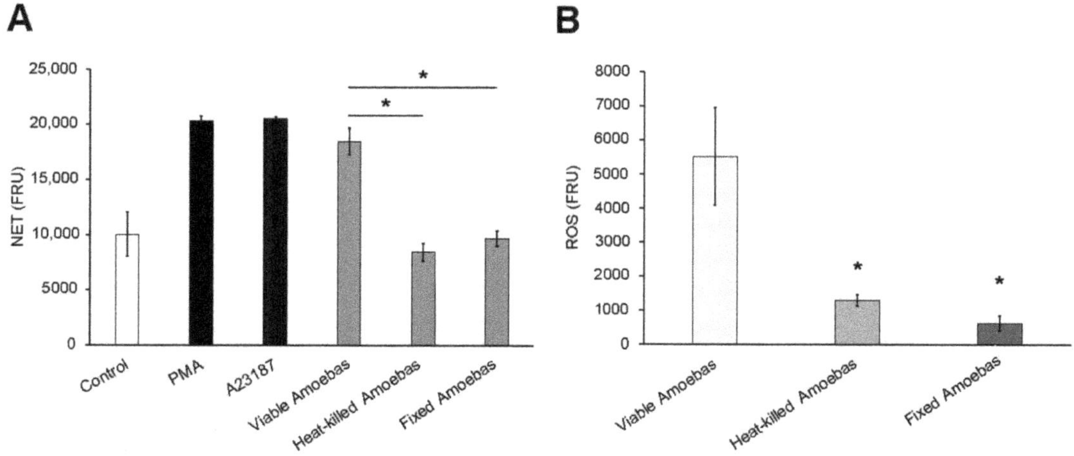

Figure 3. Dead *E. histolytica* trophozoites do not trigger NETosis and possess less ROS. (**A**) Neutrophils (1×10^5) were culture in RPMI-1640 medium supplemented with 5% FBS and added with 500 nM SYTOX® Green. Cells were stimulated with PMA (50 nM), A23187 (10 μM), viable trophozoites (5×10^3), heat-killed trophozoites (5×10^3) or formaldehyde-fixed trophozoites (5×10^3). Fluorescence was read at 4 h. (**B**) Viable, heat-killed and formaldehyde-fixed trophozoites were pretreated with H_2DCFDA for 1 h. Posteriorly, amoebas (1×10^5) were placed in a 96 well plate and fluorescence was read. NET and ROS amounts are expressed in fluorescence relative units (FRU). Values are means ± SD of three independent experiments. * $p < 0.0001$.

3.4. Amoebas Derived ROS Are Required for NETosis

Based on the above results, we evaluated the effect of reducing the amount of ROS in viable trophozoites on their capability to induce NETosis. First, we demonstrated that the pretreatment of *E. histolytica* trophozoites with the ROS scavenger pyrocatechol (50 to 200 μM) for 1.5 h resulted in a significant and dose-dependent reduction of ROS levels. Moreover, this effect persisted during the 4 h period that NETosis assay lasted (Figure 4A). The result was confirmed in H_2DCFDA-stained trophozoites. As shown in Figure 4B, the amoebic trophozoites exhibited intense green florescence in basal conditions (high ROS levels), whereas the pyrocatechol-treated amoebas exhibited a dose-dependent decrease of green fluorescence denoting a reduction of trophozoite-derived ROS. Many trophozoites pretreated with pyrocatechol 200 μM were virtually non-fluorescent (Figure 4B, lower panel, white arrows). The treatments with pyrocatechol and luminol did not affect the viability of trophozoites (Supplementary Figure S1).

Once we had demonstrated the reduction of ROS in pyrocatechol-pretreated trophozoites, we then carried out NETosis assays with these amoebas. It is noteworthy that ROS-reduced trophozoites induced a reduced NET amount compared with trophozoites pretreated with the vehicle DMSO, and the reduction was statistically significant when pyrocatechol at 100 and 200 μM was used (Figure 5A). As expected, pyrocatechol present in the culture media abolished NET release by PMA at any concentration, but it did not affect NET release by A23187 when used at doses under 100 μM. Surprisingly, A23187-induced NETosis, which is NADPH-ROS independent, was reduced with 200 μM of pyrocatechol (Figure 5A). Involvement of amoeba-derived ROS in NETosis became more evident when lower ratios of parasites per neutrophil were tested. As shown in Figure 5B, trophozoites pretreated with pyrocatechol induced less NET release than DMSO-treated parasites in a dose-dependent manner. Moreover, pretreated amoebas lost the ability to induce NETosis at lower ratios (1:50 and 1:100). It is noteworthy that catalase added to the culture media was unable to prevent NETosis induced by untreated trophozoites (Supplementary Figure S2).

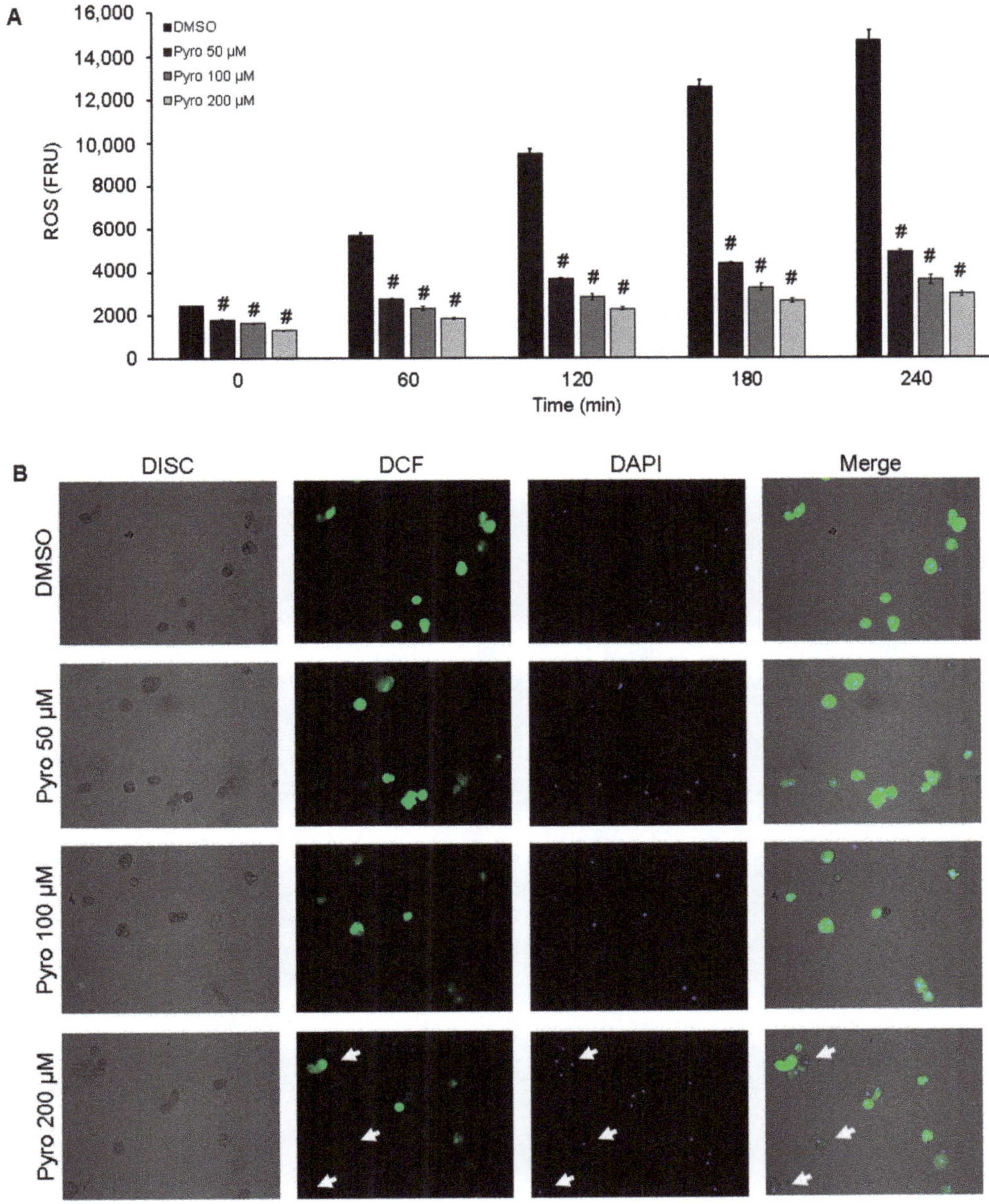

Figure 4. Pyrocatechol reduces ROS generation in viable *E. histolytica* trophozoites. Amoebic trophozoites were treated with DMSO or pyrocatechol (Pyro) at 50, 100 and 200 μM for 30 min and then H_2DCFDA (100 μM) was added. Cells were incubated for another hour and after treatment, trophozoites were resuspended in RPMI-1640 medium supplemented with 5% FBS. A total of 1×10^5 trophozoites were placed and fluorescence was read every hour during 4 h (**A**) or were fixed and counterstained with DAPI for visualization under fluorescence microscopy (**B**). NET amount is expressed in fluorescence relative units (FRU). Values are means ± SD of three independent experiments. # $p < 0.01$ with respect to the control.

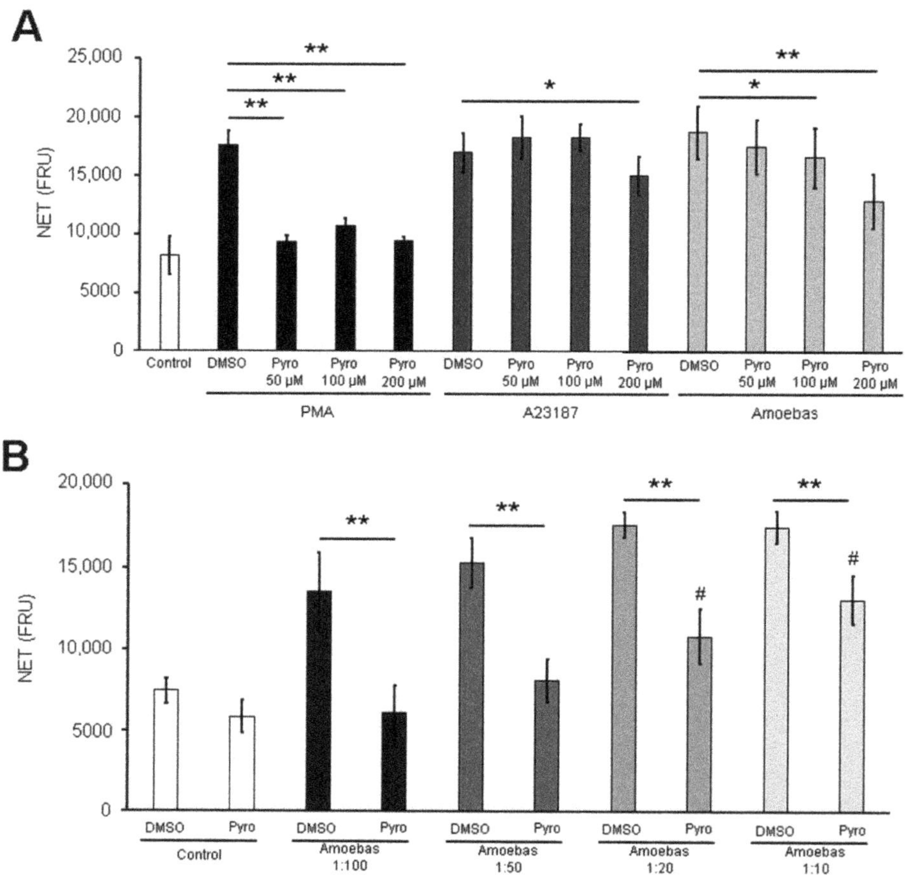

Figure 5. ROS-reduced trophozoites failed to induce NETosis on human neutrophils. (**A**) Neutrophils (1×10^5) were cultured in RPMI-1640 medium supplemented with 5% FBS, 500 nM SYTOX® Green and pyrocatechol (Pyro, 50, 100 or 200 μM or DMSO). Cells were stimulated with PMA (50 nM), A23187 (10 μM) or 5×10^3 pyrocatechol-pretreated trophozoites (according to the concentration present in the medium). Fluorescence was read after 4 h. (**B**) Neutrophils (1×10^5) were cultured in RPMI-1640 medium supplemented with 5% FBS, 500 nM SYTOX® Green and pyrocatechol (200 μM or DMSO). Cells were stimulated with trophozoites pretreated with pyrocatechol (200 μM or the vehicle DMSO) at ratios 1:100, 1:50, 1:20 or 1:10. Fluorescence was read after 4 h. NET amount is expressed in fluorescence relative units (FRU). Values are means ± SD of three independent experiments. * $p < 0.01$, ** $p < 0.0001$, # $p < 0.05$ with respect to the control.

3.5. *MPO Activity Is Detected Early during Neutrophil-Amoeba Interaction*

Trophozoites induced a rapid MPO activity, detected by luminol, that is independent of the ratio amoeba:neutrophil tested. This activity is detectable in the first minutes of interaction, reaching the maximum value at 20 min (Figure 6A–D). Then, the luminescence signal gradually decreases until it disappears after 90 min. A23187 also induced a similar kinetic of MPO activity that reached the maximum value at 10 min and then, decreased drastically until it disappeared after 100 min (Figure 6E). On the other hand, PMA induced a slower MPO activity that gradually increased during the first 20 min, then reached a steady state for 25 min (Figure 6F). Afterwards, the activity increased and newly reached the maximum value at 100 min, and then decreased to completely disappear at 200 min. It is important to notice that amoebas did not exhibit any MPO activity (Figure 6G).

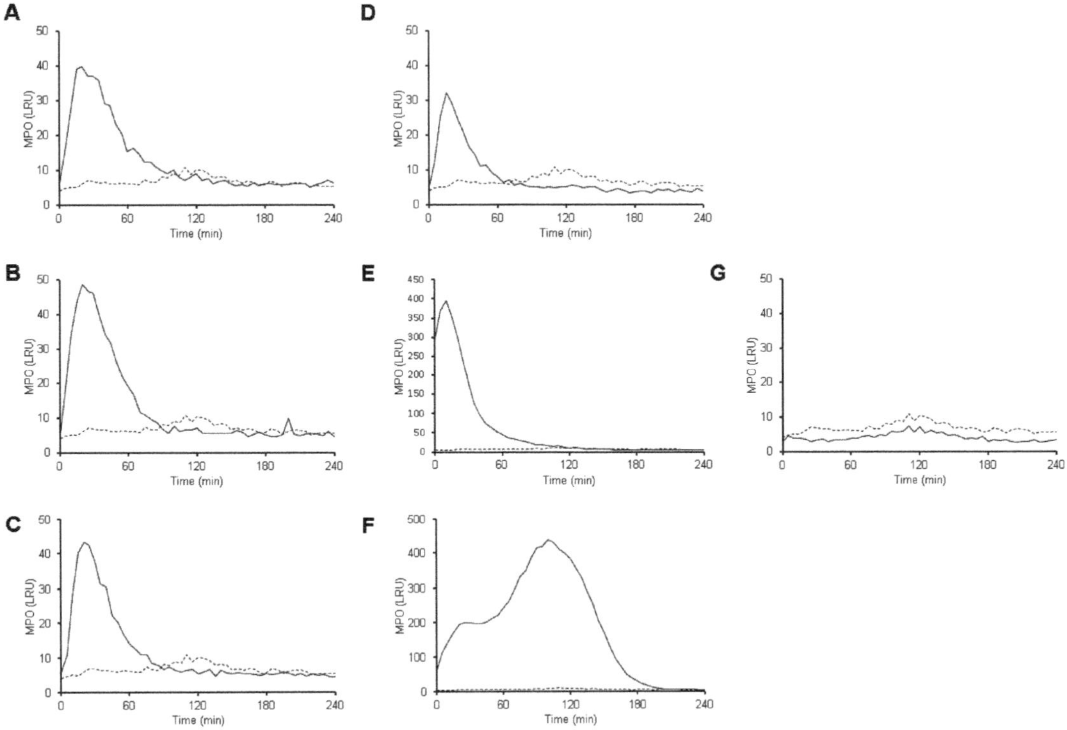

Figure 6. MPO activity is detected during interaction between neutrophils with *E. histolytica*. Neutrophils (1×10^5) were culture in RPMI-1640 medium supplemented with 5%FBS and luminol (200 µM). Cells were stimulated with viable *E. histolytica* trophozoites at ratios of 1:100 (**A**), 1:50 (**B**), 1:20 (**C**) and 1:10 (**D**), as well as 50 nM PMA (**E**), 10 µM A23187 (**F**). Amoebas alone (1×10^4) also were tested (**G**). Black line represents MPO activity (as luminescence relative units, LRU) after stimulation and dotted line represents MPO activity on neutrophils in the absence of stimuli (same for all plots). Values are means of three independent experiments.

3.6. MPO Activity Is Required for NETosis Induced by E. histolytica

Luminol, a scavenger of MPO-derived HClO, decreased NET release triggered by *E. histolytica* trophozoites in a dose-dependent manner, suggesting that HClO is involved in this process (Figure 7A). Similar results were observed in PMA and A23187-induced NETosis. When we performed this assay using different trophozoite:neutrophil ratios, luminol (200 µM) decreased NET release induced by trophozoites at 1:50, 1:20 and 1:10 ratios; nevertheless, no differences were observed at a 1:100 ratio (Figure 7B).

To confirm that this effect was due to the reduction of MPO-derived HClO from neutrophils but not a reduction of ROS from amoebas, we pretreated trophozoites with luminol for 1.5 h and then used them to induce NETosis. Surprisingly, a decrease in NET release was observed in a similar way to in the previous experiment (Figure 7C). Therefore, we estimated ROS in luminol-pretreated trophozoites and found that, unlike neutrophils, luminol strikingly caused a dose-dependent increase of ROS in amoebas, as compared to the vehicle DMSO (Figure 7D). Viability of trophozoites was not affected by the treatment with luminol throughout the experiment (Supplementary Figure S1). Taken together, these data suggest that MPO activity from neutrophils and ROS from amoebas is produced early during the contact of the two cells, which is necessary for NETosis.

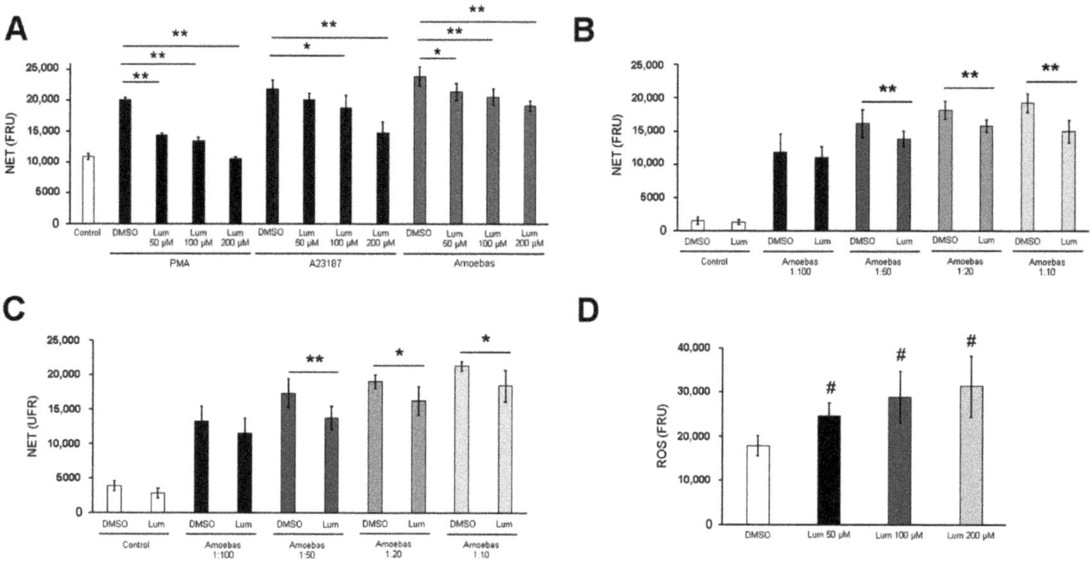

Figure 7. MPO activity is required for NETosis induced by *E. histolytica* trophozoites. (**A**) Neutrophils (1×10^5) were culture in RPMI-1640 medium supplemented with 5% FBS, 500 nM SYTOX® Green and luminol (Lum, 50, 100 or 200 μM or DMSO). Cells were stimulated with PMA (50 nM), A23187 (10μM) or 5×10^3 luminol-pretreated trophozoites (according with concentration present in the medium). Fluorescence was read after 4 h. (**B**) Neutrophils (1×10^5) were pretreated with 50, 100 and 200 μM luminol or DMSO during 30 min. Posteriorly, cells were transferred to RPMI-1640 medium added with 5% FBS and 500 nM SYTOX® Green and then stimulated with PMA (50 nM), A23187 (10 μM) or *E. histolytica* trophozoites at ratios of 1:100, 1:50, 1:20 and 1:10. Fluorescence was read at 4 h. (**C**) Neutrophils (1×10^5) were cultured in RPMI-1640 medium supplemented with 5% FBS, 500 nM SYTOX® Green and luminol (200 μM or DMSO). Cells were stimulated with trophozoites pretreated with luminol (200 μM or the vehicle DMSO) at ratios 1:100, 1:50, 1:20 or 1:10. Fluorescence was read at 4 h. (**D**) Amoebic trophozoites were treated with DMSO or luminol at 50, 100 and 200 μM for 30 min and then H_2DCFDA (100 μM) was added. Cells were incubated for another hour and after treatment, trophozoites were resuspended in RPMI-1640 medium supplemented with 5% FBS. A total of 1×10^5 trophozoites were placed and fluorescence was read. For (**A–D**) the NET or ROS amount are expressed in fluorescence relative units (FRU). Values are means ± SD of three independent experiments. * $p < 0.01$, ** $p < 0.001$, # $p < 0.0001$ with respect to the control.

3.7. MPO Is Detected on the Surface of Amoebic Trophozoites Early after Contact with Neutrophils and Its Activity Is Required for Trophozoite-Induced NETosis

The above results point out to the possibility of ROS transfer between the two cells in very early stages of contact. As ROS were not detected in neutrophils stimulated with trophozoites at a 1:20 ratio, but since MPO activity was always detected independent of the trophozoite:neutrophil ratio assayed, we explored the possibility that MPO was transferred from the neutrophil to the amoeba. As shown in Figure 8A, anti-MPO antibodies started to react with the surface of trophozoites after 5 min, covering all amoeba surfaces after 10 min of contact with neutrophils, suggesting that MPO was rapidly transferred from neutrophils to amoebas. The fluorescence detected was not due to autofluorescence or unspecific binding of the secondary antibody, as shown in trophozoites alone (Figure 8A).

To determine if the MPO activity in the surface of amoebic trophozoites is involved in NETosis, isoluminol (which does not cross the cell membrane) was used instead of luminol to scavenge MPO-derived HClO. In contrast to luminol, isoluminol reduced NETosis triggered by amoebic trophozoites at all ratios tested, showing significant differences with respect to the control (Figure 8B). Isoluminol also reduced PMA- and A23187-induced NETosis; however, NETosis was not completely abolished.

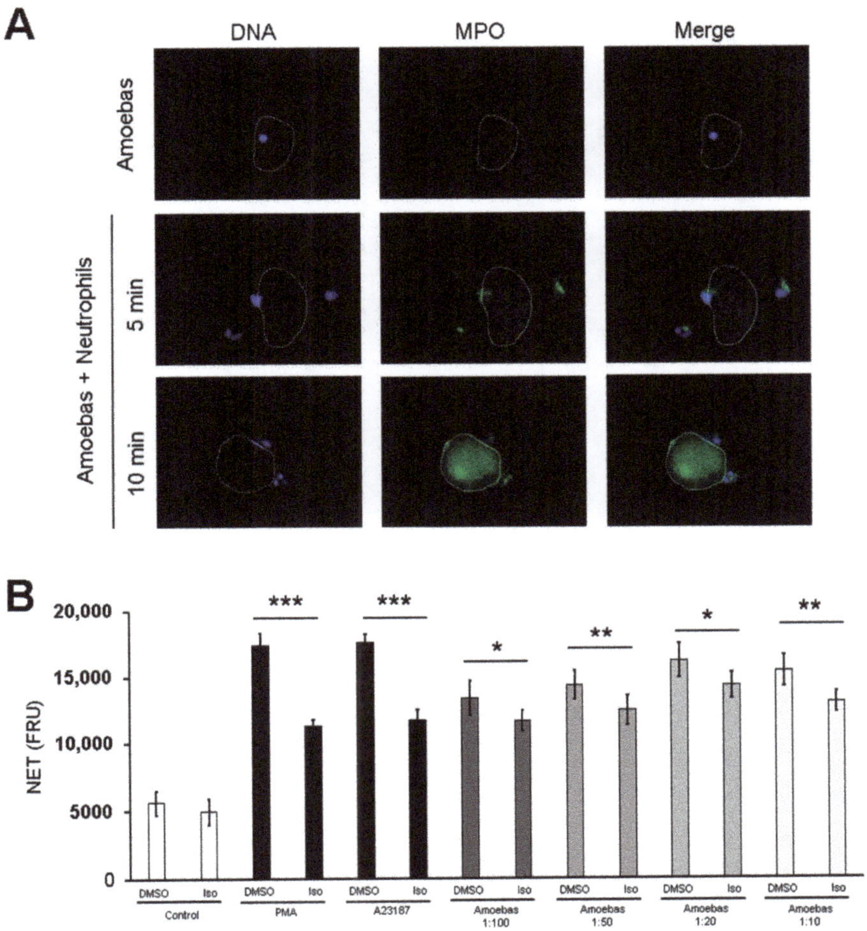

Figure 8. Extracellular MPO activity is required for NETosis induced by *E. histolytica*. (**A**) Neutrophils (2×10^5) were co-cultured with 1×10^4 *E. histolytica* trophozoites during 5 or 10 min. Cells were fixed and immunofluorescence was performed using anti-MPO antibody followed by anti-mouse IgG-FITC secondary antibody. DNA was counterstained with DAPI. Amoebas alone were used as a control. Trophozoites are indicated by doted white lines. Images were taken at 100× magnification. Scale bar 100 µm. (**B**) Neutrophils (1×10^5) were pretreated with 50, 100 and 200 µM isoluminol (Iso) or DMSO for 30 min. Posteriorly, cells were transferred to RPMI-1640 medium added with 5% FBS and 500 nM SYTOX® Green and then stimulated with PMA (50 nM), A23187 (10 µM) or *E. histolytica* trophozoites at ratios of 1:100, 1:50, 1:20 and 1:10. Fluorescence was read after 4 h. NET amount is expressed in fluorescence relative units (FRU). Values are means ± SD of three independent experiments. * $p < 0.05$, ** $p < 0.01$, *** $p < 0.001$.

3.8. E. histolytica-Induced NETosis Occurs Independently of Mitochondrial Derived ROS

We decided to explore whether mitochondrial ROS are produced during the NETosis triggered by *E. histolytica* trophozoites. As expected, PMA did not induce mitochondrial ROS, whereas the calcium ionophore A23187 induced a significant increase of these molecules. It is noteworthy that amoebas induced mitochondrial ROS in neutrophils in a dose-dependent manner (Figure 9A). To determine whether mitochondrial ROS are necessary for amoebic-induced NETosis, we used the specific scavenger mitoTEMPO, which did not affect PMA-induced NETosis. As shown in Figure 9B, NETosis was not affected by mitoTEMPO at any ratio tested, suggesting that mitochondrial ROS are not involved.

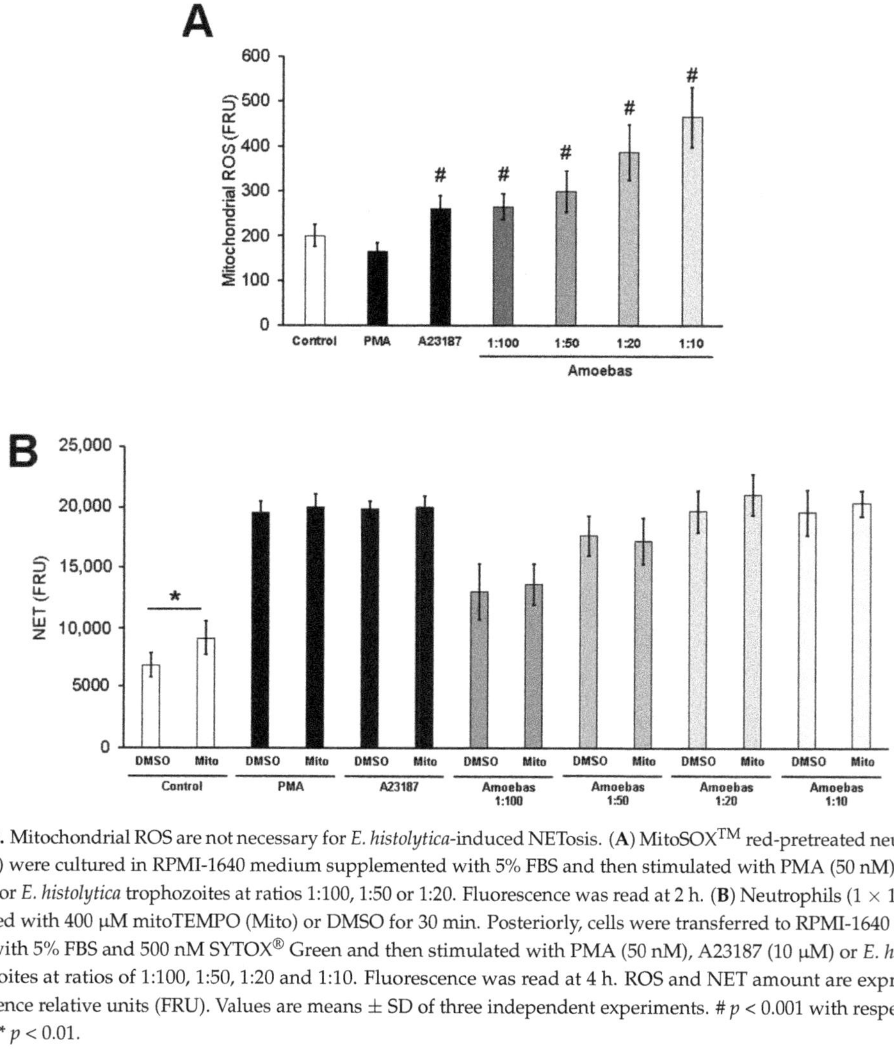

Figure 9. Mitochondrial ROS are not necessary for *E. histolytica*-induced NETosis. (**A**) MitoSOX^{TM} red-pretreated neutrophils (1×10^5) were cultured in RPMI-1640 medium supplemented with 5% FBS and then stimulated with PMA (50 nM), A23187 (10 μM) or *E. histolytica* trophozoites at ratios 1:100, 1:50 or 1:20. Fluorescence was read at 2 h. (**B**) Neutrophils (1×10^5) were pretreated with 400 μM mitoTEMPO (Mito) or DMSO for 30 min. Posteriorly, cells were transferred to RPMI-1640 medium added with 5% FBS and 500 nM SYTOX® Green and then stimulated with PMA (50 nM), A23187 (10 μM) or *E. histolytica* trophozoites at ratios of 1:100, 1:50, 1:20 and 1:10. Fluorescence was read at 4 h. ROS and NET amount are expressed in fluorescence relative units (FRU). Values are means ± SD of three independent experiments. # $p < 0.001$ with respect to the control, * $p < 0.01$.

4. Discussion

Neutrophil extracellular traps (NETs) were initially described by Brinkmann et al. [1] as a novel effector mechanism used by neutrophils to entrap and kill bacteria. Since then, many works have explored the mechanism underlying the DNA extrusion to extracellular space, process known as NETosis. Fuchs et al. [27] provided one of the first approaches to understand how NET release takes place, showing that oxidative metabolism directed by NADPH oxidase is involved. Nevertheless, the finding of calcium ionophores triggering NETosis opened the possibility that other mechanisms could also lead to the release of DNA, as they do not require the activity of an NADPH oxidase, but rather require PAD4 activity [44]. The case of NETosis induced by the *E. histolytica* trophozoites is intriguing, as we showed that the process occurs through a non-classical mechanism, independent of NADPH-ROS and PAD4 activity [41,45]. In this work, we performed a set of experiments to continue with the characterization of the amoeba-induced NETosis and found that it is dependent on ROS from *E. histolytica* trophozoites and on the activity of the MPO from neutrophils present in the surface of the parasites.

Previously, we described that amoebic trophozoites triggered NETosis on human neutrophils when co-incubated at trophozoite:neutrophil ratios of 1:20 [41,45–47]. Here, we demonstrated that lower amounts of amoebas (ratios 1:100 and 1:50) were also capable of leading NET release in a dose-dependent manner, whereas ratios higher than 1:20 did not induce more NETs. This result indicated that *E. histolytica* trophozoites are one of the most potent parasites to induce NETosis, since other protozoa require greater numbers to trigger significant DNA release. Thus, *Toxoplasma gondii* was used at MOI of 5:1, *Trypanosoma cruzi* at a 1:1 ratio and *Leishmania chagasi*, *L. major* or *L. amazonensis* at 10:1 to 1:1 ratios [12,48–50]. In contrast, we observed that 1 amoeba per 100 neutrophils is sufficient to induce NETosis. The reason is unknown but the size of the parasites, and therefore the density of NETosis triggering molecules, could be involved, since *T. gondii* and *Leishmania* forms (amastigotes and promastigotes) do not exceed 15 μm in length and *T. cruzi* trypomastigotes measure 12–30 μm, which is small compared with *E. histolytica* trophozoites measuring up to 60 μm [34,51–53]. While neutrophils can phagocyte small parasites enlisted above [54–56], they cannot phagocyte amoebas. Instead, we have observed that trophozoites engulf these leukocytes [45]. Neutrophils probably sense the pathogen size through dectin-1, to decide between phagocytosis or NETosis by sequestering of NE [57]. During phagocytosis, NE is moved to the phagolysosome compartment. In contrast, during NETosis, NE is guided to the nucleus for chromatin decondensation. This correlates with our previous observation in which NE is translocated to nuclei during the neutrophil-amoeba interaction [41]. It is conceivable that small parasites mainly drive neutrophils towards phagocytosis, whereas greater pathogens, such as amoebic trophozoites, preferentially induce NETosis.

NETosis mechanisms are generally divided into two groups: dependent on NADPH oxidase activity and independent of NADPH oxidase activity. Fuchs et al. [27] was the first to report that NADPH oxidase inhibition prevented NET release by PMA, and different authors have described the same mechanism for other stimuli [58–60]. Later, it was described that calcium ionophores trigger NETosis independently of NADPH oxidase activity, but this mechanism requires PAD4 [44]. We previously reported that *E. histolytica* trophozoites induce NETosis by a non-classical mechanism, independent of NADPH oxidase and PAD4 activities, since apocynin and GSK484, as respective inhibitors, failed to reduce the NET amount. Our previous observations also showed that amoebic trophozoites at 1:20 ratio completely suppressed the oxidative burst in neutrophils [41,46], which has also been reported by others [61]. Interestingly, in this work we found that lower numbers of amoebic trophozoites (1:50 and 1:100 ratios) did not completely abolish ROS generation in these leucocytes. This suggested that neutrophil ROS inhibition by amoebas depends on the density of parasites trespassing a threshold that may cause citrullination of proteins. Accordingly, Zhou et al. [62] showed that dysregulated calcium influx in neutrophils activates PAD4 that citrullinates the cytoplasmatic units $p47^{phox}$ and $p67^{phox}$, blocking the assemble of the NADPH oxidase complex and, in turn, preventing ROS generation. In this context, we previously reported that *E. histolytica* trophozoites trigger calcium influx on human neutrophils and when incubated at a 1:20 ratio [46] and citrullinated proteins were detected [41], which would cause the inactivation of NADPH oxidase. When smaller numbers of trophozoites are confronted, this process may not happen; however, additional studies are required on this. It is worth mentioning that PAD4-independent citrullination of proteins has also been observed during the NETosis induced by *Candida albicans* [63].

Previously, we showed that heat-killed and fixed trophozoites failed to induce NETosis, which was confirmed in this work (Figure 3A) [41]. This data suggests that some products of the *E. histolytica* trophozoites metabolism are responsible for inducing NETosis. In this context, ROS derived from pathogens have been identified as molecules that lead NET release independently of ROS produced by neutrophils [33,43]. Noteworthy, we previously showed that *Entamoeba dispar*, a non-pathogen human amoeba, does not produce ROS and does not trigger NETosis [43,47]. Since most studies suggest that NETosis requires some source of ROS that does not come from neutrophils in this case, we considered the possibility that ROS produced by *E. histolytica* trophozoites induce NETosis. Here we report that viable trophozoites produced basal ROS levels, which agrees with a previous report [64], whereas amoebas killed by heat or fixation with paraformaldehyde produced scarce ROS (Figure 3B). Even though cell death processes have been associated to an increase in ROS production [65], the low level detected by us could be explained by death-inducing agents that were used, since both reduce enzymatic activity and formaldehyde, causing protein cross-linking and the development of heat denaturalizing proteins [66].

When we treated H_2DCFDA-stained amoebas with hydrogen peroxide, they exhibited a stronger fluorescence (Supplementary Figure S3), suggesting that H_2DCFDA can be used as an indicator for hydrogen peroxide. Using this approach, we found that pretreatment of amoebas with pyrocatechol, a ROS scavenger of hydrogen peroxide [67,68], reduced the hydrogen peroxide detected in viable amoebas, which is produced as a response to detoxify oxygen through diverse enzymes including NADPH:flavin oxidoreductase (Eh43), thioredoxin reductase (TrxR), NADPH-dependent oxidoreductases (NO1/2) or Fe-superoxide dismutase (FeSOD) [64,69]. The hydrogen peroxide reduction in trophozoites by pretreatment with pyrocatechol impacted the NETosis directly, since a smaller amount of DNA was detected in the extracellular medium. Moreover, NETosis was completely abolished when amoebas were pretreated with pyrocatechol at 200 μM in 1:100 and 1:50 ratios. In accordance with this, pyrocatechol blocked NET release induced by PMA, which is triggered by ROS produced in neutrophils. All these data indicate that amoebic trophozoites, instead of neutrophils, are the source of ROS responsible for leading NETosis involving this parasite. This is the first report regarding the importance of ROS from *E. histolytica* trophozoites for NETosis.

Kenny et al. [33] proposed that hydrogen peroxide produced by *C. albicans* was able to enter to neutrophils to trigger NETosis. In our case, however, the addition of catalase (which possesses high specificity for hydrogen peroxide) [70] to the media in the cocultures amoeba-neutrophil failed to reduce NETosis. In contrast, catalase reduced significantly NET release induced by PMA, which has been linked to extracellular production of hydrogen peroxide by NADPH oxidase in the plasma membrane [27,71]. As amoebic hydrogen peroxide is important for amoeba-induced NETosis but catalase did not affect the process, we proposed that another ROS, probable derived from hydrogen peroxide released by the trophozoites but produced in the extracellular media, might be involved. This ROS may be produced by the activity of a neutrophil product released very early after the contact with amoebas. In this context, neutrophil MPO has been related to some mechanisms of NETosis [72], mainly with non-phagocytosed stimuli [73,74]. This enzyme produces hypochlorous acid (HClO) from hydrogen peroxide and chloride during oxidative burst in neutrophils [75]. Although luminol can react with other oxidants, Gross et al. [76] reported that its luminescence depends substantially on MPO activity. Here we observed that luminol-pretreated trophozoites when incubated with neutrophils exhibit MPO activity denoted by an increase in luminol signal. It is noteworthy that the MPO maximum activity was detected when the DNA extrusion started (approximately 20 min), suggesting that this could be the triggering stimulus. In addition, amoebas did not produce a luminol signal in the absence of neutrophil, indicating that the signal detected in amoebas corresponds to the activity of neutrophil´s MPO and not to other molecules produced by trophozoites or neutrophils. The role of MPO activity on amoeba-induced NETosis was confirmed by taking advantage of the luminol ability to scavenge HClO [77]. Luminol, at all concentrations tested here, reduced NET release induced by the amoebic trophozoites at 1:50, 1:20 and 1:10 ratios. As pretreatment of amoebas with luminol showed similar results but increased ROS levels in the parasite, the data together indicated that NET reduction was due to the scavenging activity of luminol on MPO derived ROS (HClO), instead of a decrease in ROS from amoebas. The reason why luminol increased the ROS of amoeba is unknown. We suspect that luminol can cause intensive stress in the trophozoites without affecting their viability, but this is a subject for further study in our laboratory. It is also interesting that luminol was unable to reduce NETosis at a 1:100 ratio and only reduced but did not abolish NETosis at higher ratios, suggesting that other mechanisms take place to compensate for NET release depending on the culture conditions. In this context, it has been reported that some pathogens such as *L. amazonensis* and *C. glabrata* can trigger NETosis by different mechanisms [50,59].

The role of mitochondrial ROS (mitROS) in amoeba-induced NETosis was also explored. Douda et al. [28] stated that ROS derived from mitochondria are required for the NETosis induced by the calcium ionophores ionomycin and A23187, results that have been replicated [78]. Since then, other NET inducers such as *Leishmania* parasites or UV light have been shown to require mitROS [50,79]. Here we observed for the first time that *E. histolytica* trophozoites induce production of mitROS on human neutrophils in a dose dependent manner. How they are generated is unknown but the recognition of amoebic LPPG by TLR2 and TLR4 [80] and the stimulation of calcium influx plus a pathological stimulus (such as PAAR detected through TLRs) could be involved, as suggested elsewhere [81,82]. Although mitROS were detected in neutrophils in contact with amoebas, the scavenger mitoTEMPO failed to reduce NET releases, suggesting that they do not participate in the process.

Immunofluorescence performed on amoeba-neutrophil cocultures exhibited that DNA from neutrophils is released as aggregated cloudy NETs in accordance with a previous classification [83]. We observed extensive areas covered by DNA that entrap trophozoites (Figure 1B). Differing from other stimuli such as LPS, monosodium urate crystals or S-nitroso-N-acetyl-D,L-penicillamine, [84–86], NETs induced by trophozoites do not exhibit an homogeneous distribution of MPO and NE; in contrast, these proteins are usually visualized as spots located in reduced areas, suggesting that at least part of the neutrophil MPO and NE proteins are released by degranulation, and bound to trophozoites before, or at the same time, as NETs [47], which can explain the scarce proteins associated with DNA. Binding of MPO to *E. histolytica* trophozoites has be described previously by Pacheco–Yépez et al. [87]. They reported that purified MPO interacts with amoebic trophozoites, leading to morphological changes and loss of viability, which were also observed by us in amoebas entrapped in NETs [47].

Finally, detection of MPO covered trophozoites during neutrophil-amoeba interaction raise a question about the origin of this enzymatic activity. Therefore, we decided to use isoluminol, a hydrophobic isomer of luminol, to explore the scavenging of extracellular HClO [88]. Isoluminol significantly reduced NET amount released in response to *E. histolytica* trophozoites at the same level as luminol, indicating that extracellular MPO activity is the responsible for NETosis. This result is in accordance with previous observations indicating that exogen HClO and hypochlorite are sufficient to induce NETosis on human neutrophils [31,89]. Although the role of MPO activity in NETosis has been controversial, with some reports indicating that is dispensable [90,91] and others indicating that is required [73,77], our results support the latter. Taken together, the mechanism that we propose for amoebic-induced NETosis is shown in Figure 10.

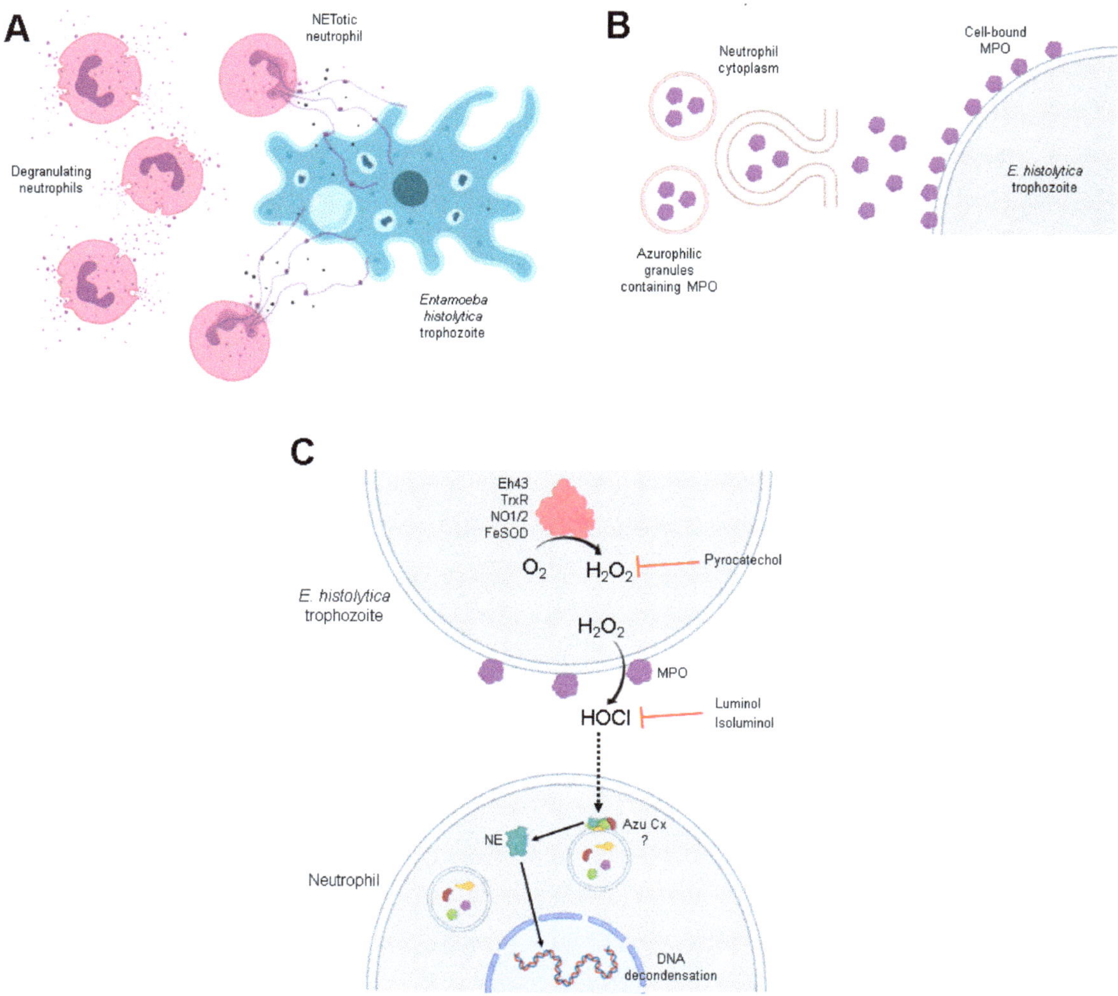

Figure 10. Mechanism proposed for NETosis induced by *E. histolytica*. (**A**) Amoebic trophozoites lead neutrophils to degranulation and NET release. (**B**) MPO derived from azurophilic cytoplasmatic granules is released and it binds to the surface of trophozoites. (**C**) During the oxidative metabolism of *E. histolytica*, H_2O_2 is generated to detoxify O_2 by diverse enzymes such as Eh43, TrxR, NO1/2 or FeSOD. H_2O_2 is rapidly converted to HClO through MPO bounded to cell surface of amoebas. HClO probably enters neutrophils by an unknown mechanism and it starts NETosis, promoting NE translocation to the nucleus. Pyrocatechol blocked NETosis scavenge H_2O_2 inside amoebas, while luminol and isoluminol react with HClO. Image was made in BioRender.com. Eh43 (NADPH:flavin oxidoreductase), TrxR (thioredoxin reductase), NO1/2 (NADPH-dependent oxidoreductases), FeSOD (Fe-superoxide dismutase), H_2O_2 (hydrogen peroxide), HOCl (hypochlorous acid), Azu Cx (azurosome complex), NE (neutrophil elastase).

5. Conclusions

In conclusion, our data show that after detecting each other, neutrophils rapidly transfer MPO to the surface of the amoeba, and that its activity could be processing the ROS of the amoebae that are necessary for the optimal induction of NETosis. In the context of parasite-neutrophil interaction, this observation is very relevant, as many parasites can inhibit respiratory burst in neutrophils but NETosis is easily induced. In general, this contribution supports the notion that parasite-induced NETosis is a very complex process, involving the rapid and active exchange of molecules between parasite and host cells.

Supplementary Materials: The following are available online at https://www.mdpi.com/article/10.3390/antiox10060974/s1, Figure S1: Pyrocatechol and luminol do not affect the viability of *E. histolytica* trophozoites, Figure S2: Catalase failed to reduce NET induced by *E. histolytica*, Figure S3: H_2DCFDA detects hydrogen peroxide in trophozoites.

Author Contributions: Conceptualization, J.C.C., C.R. and C.D.-G.; methodology, C.D.-G., J.F.J.-R. and M.N.; software, C.D.-G.; validation, C.D.-G. and J.F.J.-R.; formal analysis, S.M.-C., J.P.L., J.C.C. and C.D.-G.; investigation, C.D.-G. and J.C.C.; resources, J.C.C.; data curation, J.C.C., S.M.-C. and J.P.L.; writing, C.D.-G. and J.C.C.; visualization, J.C.C. and C.R.; supervision, J.C.C. and C.R.; project administration, J.C.C.; funding acquisition, J.C.C. All authors have read and agreed to the published version of the manuscript.

Funding: This research was funded by Consejo Nacional de Ciencia y Tecnología (CONACyT), grant numbers 284830 (to J.C.C.) and 254434 (to C.R.) and Programa de Apoyo a Proyectos de Investigación e Innovación Tecnológica (PAPIIT-UNAM), grant number IN208020 (to J.C.C.).

Institutional Review Board Statement: The study was conducted according to the guidelines of the Declaration of Helsinki, and approved by the Ethical Committee for Studies on Humans of the Instituto de Investigaciones Biomédicas, UNAM (Ethical approved number: FMED/CI/RGG/13 January 2008).

Informed Consent Statement: Informed consent was obtained from all subjects involved in the study.

Data Availability Statement: Data is contained within the article and supplementary material.

Acknowledgments: César Díaz-Godínez is a student of the Programa de Ciencias Bioquímicas, UNAM, and is a recipient of a scholarship from Consejo Nacional de Ciencia y Tecnología (CONACyT), Mexico (742064). César Díaz-Godínez thanks Alan Javier Montoya–Hernández for providing support during preparation of the manuscript.

Conflicts of Interest: The authors declare no conflict of interest.

References

1. Brinkmann, V. Neutrophil extracellular traps kill bacteria. *Science* **2004**, *303*, 1532–1535. [CrossRef]
2. Urban, C.F.; Reichard, U.; Brinkmann, V.; Zychlinsky, A. Neutrophil extracellular traps capture and kill *Candida albicans* and hyphal forms. *Cell. Microbiol.* **2006**, *8*, 668–676. [CrossRef]
3. Ramos-Kichik, V.; Mondragón-Flores, R.; Mondragón-Castelán, M.; Gonzalez-Pozos, S.; Muñiz-Hernandez, S.; Rojas-Espinosa, O.; Chacón-Salinas, R.; Estrada-Parra, S.; Estrada-García, I. Neutrophil extracellular traps are induced by *Mycobacterium tuberculosis*. *Tuberculosis* **2009**, *89*, 29–37. [CrossRef] [PubMed]
4. McCormick, A.; Heesemann, L.; Wagener, J.; Marcos, V.; Hartl, D.; Loeffler, J.; Heesemann, J.; Ebel, F. NETs formed by human neutrophils inhibit growth of the pathogenic mold *Aspergillus fumigatus*. *Microbes Infect.* **2010**, *12*, 928–936. [CrossRef]
5. Li, P.; Li, M.; Lindberg, M.R.; Kennett, M.J.; Xiong, N.; Wang, Y. PAD4 is essential for antibacterial innate immunity mediated by neutrophil extracellular traps. *J. Exp. Med.* **2010**, *207*, 1853–1862. [CrossRef] [PubMed]
6. McDonald, B.; Davis, R.P.; Kim, S.J.; Tse, M.; Esmon, C.T.; Kolaczkowska, E.; Jenne, C.N. Platelets and neutrophil extracellular traps collaborate to promote intravascular coagulation during sepsis in mice. *Blood* **2017**, *129*, 1357–1367. [CrossRef]
7. Wang, H.; Wang, C.; Zhao, M.H.; Chen, M. Neutrophil extracellular traps can activate alternative complement pathways. *Clin. Exp. Immunol.* **2015**, *181*, 518–527. [CrossRef] [PubMed]
8. Leffler, J.; Martin, M.; Gullstrand, B.; Tydén, H.; Lood, C.; Truedsson, L.; Bengtsson, A.A.; Blom, A.M. Neutrophil extracellular traps that are not degraded in systemic lupus erythematosus activate complement exacerbating the disease. *J. Immunol.* **2012**, *188*, 3522–3531. [CrossRef]
9. Vorobjeva, N.V.; Pinegin, B.V. Neutrophil extracellular traps: Mechanisms of formation and role in health and disease. *Biochemistry* **2014**, *79*, 1286–1296. [CrossRef]
10. Yang, H.; Biermann, M.H.; Brauner, J.M.; Liu, Y.; Zhao, Y.; Herrmann, M. New insights into neutrophil extracellular traps: Mechanisms of formation and role in inflammation. *Front. Immunol.* **2016**, *7*, 302. [CrossRef] [PubMed]
11. Sollberger, G.; Tilley, D.O.; Zychlinsky, A. Neutrophil extracellular traps: The biology of chromatin externalization. *Dev. Cell* **2018**, *44*, 542–553. [CrossRef]
12. Sousa-Rocha, D.; Thomaz-Tobias, M.; Diniz, L.F.A.; Souza, P.S.S.; Pinge-Filho, P.; Toledo, K.A. *Trypanosoma cruzi* and its soluble antigens induce NET release by stimulating toll-like receptors. *PLoS ONE* **2015**, *10*, e0139569. [CrossRef]

13. Wu, S.Y.; Weng, C.L.; Jheng, M.J.; Kan, H.W.; Hsieh, S.T.; Liu, F.T.; Wu-Hsieh, B.A. *Candida albicans* triggers NADPH oxidaseindependent neutrophil extracellular traps through dectin-2. *PLoS Pathog.* **2019**, *15*, e1008096. [CrossRef] [PubMed]
14. Raftery, M.J.; Lalwani, P.; Krautkrämer, E.; Peters, T.; Scharffetter-Kochanek, K.; Krüger, R.; Hofmann, J.; Seeger, K.; Krüger, D.H.; Schönrich, G. B2 Integrin mediates hantavirus-induced release of neutrophil extracellular traps. *J. Exp. Med.* **2014**, *211*, 1485–1497. [CrossRef] [PubMed]
15. Alemán, O.R.; Mora, N.; Cortes-Vieyra, R.; Uribe-Querol, E.; Rosales, C. Differential use of human neutrophil Fc γ receptors for inducing neutrophil extracellular trap formation. *J. Immunol. Res.* **2016**, *2016*, 1–17. [CrossRef]
16. Hakkim, A.; Fuchs, T.A.; Martinez, N.E.; Hess, S.; Prinz, H.; Zychlinsky, A.; Waldmann, H. Activation of the Raf-MEK-ERK pathway is required for neutrophil extracellular trap formation. *Nat. Chem. Biol.* **2011**, *7*, 75–77. [CrossRef] [PubMed]
17. Itakura, A.; McCarty, O.J.T. Pivotal role for the mTOR pathway in the formation of neutrophil extracellular traps via regulation of autophagy. *Am. J. Physiol. Cell Physiol.* **2013**, *305*, 348–354. [CrossRef] [PubMed]
18. DeSouza-Vieira, T.; Guimarães-Costa, A.; Rochael, N.C.; Lira, M.N.; Nascimento, M.T.; Lima-Gomez, P.D.S.; Mariante, R.M.; Persechini, P.M.; Saraiva, E.M. Neutrophil extracellular traps release induced by *Leishmania*: Role of PI3Kγ, ERK, PI3Kσ, PKC, and [Ca^{2+}]. *J. Leukoc. Biol.* **2016**, *100*, 801–810. [CrossRef]
19. Thiama, H.R.; Wong, S.L.; Qiu, R.; Kittisopikul, M.; Vahabikashi, A.; Goldman, A.E.; Goldman, R.D.; Wagner, D.D.; Waterman, C.M. NETosis proceeds by cytoskeleton and endomembrane disassembly and PAD4-mediated chromatin decondensation and nuclear envelope rupture. *Proc. Natl. Acad. Sci. USA* **2020**, *117*, 7326–7337. [CrossRef]
20. Neubert, E.; Meyer, D.; Rocca, F.; Günay, G.; Kwaczala-Tessmann, A.; Grandke, J.; Senger-Sander, S.; Geisler, C.; Egner, A.; Schön, M.P.; et al. Chromatin swelling drives neutrophil extracellular trap release. *Nat. Commun.* **2018**, *9*, 1–13. [CrossRef]
21. Yousefi, S.; Mihalache, C.; Kozlowski, E.; Schmid, I.; Simon, H.U. Viable neutrophils release mitochondrial DNA to form neutrophil extracellular traps. *Cell Death Differ.* **2009**, *16*, 1438–1444. [CrossRef] [PubMed]
22. Pilsczek, F.H.; Salina, D.; Poon, K.K.H.; Fahey, C.; Yipp, B.G.; Sibley, C.D.; Robbins, S.M.; Green, F.H.Y.; Surette, M.G.; Sugai, M.; et al. A novel mechanism of rapid nuclear neutrophil extracellular trap formation in response to *Staphylococcus aureus*. *J. Immunol.* **2010**, *185*, 7413–7425. [CrossRef]
23. Nauseef, W.M.; Kubes, P. Pondering neutrophil extracellular traps with healthy skepticism. *Cell Microbiol.* **2016**, *18*, 1349–1357. [CrossRef]
24. Boeltz, S.; Amini, P.; Anders, H.J.; Andrade, F.; Bilyy, R.; Chatfield, S.; Cichon, I.; Clancy, D.M.; Desai, J.; Dumych, T.; et al. To NET or not to NET: Current opinions and state of the science regarding the formation of neutrophil extracellular traps. *Cell Death Differ.* **2019**, *26*, 395–408. [CrossRef]
25. Stoiber, W.; Obermayer, A.; Steinbacher, P.; Krautgartner, W.-D. The role of reactive oxygen species (ROS) in the formation of extracellular traps (ETs) in humans. *Biomolecules* **2015**, *5*, 702–723. [CrossRef]
26. de Bont, C.M.; Koopman, W.J.H.; Boelens, W.C.; Pruijn, G.J.M. Stimulus-dependent chromatin dynamics, citrullination, calcium signalling and ROS production during NET formation. *Biochim. Biophys. Acta Mol. Cell Res.* **2018**, *1865*, 1621–1629. [CrossRef] [PubMed]
27. Fuchs, T.A.; Abed, U.; Goosmann, C.; Hurwitz, R.; Schulze, I.; Wahn, V.; Weinrauch, Y.; Brinkmann, V.; Zychlinsky, A. Novel cell death program leads to neutrophil extracellular traps. *J. Cell Biol.* **2007**, *176*, 231–241. [CrossRef] [PubMed]
28. Douda, D.N.; Khan, M.A.; Grasemann, H.; Palaniyar, N. SK3 channel and mitochondrial ROS mediate NADPH oxidase-independent NETosis induced by calcium influx. *Proc. Natl. Acad. Sci. USA* **2015**, *112*, 2817–2822. [CrossRef]
29. Pieterse, E.; Rother, N.; Yanginlar, C.; Gerretsen, J.; Boeltz, S.; Munoz, L.E.; Herrmann, M.; Pickkers, P.; Hilbrands, L.B.; Van Der Vlag, J. Cleaved N-terminal histone tails distinguish between NADPH oxidase (NOX)-dependent and NOX-independent pathways of neutrophil extracellular trap formation. *Ann. Rheum. Dis.* **2018**, *77*, 1790–1798. [CrossRef]
30. Patel, S.; Kumar, S.; Jyoti, A.; Srinag, B.S.; Keshari, R.S.; Saluja, R.; Verma, A.; Mitra, K.; Barthwal, M.K.; Krishnamurthy, H.; et al. Nitric oxide donors release extracellular traps from human neutrophils by augmenting free radical generation. *Nitric Oxide Biol. Chem.* **2010**, *22*, 226–234. [CrossRef]
31. Akong-Moore, K.; Chow, O.A.; von Köckritz-Blickwede, M.; Nizet, V. Influences of chloride and hypochlorite on neutrophil extracellular trap formation. *PLoS ONE* **2012**, *7*, e42984. [CrossRef] [PubMed]
32. McBee, M.E.; Chionh, Y.H.; Sharaf, M.L.; Ho, P.; Cai, M.W.L.; Dedon, P.C. Production of superoxide in bacteria is stress and cell state-dependent: A gating-optimized flow cytometry method that minimizes ROS measurement artifacts with fluorescent dyes. *Front. Microbiol.* **2017**, *8*, 459. [CrossRef] [PubMed]
33. Kenny, E.F.; Muth, A.; Mondal, S.; Herzig, A.; Kru, R.; Thompson, P.R.; Brinkmann, V.; Von Bernuth, H.; Zychlinsky, A. Diverse stimuli engage different neutrophil extracellular trap pathways. *eLife* **2017**, *6*, e24437. [CrossRef] [PubMed]
34. Carrero, J.C.; Reyes-López, M.; Serrano-Luna, J.; Shibayama, M.; Unzueta, J.; León-Sicairos, N.; de la Garza, M. Intestinal amoebiasis: 160 years of its first detection and still remains as a health problem in developing countries. *Int. J. Med. Microbiol.* **2020**, *310*, 151358. [CrossRef] [PubMed]
35. López, M.C.; León, C.M.; Fonseca, J.; Reyes, P.; Moncada, L.; Olivera, M.J.; Ramírez, J.D. Molecular epidemiology of *Entamoeba*: First description of *Entamoeba moshkovskii* in a rural area from central Colombia. *PLoS ONE* **2015**, *10*, e0140302. [CrossRef]
36. Al-Areeqi, M.A.; Sady, H.; Al-Mekhlafi, H.M.; Anuar, T.S.; Al-Adhroey, A.H.; Atroosh, W.M.; Dawaki, S.; Elyana, F.N.; Nasr, N.A.; Ithoi, I.; et al. First molecular epidemiology of *Entamoeba histolytica, E. dispar* and *E. moshkovskii* infections in Yemen: Different species-specific associated risk factors. *Trop. Med. Int. Health* **2017**, *22*, 493–504. [CrossRef] [PubMed]

37. Hegazi, M.A.; Patel, T.A.; El-Deek, B.S. Prevalence and characters of *Entamoeba histolytica* infection in Saudi infants and children admitted with diarrhea at 2 main hospitals at south Jeddah: A re-emerging serious infection with unusual presentation. *Braz. J. Infect. Dis.* **2013**, *17*, 32–40. [CrossRef] [PubMed]
38. Campos-Rodríguez, R.; Gutiérrez-Meza, M.; Jarillo-Luna, R.A.; Drago-Serrano, M.E.; Abarca-Rojano, E.; Ventura-Juárez, J.; Cárdenas-Jaramillo, L.M.; Pacheco-Yepez, J. A review of the proposed role of neutrophils in rodent amebic liver abscess models. *Parasite* **2016**, *6*, 14. [CrossRef]
39. Velazquez, C.; Shibayama-Salas, M.; Aguirre-Garcia, J.; Tsutsumi, V.; Calderon, J. Role of neutrophils in innate resistance to *Entamoeba histolytica* liver infection in mice. *Parasite Immunol.* **1998**, *20*, 255–262. [CrossRef]
40. Olivos-García, A.; Carrero, J.C.; Ramos, E.; Nequiz, M.; Tello, E.; Montfort, I.; Pérez-Tamayo, R. Late experimental amebic liver abscess in hamster is inhibited by cyclosporine and N-acetylcysteine. *Exp. Mol. Pathol.* **2007**, *82*, 310–315. [CrossRef]
41. Díaz-Godínez, C.; Fonseca, Z.; Néquiz, M.; Laclette, J.P.; Rosales, C. *Entamoeba histolytica* trophozoites induce a rapid non-classical NETosis mechanism independent of NOX2-derived reactive oxygen species and PAD4 activity. *Front. Cell. Infect. Microbiol.* **2018**, *8*, 1–17. [CrossRef] [PubMed]
42. García–García, E.; Uribe-Querol, E.; Rosales, C. A simple and efficient method to detect nuclear factor activation in human neutrophils by flow cytometry. *J. Vis. Exp.* **2013**, *74*, 50410. [CrossRef] [PubMed]
43. Díaz-Godínez, C.; Martínez-Flores, A.; Argüello-García, R.; Olivos-García, A.; Néquiz-Avendaño, M.; Carrero, J.C. Role of extracellular traps promoted by intestinal parasites. Relationship with virulence. In *Eukaryome Impact on Human Intestine Homeostasis and Mucosal Immunology*, 1st ed.; Guillén, N., Ed.; Springer: Cham, Switzerland, 2020; Volume 1, pp. 171–192.
44. Wang, Y.; Li, M.; Stadler, S.; Correll, S.; Li, P.; Wang, D.; Hayama, R.; Leonelli, L.; Han, H.; Grigoryev, S.A.; et al. Histone hypercitrullination mediates chromatin decondensation and neutrophil extracellular trap formation. *J. Cell Biol.* **2009**, *184*, 205–213. [CrossRef] [PubMed]
45. Ávila, E.E.; Salaiza, N.; Pulido, J.; Rodríguez, M.C.; Díaz-Godínez, C.; Laclette, J.P.; Becker, I.; Carrero, J.C. *Entamoeba histolytica* trophozoites and lipopeptidophosphoglycan trigger human neutrophil extracellular traps. *PLoS ONE* **2016**, *11*, e158979. [CrossRef]
46. Fonseca, Z.; Díaz-Godínez, C.; Mora, N.; Alemán, O.R.; Uribe-Querol, E.; Carrero, J.C.; Rosales, C. *Entamoeba histolytica* induce signaling via Raf/MEK/ERK for neutrophil extracellular trap (NET) formation. *Front. Cell. Infect. Microbiol.* **2018**, *8*, 226. [CrossRef] [PubMed]
47. Fonseca, Z.; Uribe-Querol, E.; Díaz-Godínez, C.; Carrero, J.C.; Rosales, C. Pathogenic *Entamoeba histolytica*, but not *Entamoeba dispar*, induce neutrophil extracellular trap (NET) formation. *J. Leukoc. Biol.* **2019**, *105*, 1167–1181. [CrossRef] [PubMed]
48. Abdallah, D.S.A.; Lin, C.; Ball, C.J.; King, M.R.; Duhamel, G.E.; Denkers, E.Y. *Toxoplasma gondii* triggers release of human and mouse neutrophil extracellular traps. *Infect. Immun.* **2012**, *80*, 768–777. [CrossRef]
49. Guimarães-Costa, A.B.; Nascimento, M.T.C.; Froment, G.S.; Soares, R.P.P.; Morgado, F.N.; Conceição-Silva, F.; Saraiva, E.M. *Leishmania amazonensis* promastigotes induce and are killed by neutrophil extracellular traps. *Proc. Natl. Acad. Sci. USA* **2009**, *106*, 6748–6753. [CrossRef]
50. Rochael, N.C.; Guimarães-Costa, A.B.; Nascimento, M.T.C.; Desouza-Vieira, T.S.; Oliveira, M.P.; Garciae Souza, L.F.; Oliveira, M.F.; Saraiva, E.M. Classical ROS-dependent and early/rapid ROS-independent release of neutrophil extracellular traps triggered by *Leishmania* parasites. *Sci. Rep.* **2015**, *5*, 18302. [CrossRef]
51. Leishmaniasis. Available online: https://www.cdc.gov/parasites/leishmaniasis/ (accessed on 13 January 2021).
52. Chagas Disease. Available online: https://www.cdc.gov/parasites/chagas/index.html (accessed on 13 January 2021).
53. Toxoplasmosis. Available online: https://www.cdc.gov/dpdx/toxoplasmosis/index.html (accessed on 13 January 2021).
54. Lima, T.S.; Gov, L.; Lodoen, M.B. Evasion of human neutrophil-mediated host defense during *Toxoplasma gondii* infection. *mBio* **2018**, *9*, e02027-17. [CrossRef]
55. Carlsen, E.D.; Hay, C.; Henard, C.A.; Popov, V.; Garg, N.J.; Soong, L. *Leishmania amazonensis* amastigotes trigger neutrophil activation but resist neutrophil microbicidal mechanisms. *Infect. Immun.* **2013**, *81*, 3966–3974. [CrossRef]
56. de Andrade, M.F.; de Almeida, V.D.; de Souza, L.M.S.; Paiva, D.C.C.; Andrade, C.D.M.; de Medeiros Fernandes, T.A.A. Involvement of neutrophils in Chagas disease pathology. *Parasite Immunol.* **2018**, *40*, e12593. [CrossRef]
57. Branzk, N.; Lubojemska, A.; Hardison, S.E.; Wang, Q.; Gutierrez, M.G.; Brown, G.D.; Papayannopoulos, V. Neutrophils sense microbe size and selectively release neutrophil extracellular traps in response to large pathogens. *Nat. Immunol.* **2014**, *15*, 1017–1025. [CrossRef]
58. Röhm, M.; Grimm, M.J.; D'Auria, A.C.; Almyroudis, N.G.; Segal, B.H.; Urban, C.F. NADPH oxidase promotes neutrophil extracellular trap formation in pulmonary aspergillosis. *Infect. Immun.* **2014**, *82*, 1766–1777. [CrossRef] [PubMed]
59. Johnson, C.J.; Kernien, J.F.; Hoyer, A.R.; Nett, J.E. Mechanisms involved in the triggering of neutrophil extracellular traps (NETs) by *Candida glabrata* during planktonic and biofilm growth. *Sci. Rep.* **2017**, *7*, 13065. [CrossRef]
60. Khan, M.A.; Philip, L.M.; Cheung, G.; Vadakepeedika, S.; Grasemann, H.; Sweezey, N.; Palaniyar, N. Regulating NETosis: Increasing pH promotes NADPH oxidase-dependent NETosis. *Front. Med.* **2018**, *5*, 19. [CrossRef] [PubMed]
61. Arbo, A.; Hoefsloot, M.; Ignacio-Santos, I. *Entamoeba histolytica* inhibits the respiratory burst of polymorphonuclear leukocytes. *Arch. Investig. Med.* **1990**, *21*, 57–61.
62. Zhou, Y.; An, L.L.; Chaerkady, R.; Mittereder, N.; Clarke, L.; Cohen, T.S.; Chen, B.; Hess, S.; Sims, G.P.; Mustelin, T. Evidence for a direct link between PAD4-mediated citrullination and the oxidative burst in human neutrophils. *Sci. Rep.* **2018**, *8*, 15228. [CrossRef] [PubMed]

63. Guiducci, E.; Lemberg, C.; Küng, N.; Schraner, E.; Theocharides, A.P.A.; LeibundGut-Landmann, S. *Candida albicans*-induced NETosis is independent of peptidylarginine deiminase 4. *Front. Immunol.* **2018**, *9*, 1573. [CrossRef]
64. Ramos-Martínez, E.; Olivos-García, A.; Saavedra, E.; Nequiz, M.; Sánchez, E.C.; Tello, E.; El-Hafidi, M.; Saralegui, A.; Pineda, E.; Delgado, J.; et al. *Entamoeba histolytica*: Oxygen resistance and virulence. *Int. J. Parasitol.* **2009**, *39*, 693–702. [CrossRef] [PubMed]
65. Ghosh, A.S.; Dutta, S.; Raha, S. Hydrogen peroxide-induced apoptosis-like cell death in *Entamoeba histolytica*. *Parasitol. Int.* **2010**, *59*, 166–172. [CrossRef]
66. Kast, J.; Klockenbusch, C. Optimization of formaldehyde cross-linking for protein interaction analysis of non-tagged integrin β 1. *J. Biomed. Biotechnol.* **2010**, *2010*, 927585. [CrossRef]
67. Bendary, E.; Francis, R.R.; Ali, H.M.G.; Sarwat, M.I.; El Hady, S. Antioxidant and structure–activity relationships (SARs) of some phenolic and anilines compounds. *Ann. Agric. Sci.* **2013**, *58*, 173–181. [CrossRef]
68. Nagarajan, S.; Nagarajan, R.; Kumar, J.; Salemme, A.; Togna, A.R.; Saso, L.; Bruno, F. Antioxidant activity of synthetic polymers of phenolic compounds. *Polymers* **2020**, *12*, 1646. [CrossRef]
69. Olivos-García, A.; Saavedra, E.; Nequiz, M.; Santos, F.; Luis-García, E.R.; Gudiño, M.; Pérez-Tamayo, R. The oxygen reduction pathway and heat shock stress response are both required for *Entamoeba histolytica* pathogenicity. *Curr. Genet.* **2016**, *62*, 295–300. [CrossRef] [PubMed]
70. Glorieux, C.; Calderon, P.B. Catalase, a remarkable enzyme: Targeting the oldest antioxidant enzyme to find a new cancer treatment approach. *Biol. Chem.* **2017**, *398*, 1095–1108. [CrossRef] [PubMed]
71. Vorobjeva, N.V.; Chernyak, B.V. NETosis: Molecular mechanisms, role in physiology and pathology. *Biochemistry* **2020**, *85*, 1178–1190. [CrossRef] [PubMed]
72. Kirchner, T.; Mller, S.; Klinger, M.; Solbach, W.; Laskay, T.; Behnen, M. The impact of various reactive oxygen species on the formation of neutrophil extracellular traps. *Mediat. Inflamm.* **2012**, *2012*. [CrossRef] [PubMed]
73. Parker, H.; Dragunow, M.; Hampton, M.B.; Kettle, A.J.; Winterbourn, C.C. Requirements for NADPH oxidase and myeloperoxidase in neutrophil extracellular trap formation differ depending on the stimulus. *J. Leukoc. Biol.* **2012**, *92*, 841–849. [CrossRef]
74. Parker, H.; Winterbourn, C.C. Reactive oxidants and myeloperoxidase and their involvement in neutrophil extracellular traps. *Front. Immunol.* **2012**, *3*, 424. [CrossRef]
75. Van Der Veen, B.S.; De Winther, M.P.J.; Heeringa, P. Myeloperoxidase: Molecular mechanisms of action and their relevance to human health and disease. *Antioxid. Redox Signal.* **2009**, *11*, 2899–2937. [CrossRef] [PubMed]
76. Gross, S.; Gammon, S.T.; Moss, B.L.; Rauch, D.; Harding, J.; Heinecke, J.W.; Ratner, L.; Piwnica-Worms, D. Bioluminescence imaging of myeloperoxidase activity in vivo. *Nat. Med.* **2009**, *15*, 455–461. [CrossRef]
77. Björnsdottir, H.; Welin, A.; Michaëlsson, E.; Osla, V.; Berg, S.; Christenson, K.; Sundqvist, M.; Dahlgren, C.; Karlsson, A.; Bylund, J. Neutrophil NET formation is regulated from the inside by myeloperoxidase-processed reactive oxygen species. *Free Radic. Biol. Med.* **2015**, *89*, 1024–1035. [CrossRef]
78. de Souza, C.N.; Breda, L.C.D.; Khan, M.A.; de Almeida, S.R.; Câmara, N.O.S.; Sweezey, N.; Palaniyar, N. Alkaline pH promotes NADPH oxidase-independent neutrophil extracellular trap formation: A matter of mitochondrial reactive oxygen species generation and citrullination and cleavage of histone. *Front. Immunol.* **2018**, *8*, 1849. [CrossRef] [PubMed]
79. Azzouz, D.; Khan, M.A.; Sweezey, N.; Palaniyar, N. Two-in-one: UV radiation simultaneously induces apoptosis and NETosis. *Cell Death Discov.* **2018**, *4*, 51. [CrossRef] [PubMed]
80. Maldonado-Bernal, C.; Kirschning, C.J.; Rosenstein, Y.; Rocha, L.M.; Rios-Sarabia, N.; Espinosa-Cantellano, M.; Becker, I.; Estrada, I.; Salazar-Gonzalez, R.M.; Lopez-Macias, C.; et al. The innate immune response to *Entamoeba histolytica* lipopeptidophosphoglycan is mediated by toll-like receptors 2 and 4. *Parasite Immunol* **2005**, *27*, 127–137. [CrossRef]
81. Dan Dunn, J.; Alvarez, L.A.J.; Zhang, X.; Soldati, T. Reactive oxygen species and mitochondria: A nexus of cellular homeostasis. *Redox Biol.* **2015**, *6*, 472–485. [CrossRef] [PubMed]
82. Brookes, P.S.; Yoon, Y.; Robotham, J.L.; Anders, M.W.; Sheu, S.S. Calcium, ATP, and ROS: A mitochondrial love-hate triangle. *Am. J. Physiol. Cell Physiol.* **2004**, *287*, C817–C833. [CrossRef]
83. Daniel, C.; Leppkes, M.; Muñoz, L.E.; Schley, G.; Schett, G.; Herrmann, M. Extracellular DNA traps in inflammation, injury and healing. *Nat. Rev. Nephrol.* **2019**, *15*, 559–575. [CrossRef]
84. Kaplan, M.J.; Radic, M. Neutrophil extracellular traps: Double-edged swords of innate immunity. *J. Immunol.* **2012**, *189*, 2689–2695. [CrossRef]
85. Davidsson, L.; Rudin, A.D.; Klose, F.P.S.; Buck, A.; Björkman, L.; Christenson, K.; Bylund, J. In vivo transmigrated human neutrophils are highly primed for intracellular radical production induced by monosodium urate crystals. *Int. J. Mol. Sci.* **2020**, *21*, 3750. [CrossRef]
86. Manda-Handzlik, A.; Bystrzycka, W.; Cieloch, A.; Glodkowska-Mrowka, E.; Jankowska-Steifer, E.; Heropolitanska-Pliszka, E.; Skrobot, A.; Muchowicz, A.; Ciepiela, O.; Wachowska, M.; et al. Nitric oxide and peroxynitrite trigger and enhance release of neutrophil extracellular traps. *Cell. Mol. Life Sci.* **2020**, *77*, 3059–3075. [CrossRef]
87. Pacheco–Yépez, J.; Rivera-Aguilar, V.; Barbosa-Cabrera, E.; Rojas Hernández, S.; Jarillo-Luna, R.A.; Campos-Rodríguez, R. Myeloperoxidase binds to and kills *Entamoeba histolytica* trophozoites. *Parasite Immunol.* **2011**, *33*, 255–264. [CrossRef] [PubMed]
88. Jancinová, V.; Drábiková, K.; Nosál, R.; Racková, L.; Májeková, M.; Holománová, D. The combined luminol/isoluminol chemiluminescence method for differentiating between extracellular and intracellular oxidant production by neutrophils. *Redox Rep.* **2006**, *11*, 110–116. [CrossRef] [PubMed]

89. Palmer, L.J.; Cooper, P.R.; Ling, M.R.; Wright, H.J.; Huissoon, A.; Chapple, I.L.C. Hypochlorous acid regulates neutrophil extracellular trap release in humans. *Clin. Exp. Immunol.* **2012**, *167*, 261–268. [CrossRef]
90. Papayannopoulos, V.; Metzler, K.D.; Hakkim, A.; Zychlinsky, A. Neutrophil elastase and myeloperoxidase regulate the formation of neutrophil extracellular traps. *J. Cell Biol.* **2010**, *191*, 677–691. [CrossRef] [PubMed]
91. Metzler, K.D.; Goosmann, C.; Lubojemska, A.; Zychlinsky, A.; Papayannopoulos, V. Myeloperoxidase-containing complex regulates neutrophil elastase release and actin dynamics during NETosis. *Cell Rep.* **2014**, *8*, 883–896. [CrossRef] [PubMed]

antioxidants MDPI

Article

Iron Uptake Controls *Trypanosoma cruzi* Metabolic Shift and Cell Proliferation

Claudia F. Dick [1,2,*], Carolina L. Alcantara [1,2], Luiz F. Carvalho-Kelly [3], Marco Antonio Lacerda-Abreu [3], Narcisa L. Cunha-e-Silva [1,2], José R. Meyer-Fernandes [3] and Adalberto Vieyra [1,2,4]

1 Instituto de Biofísica Carlos Chagas Filho, Universidade Federal do Rio de Janeiro, Rio de Janeiro 21941-902, RJ, Brazil; alcantara@biof.ufrj.br (C.L.A.); narcisa@biof.ufrj.br (N.L.C.-e.-S.); avieyra@biof.ufrj.br (A.V.)
2 Centro Nacional de Biologia Estrutural e Bioimagem, Universidade Federal do Rio de Janeiro, Rio de Janeiro 21941-902, RJ, Brazil
3 Instituto de Bioquímica Médica Leopoldo de Meis, Universidade Federal do Rio de Janeiro, Rio de Janeiro 21941-902, RJ, Brazil; lfkelly@bioqmed.ufrj.br (L.F.C.-K.); mantonio.abreu@bioqmed.ufrj.br (M.A.L.-A.); meyer@bioqmed.ufrj.br (J.R.M.-F.)
4 Programa de Pós-Graduação em Biomedicina Translacional /BIOTRANS, Universidade do Grande Rio, Duque de Caxias 25071-202, RJ, Brazil
* Correspondence: cfdick@bioqmed.ufrj.br; Tel.: +55-2139386781

Abstract: (1) Background: Ionic transport in *Trypanosoma cruzi* is the object of intense studies. *T. cruzi* expresses a Fe-reductase (TcFR) and a Fe transporter (TcIT). We investigated the effect of Fe depletion and Fe supplementation on different structures and functions of *T. cruzi* epimastigotes in culture. (2) Methods: We investigated growth and metacyclogenesis, variations of intracellular Fe, endocytosis of transferrin, hemoglobin, and albumin by cell cytometry, structural changes of organelles by transmission electron microscopy, O_2 consumption by oximetry, mitochondrial membrane potential measuring JC-1 fluorescence at different wavelengths, intracellular ATP by bioluminescence, succinate-cytochrome c oxidoreductase following reduction of ferricytochrome c, production of H_2O_2 following oxidation of the Amplex® red probe, superoxide dismutase (SOD) activity following the reduction of nitroblue tetrazolium, expression of SOD, elements of the protein kinase A (PKA) signaling, TcFR and TcIT by quantitative PCR, PKA activity by luminescence, glyceraldehyde-3-phosphate dehydrogenase abundance and activity by Western blotting and NAD^+ reduction, and glucokinase activity recording $NADP^+$ reduction. (3) Results: Fe depletion increased oxidative stress, inhibited mitochondrial function and ATP formation, increased lipid accumulation in the reservosomes, and inhibited differentiation toward trypomastigotes, with the simultaneous metabolic shift from respiration to glycolysis. (4) Conclusion: The processes modulated for ionic Fe provide energy for the *T. cruzi* life cycle and the propagation of Chagas disease.

Keywords: trypanosomatids; growth and differentiation; parasite oxidative stress; mitochondrial function; ATP synthesis; parasite lipid content

Citation: Dick, C.F.; Alcantara, C.L.; Carvalho-Kelly, L.F.; Lacerda-Abreu, M.A.; Cunha-e-Silva, N.L.; Meyer-Fernandes, J.R.; Vieyra, A. Iron Uptake Controls *Trypanosoma cruzi* Metabolic Shift and Cell Proliferation. *Antioxidants* **2023**, *12*, 984. https://doi.org/10.3390/antiox12050984

Academic Editor: Serge Ankri

Received: 17 March 2023
Revised: 18 April 2023
Accepted: 20 April 2023
Published: 22 April 2023

1. Introduction

The etiological agent of Chagas disease, *Trypanosoma cruzi*, has a complicated life cycle that alternates between intermediate invertebrates and definitive mammal hosts [1], and Chagas disease is considered a neglected disease worldwide [2,3].

Iron (Fe) is one nutritional element that controls *T. cruzi* growth and differentiation during its life cycle since it is a necessary micronutrient for all forms of life and a cofactor of many enzymes in a considerable number of metabolic pathways [4]. Fe is also hazardous because of its potential to accelerate the creation of reactive oxygen species (ROS), and all biological systems have evolved mechanisms for managing Fe intake, metabolism, and storage [5].

Fe is essentially physiologically inaccessible due to the limited solubility of its thermodynamically stable +3 oxidation state in the presence of O_2 at neutral pH [6,7]. The concentration of free Fe in the environment ranges between 10^{-9} and 10^{-18} M, which is lower than the concentration necessary for microbial development [8]. Fe is required for DNA synthesis [9,10], energy generation [11], and oxidative stress in trypanosomes [12]. Furthermore, mammalian hosts sequester free Fe into proteins such as transferrin and lactoferrin [13–15], resulting in a free Fe concentration in serum of roughly 10^{-24} M [13–15]. Thus, there are three essential sources of Fe in mammals' bodies that a pathogenic microbe may use: (i) transferrin, (ii) ferritin, and (iii) heme-containing proteins like hemoglobin [16].

Trypanosoma cruzi requires iron (Fe) for growth, in vitro proliferation of epimastigotes forms (mobilizing heminic or non-heminic Fe), and pathogenicity in mice [17]. *Trypanosoma cruzi* can hijack Fe-proteins from mammalian hosts. In culture, adding deferoxamine, a Fe chelator, or transferrin-free serum can reduce amastigotes cell multiplication, demonstrating that Fe is an essential nutrient [18]. This parasite has evolved human transferrin receptors that bind exogenous transferrin. Acid treatment does not remove transferrin attached to amastigote cells, indicating that this transferrin may be internalized and used [18]. Transferrin is taken up by the cytostome, a specialized structure consisting of a profound membrane invagination in the anterior area near the flagellar pocket [19]. *Trypanosoma cruzi* also uses heme as a Fe source; it can boost *T. cruzi* growth in culture in a dose-dependent way [20]. Furthermore, *T. cruzi* epimastigotes internalize heme/porphyrin through a process that might be mediated by an ABC transporter protein [21]. However, because no heme oxidase gene is indicated on the *T. cruzi* genome, the first heminic ring hydrolysis for Fe release is the limiting step for pathogenic trypanosomatids using heme [5].

Since Fe is present in aerobic conditions as Fe^{3+}, it must be converted to Fe^{2+} by the Fe-reductase enzyme [22] to be transported across the plasma membrane. There is much evidence that Fe^{3+} reduction is frequently linked to Fe^{2+} transport in bacteria, yeast, plant, and animal cells [23]. As a result, the identification of Fe-reductase activity in *Leishmania chagasi* [24], *L. amazonensis* [22], and, subsequently, *T. cruzi* [25] was a strong signal of the presence of a Fe^{2+} transport mechanism in trypanosomatids.

The discovery of Fe-reductase activity in trypanosomatids reinforced the hypothesis of a two-step Fe transport mechanism: first, the reduction of Fe^{3+} to Fe^{2+}, followed by the absorption of Fe^{2+} by specialized transporters [26]. In addition, the finding of a Fe transporter (LIT, [27]) in the plasma membrane of *L. amazonensis*, which belongs to the zinc and iron transporter family (ZIP family), gave early support to this idea. ZIP family members are said to be capable of transporting Zn^{2+}; however, some members of this family can also transport Fe^{2+} [28]. Recently, the discovery of the TcIT Fe transporter in *T. cruzi* [29] supports the hypothesis of a functional link between TcFR [25] and TcIT for Fe^{2+} uptake in this parasite.

Due to its low redox potential, Fe is a suitable element for redox catalysis processes [30,31], acting as an electron donor and receptor and being able to catalyze the formation of Once internalized, free Fe must be stored or processed as it enters the cytosol to avoid the generation of ROC. Corrêa et al. [32] discovered Fe in the acidocalcisomes of *T. cruzi* blood trypomastigotes. A putative Fe transporter in the form of a metal ion in the acidocalcisome of *T. cruzi* supports the hypothesis of Fe storage in this organelle [33]. This metabolic element has lately gained prominence due to the discovery that Fe mobilization and oxidative stress play a critical role in the parasite's persistence and survival in the tissues of the mammalian host—a role that had not previously been proven [34].

Although Fe is a critical micronutrient for trypanosomatids, as previously stated [9–12], there are still many unknowns about its involvement in the life cycle and pathogenicity of these organisms, as well as the ways in which it is acquired and used. The main aim of the present work was to investigate whether exogenous ionic Fe modulates *T. cruzi's* redox status and metabolic pathways, as well as the proliferation and differentiation of the parasite. The study revealed molecular mechanisms and intracellular processes modulated by exogenous ionic Fe that had not previously been reported.

2. Materials and Methods

2.1. Epimastigote Growth and Metacyclogenesis

Trypanosoma cruzi (Dm28c strain) epimastigotes were grown in stationary phase at 28 °C in Brain Heart Infusion (BHI) medium supplemented with 10% FBS, 30 μM hemin, and 1% penicillin-streptomycin (P/S) cocktail (referred to hereafter as Regular Media, RM). Iron-Depleted Media (IDM) was prepared using an iron-free BHI medium, as described by Dick et al. [25]. Briefly, BHI medium without hemin addition was treated with Chelex (5 g/100 mL) for 1 h at room temperature and sterilized by filtration using 0.22 μm pore-size filters. To this Iron-Free BHI medium was added 1% P/S cocktail and 10% Iron-Free FBS. The iron-free FBS was prepared by adding 10 mM ascorbic acid for 6–7 h at 37 °C until the optical density at 405 nm had decreased by 50%. Then, the solution was supplied with 5 g of Chelex resin per 100 mL and incubated at room temperature under stirring at 50 rpm for 3–4 h, filtered to remove the resin, and dialyzed (with a cutoff of 2000 Da) against 4 l of cold, sterile PBS for 6 h, changing the solution every 2 h. The iron-free FBS was also sterilized by filtration using 0.22 μm pore-size filters and stored at −20 °C. Iron-Free FBS Iron-Depleted Media with Fe (IDM + Fe) was made the same way as IDM but with 8 μM Fe-citrate added. The parasites were inoculated (10^6 cells/mL) into the BHI medium on the sixth day of growth to test epimastigotes proliferation (RM, IDM, or IDM + Fe). Every day, cell proliferation was measured by counting the number of cells in a hemocytometer. Dm28c is a strain that differentiates "in vitro" and was previously used in our previous studies that demonstrated the existence of a Fe-reductase and a Fe-transporter in *T. cruzi* [25,29]. This strain is deposited in the Collection of Trypanosoma from Wild and Domestic Mammals and Vectors (COLTRYP), Oswaldo Cruz Foundation, Rio de Janeiro, Brazil.

Metacyclogenesis was induced according to Koeller et al. [35]. Epimastigotes in the transition from the logarithmic to the stationary phase were adjusted to 5×10^8 parasites/mL in triatomine artificial urine (TAU) medium (190 mM NaCl, 17 mM KCl, 2 mM $MgCl_2$, 2 mM $CaCl_2$, 0.035% (*w*/*v*) $NaHCO_3$, and 8 mM phosphate buffer at pH 6.0). After 2 h at 28 °C, the cultures were diluted 100-fold in 10 mL TAU medium supplemented with 10 mM proline and 250 mM glucose (TAU-P) plus 500 g/mL G418 (Sigma-Aldrich, Saint Louis, MO, USA) and transferred to T25 flasks—lying at a 45° angle to increase the area in contact with O_2—and kept at 28 °C to promote metacyclogenesis. Following 3–5 days, parasites were counted using hemocytometry, and the proportion of metacyclic epimastigotes (Tryp) was determined using their morphology after Giemsa staining.

2.2. Intracellular Fe Concentration Determination

A colorimetric test based on ferrozine was used to assess the quantity of intracellular ionic Fe accumulated under different circumstances, as described before [29]. Briefly, suspensions containing 10^8 parasites were obtained from various cultures and washed three times with PBS pretreated with 5 g/100 mL Chelex resin (Sigma-Aldrich). The cells were lysed with 100 μL of 50 mM NaOH, and then 100 μL of 10 mM HCl was added; the release of ionic Fe bound to intracellular structures was induced by adding 100 μL of a mixture containing 1.4 M HCl and 4.5% (*w*/*v*) $KMnO_4$ (1:1) to the cell lysate, followed by incubation at 60 °C for 2 h. Then, 30 μL Fe detection reagent was added (6.5 mM ferrozine, 6.5 mM neocuproin, 2.5 M ammonium acetate, and 1 M ascorbic acid). The sample's absorbance at 550 nm was measured after 30 min of incubation at room temperature. A standard curve with known $FeCl_3$ concentrations (0–75 μM; [36]) (Merck, Darmstadt, Germany) was used to calculate the Fe content.

2.3. Real-Time-PCR

Total *T. cruzi* RNA was extracted using a Direct-zol RNA Miniprep Kit (Zymo Research, Orange, CA, USA) from epimastigotes kept at RM, IDM, or IDM + Fe for 5 days (as indicated in the figure legends). The high-capacity cDNA reverse transcription kit was used to reverse-transcribe whole RNA (Thermo Fisher Scientific, Waltham, MA, USA). For RT-PCR, 100 ng/μL cDNA per well (15 μL total volume) was utilized, coupled with a 5 μM primer mix and a 7 μL PowerUp SYBR green master solution (Thermo Fisher Scientific). Primers were designed using the Primer3 software, with predicted amplicon sizes of 100 pb each [37]. The primers for amplification are shown in Table 1. Gene expression data were normalized to an endogenous reference, β-tubulin. The expression ratios were determined using the threshold cycle (ΔΔCT) [38].

Table 1. Primer sequences for nine genes analyzed.

Primers Name	Forward Primer	Reverse Primer
T. cruzi iron transporter	TCTGGTCGCTTCTCTTCTCG	TAAAGACTCCGGCACACAGT
T. cruzi ferric reductase	GTGGTTTGTAGACCGGCTGT	GTGCCATTGCAAGAGAGACA
T. cruzi heme-regulated inhibitor	CATTGTGGAGGCGTTGGAAA	TGGAAGAGCACCGTGAAGAT
T. cruzi eukaryotic initiation factor 2α	CCGTTTAACGTTCCCTTTGA	GTCCCAGCTCGTTACTCCAA
T. cruzi protein kinase A	CCGGGTGTACTTTGTGTTGG	CGCAAACCCAAAGTCAGTCA
T. cruzi glyceraldehyde-3-phosphate dehydrogenase	GCAAGCTTGGTGTGGAGTAC	CTCACTGGGGTTGTACTCGT
T. cruzi superoxide dismutase	GTCGGATATTGTGTTGGGCC	CCCTGTACCACGGAAACTCT
T. cruzi ascorbate peroxidase	CACGACAAGTACGGCTTTGA	CTTCGCGATGGAACGATATT
T. cruzi tubulin	AAGCGCACGATTCAGTTTGT	CTCCATACCCTCACCAACGT

2.4. Endocytosis Assay

Epimastigotes were submitted to endocytosis with 30 μg/mL of transferrin-FITC, hemoglobin-FITC, or BSA-FITC in Roswell Park Memorial Institute (RPMI) medium for 30 min at 28 °C. The parasites were fixed with 4% (*v*/*v*) paraformaldehyde in phosphate-buffered saline (PBS, pH 7.2) for 1 h for flow cytometry and imaging using the fluorescence microscope AxioObserver (Zeiss, Oberkochen, Germany) after staining with DAPI.

2.5. Cell Cytometry

After the endocytosis assay, tracer uptake was measured on a BD Accuri C6 flow cytometer (Becton Dickinson Bioscience, BDB, San José, CA, USA), counting 10,000 events at the FL2 channel. The data were analyzed by the BD Accuri C6 software. This analysis was performed in three independent experiments.

2.6. Transmission Electron Microscopy

Samples were fixed with 2.5% (*v*/*v*) glutaraldehyde in 0.1 M cacodylate buffer (pH 7.2) for 1 h at room temperature. After a wash in cacodylate buffer, cells were post-fixed using an osmium-thiocarbohydrazide-osmium (OTO) protocol as already described [39]. Samples were washed in water, dehydrated in an acetone series, and embedded in epoxy resin (EMBED 812 resins, EMS, Hatfield, PA, USA). Ultrathin sections were cut with a UC7 ultramicrotome (Leica, Wetzlar, Germany), stained with 5% (*w*/*v*) uranyl acetate and lead citrate, and observed in an HT7800 transmission electron microscope (Hitachi, Tokyo, Japan) operating at 100 kV.

2.7. Protein Kinase A (PKA) Activity

Epimastigotes cells (5×10^7 cells/mL) were washed twice in ice-cold phosphate buffer saline (PBS, pH 7.2) and lysed in 0.5 mL radioimmunoprecipitation assay buffer (RIPA buffer) for 30 min. PKA activity was assayed in the presence of 4 mM Hepes-Tris (pH 7.0), 0.4 mM $MgCl_2$, 1 mM $CaCl_2$, 1 μM ATP, and 50 μg of lysed cells in a final volume of 50 μL, in the presence or absence of 5 μM 3-isobutyl-1-methylxanthine (IBMX, a PKA activator), in MTS-11C mini tubes (Axygen Scientific, Union City, CA, USA). The reaction was triggered by adding 50 μL of the Kinase-Glo luminescent kit, and after 10 min at 37 °C, the samples were placed in a GloMax Multi JR detection system (Promega Corporation, Fitchburg, WI, USA). PKA activity was quantified as the difference between the reading in the presence of IBMX and in the absence of the activator.

2.8. High-Resolution Respirometry in Different Respiratory States

Oxygen consumption of intact epimastigotes (5 × 10^7 parasites/chamber) was measured using an O2k-system high-resolution oxygraph (Oxygraph-2K; Oroboros Instruments, Innsbruck, Austria) at 28 °C with continuous stirring. The cells were suspended in a 2 mL respiration solution containing 100 mM sucrose, 50 mM KCl, and 50 mM Tris–HCl (pH 7.2), and 50 µM digitonin was added to permeabilize the parasites. Following that, 10 mM succinate and 200 µM ADP were added. Uncoupled respiration was induced with 3 µM carbonyl cyanide 4-(trifluoromethoxy)phenylhydrazone (FCCP), and then blocked with 2.5 µg/mL antimycin A to assess residual O_2 consumption [29]. Oxygen concentrations and O_2 consumption were recorded using DatLab software coupled to Oxygraph-2K.

2.9. Succinate-Cytochrome C Oxidoreductase Activity

The activity of succinate-cytochrome c oxidoreductase (complex II/III) was determined by the increase in absorbance due to the reduction of ferricytochrome c at 550 nm [40,41]. Frozen-thawed parasite homogenates (50 µg) were incubated with 25 mM potassium phosphate (pH 7.4), 10 mM succinate, 1 mM KCN, and 5 mM $MgCl_2$ for 20 min to allow for complete activation of succinate dehydrogenase, after which the reaction was initiated with 50 µM horse heart cytochrome c and monitored for 2 min. Protein concentrations were determined by the Bradford method, using bovine serum albumin as a standard [42].

2.10. Mitochondrial Membrane Potential

Mitochondrial membrane potential was analyzed in the *T. cruzi* epimastigotes using the MitoProbe JC-1 assay kit (Molecular Probes; Thermo Fisher Scientific). Epimastigotes (1 × 10^7 cells/mL) were loaded with 10 µM 5,5′,6,6′-tetrachloro-1,1′3,3′-tetraethylimidacarbocyanine iodide (JC-1) and incubated for 40 min at room temperature. The fluorescence intensity ratio of red (540 nm excitation and 590 nm emission) to green (490 nm excitation and 540 nm emission) was measured using a multi-well fluorescence reader [43].

2.11. Intracellular ATP Quantification

Intracellular ATP (ATP_i) was measured using an ATP bioluminescent somatic cell test kit (Sigma-Aldrich). In brief, epimastigotes (1 × 10^7 parasites per tube) were incubated in a solution containing 100 mM sucrose, 50 mM KCl, and 50 mM Tris-HCl in 0.1 mL (pH 7.2 adjusted with HCl). Cellular extracts were prepared by combining them with 0.1 mL of somatic cell ATP-releasing reagent and then chilling the mixture for 1 min. The mixture was transferred to MTS-11C mini tubes containing 0.1 mL ATP assay mix (*v:v*; Axygen) and swirled for 10 s at room temperature. A GloMax Multi JR detection system (Promega) measured the overall quantity of light emitted. In each experiment, the total intracellular ATP concentration per 10^7 cells was determined using a standard ATP curve [29].

2.12. Glyceraldehyde-3-Phosphate Dehydrogenase (GAPDH) Activity

The reduction of NAD^+ to NADH was used to assess the conversion of glyceraldehyde-3-phosphate to 1,3-bisphosphoglicerate, as described before [44], with slight modifications. Total epimastigote lysates were incubated for 15 min at 37 °C in reaction media containing 100 mM Triethalonamine:HCl buffer (pH 7.5) containing 1.0 mM EDTA, 5 mM $MgSO_4$, 1.0 mM dithiothreitol, 1.5 mM NAD^+, and 30 mM KH_2AsO_4 [45,46]. The reaction was started using 2 mM glyceraldehyde-3-phosphate, and the absorbance at 340 nm was measured every 1 min for 5 min. Total NADH generation was determined using the NADH standard curve.

2.13. Glucokinase Activity

Cellular extracts of epimastigotes were incubated in a reaction buffer containing 20 mM Tris-HCl (pH 7.4), 5 mM $MgCl_2$, 1 mM glucose, 1 unit/mL glucose-6-phosphate dehydrogenase (G6PDH) (*Leuconostoc mesenteroides*), 0.1% Triton X-100, 1 mM NaF, 5 mM NaN_3, 1 mM ATP, and 50–100 µg/mL protein [47]. After 3 min of incubation, the reactions were started by the addition of 0.5 mM β-$NADP^+$, and quantified spectrophotometrically following the reduction of β-$NADP^+$ to β-NADPH (λ = 340 nm) for 10 min. Total NADPH generation was determined using the NADPH standard curve.

2.14. Western Blotting

For western blotting detection, epimastigotes were lysed with 1 mL RIPA buffer supplemented with 1 mM phenylmethanesulfonyl fluoride and 5 mM leupeptin, for 30 min at 4 °C. Then, homogenates were centrifuged at 16,000 rpm for 10 min. Aliquots of the supernatants (containing 100 μg total protein) were separated by 12% SDS-PAGE and transferred to nitrocellulose membranes (Merck Millipore, Burlington, MA, USA), which were blocked with 5% milk in PBS plus 0.1% (*w*/*v*) Tween 20, probed overnight at 4 °C with the primary mouse anti-GAPDH antibody (1:500, Sigma-Aldrich), and detected using horseradish peroxidase (HRP)-conjugated anti-mouse IgG secondary antibody (1:5000, Santa Cruz Biotechnology, Dallas, TX, USA). The loading control was probed with a primary rabbit anti-tubulin antibody (1:500, Sigma-Aldrich) and detected using an HRP-conjugated anti-rabbit IgG secondary antibody (1:10,000, Santa Cruz Biotechnology). Luminescence was detected using an ImageQuant LAS 4000 digital imaging system (GE Healthcare Life Sciences, Amersham, UK) after the reaction with LuminataTM Forte Western HRP Substrate (Millipore, Billerica, MA, USA). Densitometric analysis was performed using ImageJ software version 1.50i (NIH Image, Bethesda, MD, USA) with background correction.

2.15. Superoxide Dismutase (SOD) Activity

Total SOD activity was assessed as previously described [29], based on SOD inhibiting the reduction of nitro blue tetrazolium (NBT) by $O_2^{\bullet -}$. Epimastigote cells were harvested by centrifugation, washed three times in cold PBS, and disrupted by freeze-thaw. Centrifugation as described above was used to collect epimastigote cells, which were washed three times in cold PBS and disrupted by freeze-thaw. The Bradford method [42] was used to determine the protein content in the whole homogenate. In a final volume of 200 μL, the homogenates (using known amounts of protein in the range of 10–50 μg) were incubated in a reaction medium containing 45 mM potassium phosphate buffer (pH 7.8), 6.5 mM EDTA, and 50 mM NBT. The reaction began with the addition of 2 mM riboflavin. The sample's absorbance at 560 nm was measured after 15 min in a lightbox. SOD activity was expressed as the quantity of enzyme-blocking NBT reduction by 50% for each amount of protein.

2.16. Amplex® Red Peroxidase Assay

The rate of H_2O_2 reduction was assayed by the production of H_2O_2 in epimastigotes to H_2O, which is stoichiometrically coupled (1:1) to the simultaneous oxidation of the non-fluorescent Amplex® Red probe to the fluorescent resorufin [48]. Briefly, 10^7 parasites/mL were incubated with 5 mM Tris-HCl (pH 7.4), 1.7 μM Amplex® Red (Invitrogen, Carlsbad, CA, USA), and 6.7 U/mL horseradish peroxidase (Sigma-Aldrich) in a final volume of 100 μL, for 30 min at room temperature. Fluorescence evolution was observed at excitation/emission wavelengths of 563/587 nm. H_2O_2 concentration was determined using a standard curve.

2.17. Statistical Analysis

The data are provided as mean ± standard error of the mean (SEM). The unpaired Student's *t*-test was used to compare two means. When comparing more than two means, a one-way ANOVA with Tukey's test was applied, as specified in the text or the figure captions. The normal distribution was assessed before each ANOVA analysis. When results were expressed as a percentage of the RM group, the SEM was obtained from the absolute values. A comparison of the absolute results was carried out using the parametric test. Significance was set at $p < 0.05$. Except when otherwise indicated, different lowercase letters in superscripts indicate statistical differences among means in the same line of tables. Asterisks (also indicating $p < 0.05$) were used in figures and for comparing values from different lines in figures. GraphPad Prism 7.0 was used for statistical analysis and the preparation of figures (GraphPad Software, San Diego, CA, USA).

3. Results

3.1. *Trypanosoma cruzi Epimastigotes Depend on Exogenous Fe Content for Fe Proliferation*

The proliferation of *T. cruzi* epimastigotes during the exponential growth phase (Figure 1) depends on Fe in a concentration-dependent manner. In Fe-depleted medium (IDM, black circles), the maximum number of protozoa after 7 days decreased from 4.4×10^7 to 1.0×10^7 parasites/mL, compared to epimastigotes maintained in regular medium (RM, empty circles). The proliferative capacity of the cells was recovered with 8 µM Fe citrate added to the Fe-depleted medium (IDM + Fe, gray circles): the number of protozoa after 7 days increased from 1.0×10^7 to 5.1×10^7 parasites/mL, a condition similar to the control. In addition, Fe citrate supplementation was enough to restore intracellular Fe concentration compared to RM, while cells maintained at IDM presented low intracellular Fe content (Table 2). Although in IDM and in IDM + Fe there was an increase in the expression of Fe reductase (TcFR), the expression of the Fe transporter (TcIT) in cells maintained in IDM + Fe remained at levels similar to the RM condition (Table 2).

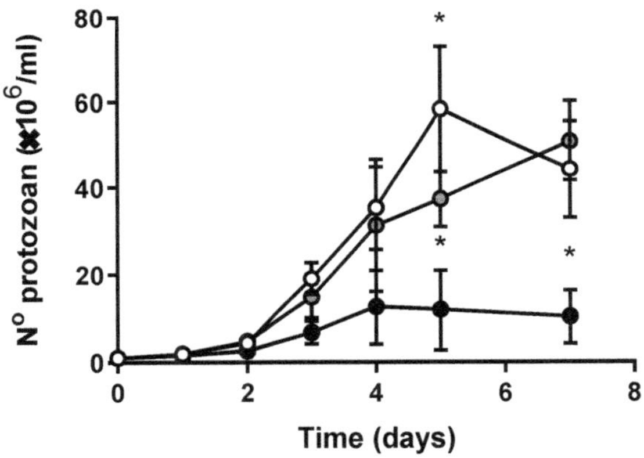

Figure 1. The effect of exogenous Fe on *T. cruzi* growth. *Trypanosoma cruzi* epimastigotes were harvested, washed twice, seeded in fresh medium, and grown for the indicated times in Regular Media (RM: Brain Heart Infusion medium (BHI) supplemented with 30 µM hemin and 10% Fetal Bovine Serum (FBS)) (white circles), Iron-Depleted Media (IDM: BHI without hemin supplementation and treated with Chelex for Fe depletion, supplied with 10% Iron-depleted FBS) (black circles), or in Iron-Depleted Media supplemented with Fe-citrate (IDM + Fe: IDM described before, plus 8 µM Fe-citrate) (gray circles), (n = 6). Using one-way ANOVA with Tukey's test, we assessed differences between mean values at time-matched determinations; * $p < 0.05$ with respect to RM.

Table 2. Intracellular Fe content and quantification of the TcFR and TcIT transcripts *.

	RM	IDM	IDM + Fe
Fe content	22.07 ± 0.78 [a]	11.01 ± 2.42 [b]	26.59 ± 3.31 [a]
TcFR transcript	0.96 ± 0.05 [a]	1.59 ± 0.11 [b]	2.14 ± 0.16 [b]
TcIT transcript	0.85 ± 0.05 [a]	1.72 ± 0.24 [b]	0.72 ± 0.12 [a]

* *Trypanosoma cruzi* epimastigotes were maintained in Regular Media (RM), Iron-Depleted Media (IDM), or Iron-Depleted Media plus 8 µM Fe-citrate (IDM + Fe). Intracellular Fe content determination ($n = 5$) and quantitative PCR for TcFR ($n = 5$) or TcIT ($n = 5$) (using 100 ng cDNA) were carried out in epimastigotes in a mid-log phase when maintained in the different media. Using one-way ANOVA with Tukey's test, we assessed differences between mean values. Different lower-case letters as superscripts in the same line indicate different mean values ($p < 0.05$).

3.2. Exogenous Fe Selectively Modulates the Endocytic Capacity of Epimastigote Cells

Figure 2 shows that ionic Fe modulates both the function and structure of the endocytic pathway in *T. cruzi* epimastigotes. Figure 2A (see also representative cytometric analyses in Supplementary Figure S1) demonstrates that parasites from IDM and IDM + Fe upregulate transferrin (Tf) uptake compared to control parasites grown in RM. However, hemoglobin (HB) uptake is upregulated only in IDM, compared to RM and IDM + Fe, indicating that free Fe in the culture medium controls hemoglobin uptake, in contrast with that observed with the transferrin uptake mechanism. Fe depletion stimulates bovine serum albumin (BSA) uptake in parasites grown in IDM, and the stimulation persists in IDM + Fe, indicating that this endocytic event is higher in both conditions. The fluorescence images in Figure 2B visually represent what we found in the cell cytometry data. Parasites upregulate the uptake of the endocytic tracers in IDM and IDM + Fe conditions.

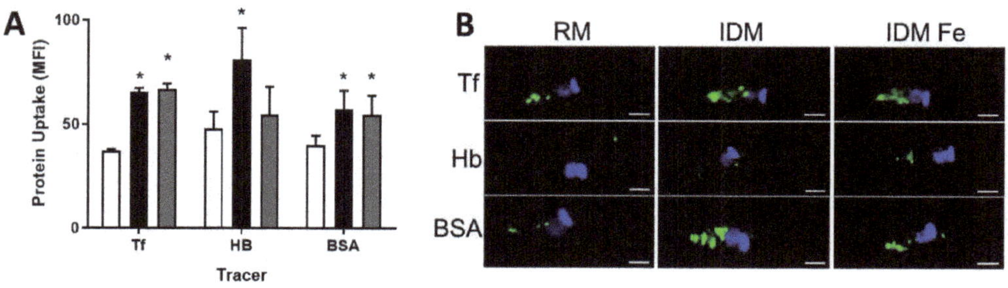

Figure 2. Fe exogenous content alters endocytosis in epimastigotes. Epimastigotes were incubated in RPMI medium containing transferrin–FITC (Tf), hemoglobin–FITC (HB), or BSA–FITC (BSA) for 30 min at 28 °C. (**A**) Endocytosis assay evaluated by flow cytometry in epimastigotes maintained at RM (white bars), IDM (black bars), or IDM + Fe (gray bars) (n = 4 from different cultures); * $p < 0.05$; (**B**) immunofluorescence of epimastigotes at different culture conditions and endocytic tracers incubation. Endocytic content staining with fluorescence macromolecules transferrin–FITC (Tf), hemoglobin–FITC (HB), or BSA–FITC (BSA) (green). Nuclei staining with DAPI (blue). Bars: 10 μm.

Ultrastructural analysis of epimastigotes grown in the three different culture media revealed opposite trends depending on the organelles analyzed (Figure 3). While the general morphological aspect remained unmodified in IDM and IDM + Fe when compared with the parasites maintained in RM, important alterations can be seen in the reservosomes, lysosome-like organelles that are also involved in the storage of exogenous lipids, especially cholesterol, and enzymes involved in lipid synthesis from acetyl-CoA [49–52]. The homogeneous content of the reservosomes encountered in RM epimastigotes (upper line) contrasts with that seen in the other two groups. The depletion of Fe (middle line) provoked a marked increase in lipid accumulation, which was not reversed by the supplementation with Fe citrate (bottom line), indicating significant metabolic modifications.

3.3. Free Iron Supplementation-Induced Modifications in the Heme-Regulated Eukaryotic Initiation Factor 2α Kinase-Signaling Pathway

Facing the epimastigotes proliferation results presented in Figure 1 and considering that Fe depletion downregulates cytosolic and mitochondrial ribosomal proteins [53] via the heme-regulated inhibitor (HRI) of the translation through phosphorylation of the eukaryotic initiation factor 2α (eIF2α), we investigated two key elements of this signaling pathway (Table 3). Removal of heme and free Fe (IDM) increased the expression of HRI by approximately 90% and decreased by 50% the expression of eIF2α. In the case of HRI, the supplementation with free Fe restored the levels of HRI and upregulated those of eIF2α.

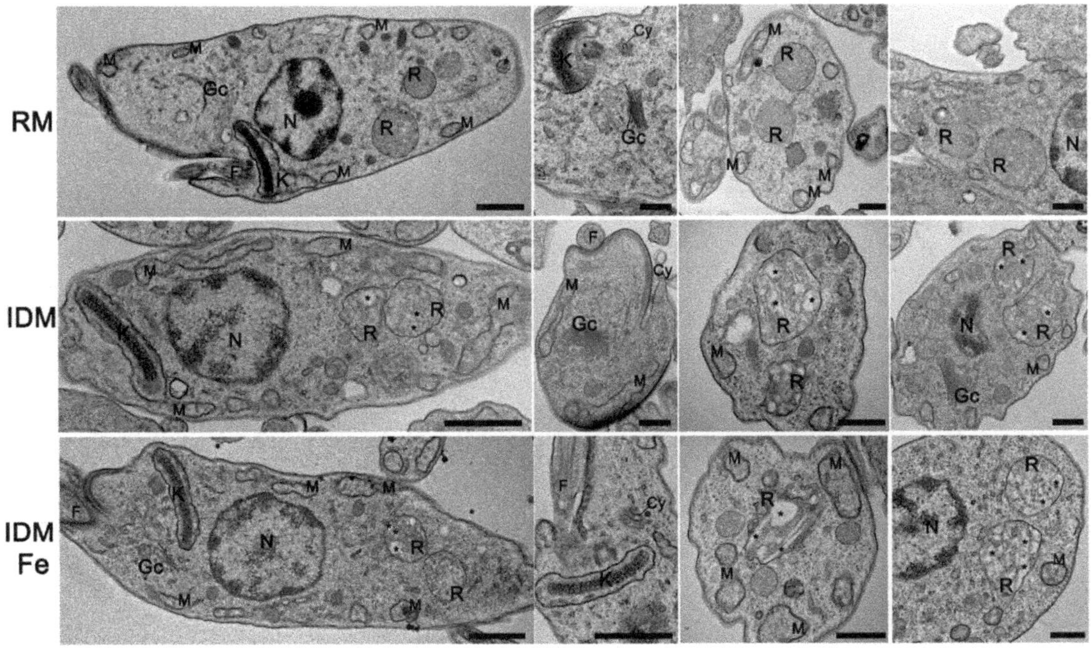

Figure 3. Ultrastructural changes in *T. cruzi* parasites submitted to different exogenous Fe contents. Epimastigotes were processed and observed by transmission electron microscopy. (Upper line) Epimastigotes maintained in RM show normal morphology of major cellular organelles such as the kinetoplast (K), nucleus (N), golgi complex (Gc), cytostome-cytopharynx complex (Cy), reservosomes (R), mitochondria (M), and flagellum (F). (Middle line) Epimastigotes were maintained in IDM. (Bottom line) Epimastigotes maintained in IDM + Fe. Epimastigotes maintained at IDM or IDM + Fe media present typical organelles morphology and positioning compared to those maintained at RM, except for reservosomes, which present a higher content of lipid inclusions (asterisks) compared to the ones in RM. Bars: 500 nm. Five culture samples were examined.

Table 3. Influence of PKA pathway on the response to exogenous ionic Fe in *T. cruzi* epimastigotes *.

	RM	IDM	IDM + Fe
TcHRI transcript	1.04 ± 0.02^{a}	1.70 ± 0.11^{b}	1.44 ± 0.25^{a}
eIF2α transcript	1.00 ± 0.07^{a}	0.41 ± 0.05^{b}	1.45 ± 0.05^{b}
TcPKA transcript	1.04 ± 0.04^{a}	0.78 ± 0.07^{a}	1.57 ± 0.16^{b}
PKA activity	0.98 ± 0.02^{a}	0.29 ± 0.08^{b}	2.12 ± 0.24^{a}

* Quantification of the TcHRI (n = 5), eIF2α (n = 5), and TcPKA (n = 5) transcripts in *T. cruzi* epimastigotes. Quantitative PCR was done using 100 ng of cDNA from epimastigotes in a mid-log phase when maintained in RM, IDM, or IDM + Fe. PKA activity was measured, as described in Materials and Methods (n = 5), in epimastigotes maintained in RM, IDM, or IDM + Fe. In all cases, differences were assessed using one-way ANOVA with Tukey's test. Different lower-case letters as superscripts in the same line indicate different mean values.

The faster response mediated by the HRI→eIF2α is integrated with the PKA pathway in several eukaryotic organisms [54], and the PKA pathway participates in the sensing of and response to nutrients, including Fe [55]. Table 3 also shows a decrease in PKA expression in epimastigotes grown in IDM and upregulation in IDM + Fe when both conditions are compared to RM. The impact of Fe depletion/supplementation was better seen with PKA activity: a pronounced decrease in parasites from IDM and an upregulation of more than 100% in IDM + Fe conditions (Table 3).

3.4. Exogenous Fe Alters Mitochondrial Respiratory Rates in T. cruzi Epimastigotes

As mentioned above, mitochondrial proteins decrease when the HRI→eIF2α pathway is dysfunctional [53], as encountered in the case of Fe depletion (Table 3). Fe concentration in the culture altered mitochondrial O_2 consumption by *T. cruzi*, as observed in Table 4. The digitonin permeabilization of parasites did not alter the basal respiration profile. There was a difference in the O_2 consumption after adding succinate (the consumption of O_2 in the LEAK state) to IDM compared to RM and IDM + Fe (line 1). The O_2 consumption in the presence of succinate without further additions matched the succinate-cytochrome c oxidoreductase activity (line 5). However, after additions of ADP (oxidative phosphorylation state-OXPHOS) (line 2) and H^+ ionophore FCCP (electron transfer system uncoupled from phosphorylation-ETS) (line 3), there was a significant and similar decrease in O_2 consumption in both IDM and IDM + Fe. After adding antimycin A to inhibit the mitochondrial complex III, the residual respiration (ROX) was almost completely abolished under all conditions (line 4). The impact of Fe depletion (even after Fe citrate supplementation) on mitochondrial internal membrane potential ($\Delta\Psi_m$) can be seen in line 6. The ratio between the safranin A fluorescence in the absence and presence of FCCP (F/F_{FCCP}) increased by 600% in parasites grown in IDM and attained an intermediary level ~200% higher in IDM + Fe conditions.

Table 4. Removal of exogenous Fe leads to mitochondrial alterations in epimastigotes *.

Line	Determination	Additions	RM	IDM	IDM + Fe
	Mitochondrial respiration state				
1	LEAK	Succinate	0.87 ± 0.13 [a]	0.38 ± 0.06 [b]	0.66 ± 0.09 [a]
2	OSPHOS	ADP	1.51 ± 0.16 [a]	0.44 ± 0.07 [b]	0.73 ± 0.09 [b]
3	ETS	FCCP	2.16 ± 0.26 [a]	0.54 ± 0.10 [b]	0.92 ± 0.10 [b]
4	ROX	Antimycin A	0.33 ± 0.02	0.23 ± 0.01	0.12 ± 0.02
5	Succinate cytochrome c oxidoreductase activity	Succinate	103.5 ± 5.7 [a]	41.1 ± 7.9 [b]	93.9 ± 14.4 [a]
6	$\Delta\Psi_m$ (F/F_{FCCP})	Succinate + FCCP	1.35 ± 0.03 [a]	9.23 ± 0.57 [b]	2.96 ± 0.59 [c]
7	ATP_i	No inhibitors	24.24 ± 0.96 [a]	14.28 ± 1.52 [b]	12.51 ± 2.09 [b]
8		Oligomycin	10.63 ± 0.57 [a]	7.79 ± 0.94 [a]	8.98 ± 2.25 [a]
9		Iodoacetamide	11.74 ± 2.45 [a]	6.42 ± 1.14 [b]	6.28 ± 0.54 [b]

* *Trypanosoma cruzi* epimastigotes maintained in RM, IDM, or IDM + Fe media were tested for respiration of intact epimastigotes. The mitochondrial respiration, the succinate cytochrome c oxidoreductase activity, and the ATP_i are expressed in pmol of O_2 consumed per million of cells, % of the activity in RM, and nmol of intracellular ATP per 10^7 cells, respectively. Epimastigotes were digitonin-permeabilized, as described in Nogueira et al. [40], to measure mitochondrial respiration after the successive additions of succinate, ADP, FCCP, and antimycin A ($n = 3$) (lines 1–4). Succinate-cytochrome c oxidoreductase activity was measured as the rate of ferricytochrome c reduction upon adding succinate in epimastigotes ($n = 3$) (line 5). The F/F_{FCCP} ratio was used to estimate the $\Delta\Psi_m$, where F is the mean fluorescence intensity in the absence of the uncoupler FCCP and F/F_{FCCP} is the mean fluorescence in the presence of FCCP ($n = 3$) (line 6). Intracellular ATP in *T. cruzi* epimastigotes was measured in the absence or presence of oligomycin or iodoacetamide ($n = 3$) (lines 7–9). We assessed differences between mean values using a one-way ANOVA with Tukey's test. Different lower-case letters as superscripts indicate statistical differences among the mean values in the same line ($p < 0.05$). Asterisks in lines 8 and 9 indicate a statistical difference ($p < 0.05$) with respect to RM in the absence of inhibitors.

Parasites maintained in the RM medium presented intracellular ATP content (ATP_i) more than twice as high when compared to parasites grown in IDM or IDM + Fe media (Table 4, line 7). The addition of oligomycin to block the F_oF_1-ATP synthase [56] decreased ATP_i content by more than 50% with respect to the RM group without inhibitors (line 8), indicating that, under these conditions, the cellular ATP supply comes from mitochondrial activity. Moreover, the addition of iodoacetamide (line 9), an inhibitor of glyceraldehyde 3-phosphate dehydrogenase (GAPDH): (i) reduced ATP_i in parasites grown in RM to levels that are similar to those encountered in IDM and IDM + Fe conditions in the absence or presence of oligomycin; (ii) provoked a further 50% decrease in ATP_i from IDM and IDM + Fe parasites when compared to that grown in RM, indicating that the glycolytic pathway is responsible for ATP production in IDM and IDM + Fe conditions.

3.5. Upregulation of Glycolytic Enzymes in Parasites Grown in Fe-Depleted Media

The activity and abundance of two glycolytic enzymes were evaluated in the experiments depicted in Figure 4. Epimastigotes grown in IDM and IDM + Fe presented, respectively, with GAPDH activities two and three times higher than those grown in RM (Figure 4A), an increase that matches the higher expression of the enzyme (Figure 4B). The abundance of the enzyme was also investigated; the levels increased by more than 100% in IDM + Fe parasites compared with the RM group without modification in the IDM group (Figure 4C). The shift toward a glycolytic metabolic profile was confirmed by the more than 200% increase in the activity of glucokinase (Figure 4D), the enzyme that connects the pentose phosphate pathway and the glycolytic pathway, which are the two major pathways for glucose metabolism in *T. cruzi* [57].

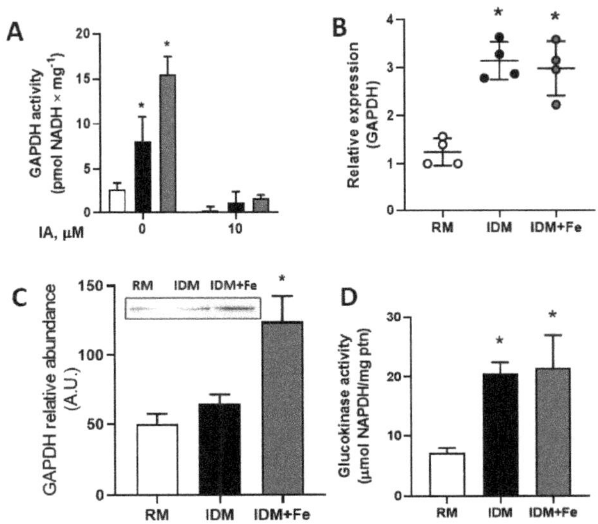

Figure 4. Exogenous Fe leads to glycolytic changes in epimastigotes. (**A**) GAPDH activity in the absence or presence of iodoacetamide (IAA, 10 µM), as indicated on the abscissa, in epimastigotes in a mid-log phase when maintained in RM (empty bar), IDM (black bar), or IDM + Fe (gray bar). The absorbance due to the formation of NADH was monitored at 340 nm ($n = 4$); (**B**) quantification of the GAPDH transcript in *T. cruzi* epimastigotes. Quantitative PCR was done using 100 ng cDNA from epimastigotes maintained at RM (empty circles), IDM (black circles), or IDM + Fe (gray circles), ($n = 4$); (**C**) densitometric analysis of Western blot results ($n = 4$), with the inset showing representative Western blotting analysis. Membranes were probed with primary mouse anti-GAPDH antibody (1:500) as described in the Materials and Methods section; (**D**) glucokinase activity was measured as described in Materials and Methods in epimastigotes maintained at RM (empty bar), IDM (black bar), or IDM + Fe (gray bar). The absorbance due to the formation of NADPH was monitored at 340 nm ($n = 4$); * $p < 0.05$ in all cases with respect to RM. We used a one-way ANOVA with Tukey's test to assess differences between mean values.

3.6. Modifications in Redox Signaling Induced by Depletion of Fe in Culture Medium

The mutual regulation of hexokinases and redox signaling comprises important events in parasites that cause severe diseases, e.g., *T. cruzi* [58]. This is the reason why we decided to investigate two enzymes involved in *T. cruzi* redox signaling and see the influence of Fe depletion: (i) superoxide dismutase (TcSOD), which has Fe as a cofactor in *T. cruzi* [59], and (ii) ascorbate peroxidase (TcAPX) [60], an enzyme that is absent in the human host [61], which metabolizes H_2O_2 to H_2O. We measured TcSOD activity by following two strategies. First, by evaluating the formation of nitroblue tetrazolium formazan (NBT formazan) after oxidation of xanthine to uric acid and generation of $O_2^{\bullet-}$, a reaction in which the higher superoxide dismutase activity is, the lower NBT formazan formation will be. Second, measuring the H_2O_2 formed through different processes, including the premature transfer of electrons to O_2 in the *T. cruzi* mitochondrion. Table 5 shows that the SOD activity associated with the $O_2^{\bullet-}$ formed during the oxidation of xanthine increased by approximately 200% in parasites grown in IDM regardless of supplementation with Fe. Moreover, when the total H_2O_2 production was assayed, a significant decrease of about 40% was found in IDM parasites, which was suppressed in the IDM + Fe group. Finally, the expression profile of TcSOD was the mirror of the activity, whereas that of TcAPX matched the activity, i.e., decreased in the parasites grown in IDM, with recovered levels in the IDM + Fe group.

Table 5. Effect of exogenous Fe on redox signaling *.

	RM	IDM	IDM + Fe
SOD activity	7.98 ± 2.26 [a]	21.47 ± 3.08 [b]	23.24 ± 3.94 [b]
Production of H_2O_2	55.01 ± 0.75 [a]	33.59 ± 6.33 [b]	60.50 ± 5.34 [a]
TcSOD transcript	1.01 ± 0.01 [a]	0.45 ± 0.09 [b]	0.44 ± 0.06 [b]
TcAPX	1.17 ± 0.10 [a]	0.14 ± 0.04 [b]	1.29 ± 0.20 [a]

* SOD activity ($n = 4$) and production of H_2O_2 ($n = 3$) were assessed in living epimastigotes maintained in RM, IDM, or IDM + Fe. The quantification of the TcSOD ($n = 5$) and of the TcAPX transcripts ($n = 5$) in *T. cruzi* epimastigotes was done using 100 ng cDNA from epimastigotes grown in each culture medium. We used one-way ANOVA with Tukey's test to assess differences between mean values; Different lower-case letters as superscripts indicate statistical differences among the mean values in the same line ($p < 0.05$).

3.7. Differentiation of Epimastigotes Is Lower in Fe-Depleted Medium and Is Recovered after Fe Supplementation

Differentiation is a crucial step in the life cycle of *T. cruzi* because it ensures the success of infection and because the redox status of the parasitic environment is an important modulator [62]. Figure 5 demonstrates that, in culture, differentiation is critically dependent on the presence of Fe in the medium, and this dependence is more evident at initial times. While differentiation was barely detectable along the first day in IDM (filled circles), it exhibited a linear behavior in RM (empty circles) and depicted a burst in IDM + Fe (gray circles). Remarkably, the initial differentiation rate slop doubled in this group regarding RM parasites, but it seemed to tend toward a lower plateau from the 3rd day of culture.

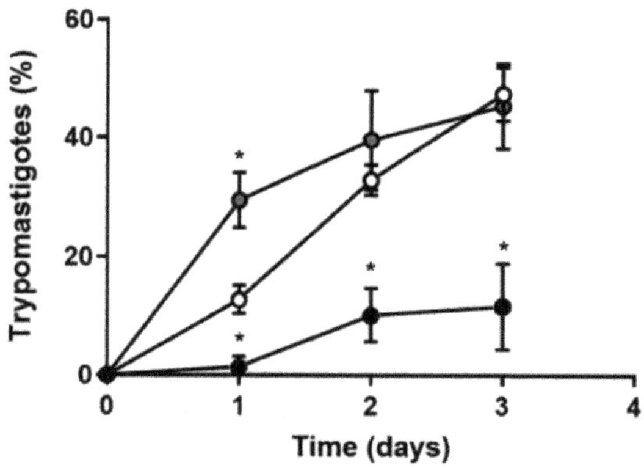

Figure 5. Fe depletion arrests *T. cruzi* differentiation to trypomastigotes. Epimastigotes cultivated in RM (empty circles), IDM (black circles), or IDM + Fe (gray circles) were submitted to in vitro differentiation media TAU. Each day, epimastigotes and trypomastigotes were differentially counted to determine the percentage of trypomastigotes in culture ($n = 4$); * $p < 0.05$ comparing time-matched determinations with respect to RM. Using one-way ANOVA with Tukey's test, we assessed differences between mean values.

4. Discussion

The central findings in the present study reveal a key role of medium ionic Fe in the proliferation of *T. cruzi* epimastigotes, with Fe depletion promoting increased oxidative stress, selective modifications in the intracellular ATP content, alterations in the HRI→eIF2α and PKA signaling pathways, increased lipid accumulation in the reservosomes, decreased mitochondrial function, and inhibition of differentiation toward trypomastigotes, in a metabolic condition shifted from respiration to glycolysis. This ensemble of results points to a pleiotropic function of ionic Fe in connected processes and pathways in *T. cruzi* epimastigotes. Using the Dm28c strain, which differentiates from trypomastigotes, allowed us to investigate the influence of exogenous Fe on the evolution of epimastigotes to trypomastigotes and, therefore, on a vital step of the parasite's life cycle.

The Fe depletion-induced lower intracellular content of Fe (Table 2). Notably, in the IDM + Fe medium, the expression of TcFR increased without a parallel increase in TcIT transcription. Even though TcFR and TcIT are coupled in the process of Fe uptake by the parasite, their expression is differentially modulated by the intracellular Fe content. Free Fe could regulate the TcIT transcription, so in the IDM + Fe medium, TcIT is downregulated relative to the TcIT transcript in IDM. Differently, TcFR uses Fe-containing proteins for the reaction $Fe^{3+} \rightarrow Fe^{2+}$, and since these proteins (hemin and transferrin) were neither added to IDM nor to IDM + Fe, TcFR is upregulated in both cases (Table 2).

The decreased succinate-cytochrome c oxidoreductase activity, the dropped O_2 consumption in the presence of normal partial pressure of O_2, and the lower intracellular ATP (Table 4) are indicative of mitochondrial damage, as proposed several years ago [63]. The impairment of the mitochondrial function seems to be functional because the ultrastructure of the organelle is preserved (Figure 3). Therefore, it could be hypothesized that Fe starvation promotes dysfunction at the level of the iron-sulfur clusters in the heterodimeric $SDH2_N$:$SDH2_C$ subunit described in the mitochondrial complex II from *T. cruzi* [64], as well as in the FoF_1-ATPase [65], a possibility that emerges from the accentuated inhibition of respiration in the presence of ADP (the oxphos state) and the increased $\Delta\Psi_m$ (Table 4). Although succinate-cytochrome c oxidoreductase activity is restored by Fe-citrate supplementation (Table 4), mitochondrial function impairment seems linked to Fe-protein depletion, especially hemin. It has been demonstrated that, in epimastigotes, heme changes mitochondrial physiology [40]. NADH-ubiquinone oxidoreductase gene (0.8-fold) and succinate dehydrogenase (1.40-fold) are upregulated in the presence of heme. Besides, heme influences *T. cruzi* epimastigote energy metabolism. The contribution to ATP synthesis may depend on glycosomal fermentation, which provides energy support for the parasite's growth, an establishment inside the vector [66], and differentiation into trypomastigotes (Figure 5).

The proposal that Fe and hemin depletion promotes a shift from oxidative metabolism to a glycolytic one is reinforced by the increased GADPH expression and activity (Figure 4A,B) and the increased glucokinase activity (Figure 4D). The upregulated glycolytic and pentose phosphate pathways, which are considered central for the glucose metabolism in *T. cruzi* [57], likely provide acetyl-CoA and NADPH, respectively, for the proposed increase of fatty acid synthesis accumulation within the reservosomes (Figure 3). As mentioned earlier, these organelles present a varied repertoire of enzymes that catalyze different lipid metabolism pathways [50], and for this reason, lipid accumulation in the reservosomes deserves special discussion in the context of the other enzyme modifications and proliferation.

The impairment of the IDM in the life cycle of epimastigotes could be linked to the alterations encountered in HRI→eIF2α and PKA signaling. The increased HRI, which phosphorylates eIF2α [54], together with the downregulation of eIF2α itself (Table 3), likely culminate in repressed gene expression and overall protein translation, thus compromising the evolution of the parasite in the Fe-deprived medium. The pronounced downregulation of PKA activity (and expression) could be associated with the downregulation of lipase activity and fatty acid release and oxidation. It may be that decreased PKA activity in

IDM parasites (Table 3) results in the inhibition of a PKA-modulated lipase and functional immobilization of the lipids in the reservosomes. This proposal is supported by the fact that PKA recovery and expression increased to levels even higher than in RM conditions after Fe supplementation (Table 3), which ensures lipid turnover and recovery of parasite evolution as discussed above. As additional support for this view, it is noteworthy that the genetic inhibition of PKA is lethal for *T. cruzi* [67].

Lipolysis in *T. cruzi* is associated with glucose metabolism [57]. For this reason, the upregulation of central enzymes of the glycolytic pathway, GADPH, and glucokinase, in both IDM and IDM + Fe media (Figure 4), leads us to hypothesize that the absence of hemin is central to the upregulation of glycolysis and the pentose phosphate pathway, but that replenishing of ionic Fe is responsible for the possible stimulation of lipid hydrolysis, glycerol release, formation of glycerol 3-phosphate catalyzed by a Tc-glycerol kinase [68], and further feeding of the glycolytic pathway. The other metabolic branch after Fe-stimulated lipid turnover, the β-oxidation of fatty acids [69], can feed the acetyl-CoA pool in the epimastigotes cell, further stimulating the formation of ATP via its condensation with succinate, synthesis of succinyl-CoA, and recycling of succinate with the release of CoA, as proposed in genomic studies carried out in *T. cruzi* [57] and earlier demonstrated in *T. brucei* [70].

Ionic Fe and heme depletion lead to a downregulation of total FeSOD, regardless of FeSOD origin. It is possible that SOD activity is higher in epimastigotes maintained at Fe/heme or heme depletion, demonstrating a compensating mechanism, probably due to higher activity of cytosolic and mitochondrial FeSOD (FeSODB and FeSODA). While FeSODB has a crucial role in the defense of parasites against $O_2^{\bullet-}$ [71], FeSODA is related to mitochondrial redox balance and generates the signaling molecule for amastigote differentiation, H_2O_2 [72]. A downregulation of H_2O_2 levels in Fe depletion conditions probably deregulates parasite differentiation. These low H_2O_2 levels could be due to a non-enzymatic system besides the glutathione ascorbate cycle. Recently, it was demonstrated that *T. cruzi* trypomastigotes employ ROS as a signaling molecule to differentiate, whereas epimastigotes use ROS to proliferate rather than differentiate [72].

Finally, Figure 6 presents a hypothetical mechanistic model regarding the overall mechanisms occurring during Fe depletion or Fe supplementation. Although the metabolic shift occurs in both cases, the ROS formation and pathway signaling present slight differences that culminate in differentiation/proliferation impairment (Figure 6A), which is restored by Fe supplementation (Figure 6B). In conclusion, although heme (or Fe-containing proteins) is essential for a functional mitochondrial metabolism, exogenous Fe is required for proper signaling to control parasite proliferation and H_2O_2 formation, which stimulate parasite differentiation, thus interfering with parasite virulence. These related mechanisms and processes modulated by exogenous ionic Fe have implications for human health because, by providing energy support for parasite growth and differentiation, they ensure the continuity of the *T. cruzi* life cycle and the propagation of Chagas disease.

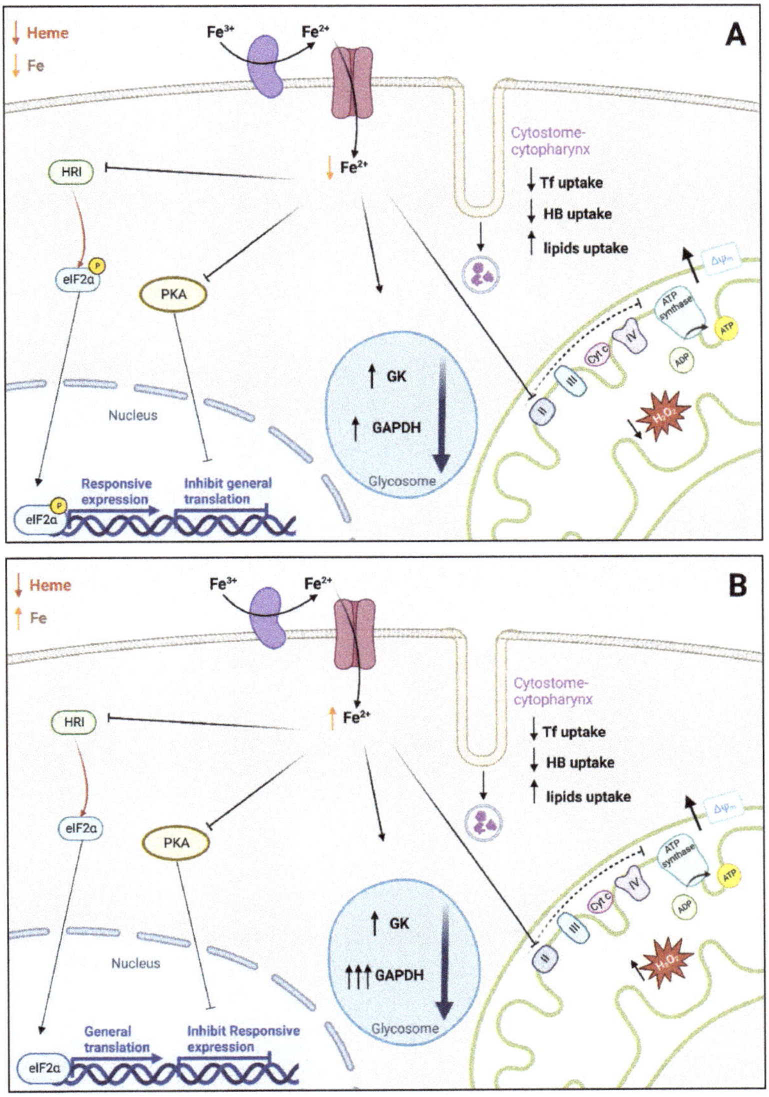

Figure 6. A proposed model describing the main events occurring on *T. cruzi* submitted to (**A**) Fe depletion or (**B**) Fe supplementation in low heme concentration (left outside part of the figure), which culminate in decreased or restored intracellular Fe^{2+}, as demonstrated by the opposite direction of the orange arrow near the upper part of the figure. The decrease in intracellular Fe in (**A**) was 50% in parasites grown in IDM (Table 2). The vesicles near the cell membrane (cytostome-cytopharynx region) indicate the selective modifications in the endocytic protein and lipid uptake. The left part of the arrow ensembles represents the modifications of the HRI→eIF2α and PKA signaling pathways and the different modifications in expression and translation. The blue circles in the middle of the panels present the quantitative differences in GAPDH abundance and activity. The blunt black arrows (in the right part of both panels) point to the impaired mitochondrial function provoked by Fe depletion, which is not sufficiently restored by Fe supplementation. The red stars in the lower right corner depict the opposite effects of Fe depletion and Fe supplementation on H_2O_2 formation. The figure was designed using BioRender.com.

Supplementary Materials: The following supporting information can be downloaded at: https://www.mdpi.com/article/10.3390/antiox12050984/s1. Table S1: Primer Information. Figure S1: Cytometry demonstrates selective modifications of protein uptake induced by ionic Fe depletion.

Author Contributions: Conceptualization, C.F.D. and A.V.; methodology, C.F.D., L.F.C.-K., C.L.A. and M.A.L.-A.; validation, C.F.D., L.F.C.-K., C.L.A. and M.A.L.-A.; formal analysis, C.F.D., C.L.A. and A.V.; investigation, C.F.D., L.F.C.-K., C.L.A. and M.A.L.-A.; resources, J.R.M.-F., N.L.C.-e.-S. and A.V.; writing—original draft preparation, C.F.D. and A.V.; writing—review and editing, C.F.D., C.L.A., N.L.C.-e.-S., J.R.M.-F. and A.V.; visualization, C.F.D., L.F.C.-K., C.L.A. and M.A.L.-A.; supervision, J.R.M.-F. and A.V.; project administration, C.F.D. and A.V.; funding acquisition, J.R.M.-F. and A.V. All authors have read and agreed to the published version of the manuscript.

Funding: This research was funded by the Brazilian National Research Council/CNPq to J.R.M.-F. (grant # 401134/2014-8), and A.V. (grant # 307605/2015-9). The support of the Rio de Janeiro State Research Foundation/FAPERJ to J.R.M.-F. (grant # E-26/201.300/2014), N.L.C.-e.-S. (grant # E-26/210.449/2019), and A.V. (grant # E-26/2012.963/2017) is also acknowledged.

Institutional Review Board Statement: Not applicable.

Informed Consent Statement: Not applicable.

Data Availability Statement: The original contributions presented in the study are included in the article and the supporting material. Further inquiries can be directed to the corresponding author.

Conflicts of Interest: The authors declare no conflict of interest.

Abbreviations

ANOVA	Analysis of variance
BHI medium	Brain heart infusion medium
BSA	Bovine serum albumin
DAPI	Diamidino-2-phenylindole
eIF2α	Eukaryotic initiation factor 2α
ETS	Mitochondrial respiration uncoupled from phosphorylation
FCCP	Carbonyl cyanide 4-(trifluoromethoxy)phenylhydrazone
Fe	Iron
FITC	Fluorescein isothiocyanate
G418	Geneticin (polypeptide synthesis blocker)
G6PDH	Glucose-6-phosphate dehydrogenase
HB	Hemoglobin
HRI	Heme-regulated inhibitor
HRP	Horseradish peroxidase
IAA	Iodoacetamide
IBMX	3-isobutyl-1-methylxanthine
IDM	Iron-depleted medium
IDM + Fe	Iron-depleted medium plus 8 μM Fe citrate
JC-1	5,5′,6,6′-tetrachloro-1,1′,3,3′-tetraethyl-imidacarbocyanine iodide
LEAK	Mitochondrial respiration in the presence of respiratory substrate only
LIT	Leishmania amazonensis iron transporter family
NBT	Nitro blue tetrazolium
$O_2^{\bullet-}$	Anion superoxide
$OH^{\bullet}$	Hydroxyl radical
OTO	Osmium-thiocarbohydrazide-osmium
OXPHOS	Oxidative phosphorylation state (mitochondrial respiration in the presence of ADP)
PBS	Phosphate buffer saline
PKA	cyclic AMP-dependent protein kinase
RIPA buffer	Radioimmunoprecipitation assay buffer
RM	Regular medium containing Fe and hemin
ROS	Reactive O_2 species
ROX	Residual mitochondrial respiration after inhibition of Complex III

RPMI medium	Roswell Park Memorial Institute medium
RT-PCR	Reverse transcription polymerase chain reaction
SEM	Standard error of the mean
TAU	Triatomine artificial urine
TcAPX	*T. cruzi* ascorbate peroxidase
TcFR	*T. cruzi* Fe reductase
TcIT	*T. cruzi* Fe transporter
TcSOD	*T. cruzi* superoxide dismutase
Tf	Transferrin
Tryp	Metacyclic epimastigotes
ZIP Family	Zinc and iron transporter family

References

1. Figueiredo, R.C.; Rosa, D.S.; Soares, M.J. Differentiation of *Trypanosoma cruzi* epimastigotes: Metacyclogenesis and adhesion to substrate are triggered by nutritional stress. *J. Parasitol.* **2000**, *86*, 1213–1218. [CrossRef]
2. Rassi, A., Jr.; Rassi, A.; Marin-Neto, J.A. Chagas disease. *Lancet* **2010**, *375*, 1388–1402. [CrossRef]
3. Franco-Paredes, C.; Von, A.; Hidron, A.; Rodríguez-Morales, A.J.; Tellez, I.; Barragán, M.; Jones, D.; Náquira, C.G.; Mendez, J. Chagas disease: An impediment in achieving the Millennium Development Goals in Latin America. *BMC Int. Health Hum.* **2007**, *7*, 7. [CrossRef]
4. Dlouhy, A.C.; Outten, C.E. The iron metallome in eukaryotic organisms. *Metallomics* **2013**, *12*, 241–278. [CrossRef]
5. Taylor, M.C.; Kelly, J.M. Iron metabolism in trypanosomatids, and its crucial role in infection. *Parasitology* **2010**, *137*, 899–917. [CrossRef] [PubMed]
6. Sunda, W.G.; Huntsman, S.A. High iron requirement for growth, photosynthesis, and low-light acclimation in the coastal cyanobacterium *Synechococcus bacillaris*. *Front. Microbiol.* **2015**, *6*, 561. [CrossRef] [PubMed]
7. Benjamin, J.A.; Desnoyers, G.; Morissette, A.; Salvail, H.; Massé, E. Dealing with oxidative stress and iron starvation in microorganisms: An overview. *Can. J. Physiol. Pharmacol.* **2010**, *88*, 264–272. [CrossRef]
8. Miethke, M.; Marahiel, M.A. Siderophore-based iron acquisition and pathogen control. *Microbiol. Mol. Biol. Rev.* **2007**, *71*, 413–451. [CrossRef]
9. Dormeyer, M.; Schöneck, R.; Dittmar, G.A.; Krauth-Siegel, R.L. Cloning, sequencing and expression of ribonucleotide reductase R2 from *Trypanosoma brucei*. *FEBS Lett.* **1997**, *414*, 449–453. [CrossRef]
10. Hofer, A.; Schmidt, P.P.; Gräslund, A.; Thelander, L. Cloning and characterization of the R1 and R2 subunits of ribonucleotide reductase from *Trypanosoma brucei*. *Proc. Natl. Acad. Sci. USA* **1997**, *94*, 6959–6964. [CrossRef] [PubMed]
11. Fairlamb, A.H.; Bowman, I.B. *Trypanosoma brucei*: Suramin and other trypanocidal compounds' effects on sn-glycerol-3-phosphate oxidase. *Exp. Parasitol.* **1997**, *43*, 353–361. [CrossRef] [PubMed]
12. Le Trant, N.; Meshnick, S.R.; Kitchener, K.; Eaton, J.W.; Cerami, A. Iron-containing superoxide dismutase from *Crithidia fasciculata*. Purification, characterization, and similarity to leishmanial and trypanosomal enzymes. *J. Biol. Chem.* **1983**, *258*, 125–130. [CrossRef] [PubMed]
13. Barasch, J.; Mori, K. Cell biology: Iron thievery. *Nature* **2004**, *432*, 811–813. [CrossRef] [PubMed]
14. Schaible, U.E.; Kaufmann, S.H. Iron and microbial infection. *Nat. Rev. Microbiol.* **2004**, *2*, 946–953. [CrossRef]
15. Braun, V. Bacterial iron transport related to virulence. *Contrib. Microbiol.* **2005**, *12*, 210–233. [CrossRef]
16. Ratledge, C. Iron metabolism and infection. *Food Nutr. Bull.* **2007**, *28*, S515–S523. [CrossRef]
17. Lalonde, R.G.; Holbein, B.E. Role of iron in *Trypanosoma cruzi* infection of mice. *J. Clin. Investig.* **1984**, *73*, 470–476. [CrossRef]
18. Lima, M.F.; Villalta, F. *Trypanosoma cruzi* receptors for human transferrin and their role. *Mol. Biochem. Parasitol.* **1990**, *38*, 245–252. [CrossRef]
19. Porto-Carreiro, I.; Attias, M.; Miranda, K.; De Souza, W.; Cunha-e-Silva, N. *Trypanosoma cruzi* epimastigote endocytic pathway: Cargo enters the cytostome and passes through an early endosomal network before storage in reservosomes. *Eur. J. Cell Biol.* **2000**, *79*, 858–869. [CrossRef]
20. Lara, F.A.; Sant'anna, C.; Lemos, D.; Laranja, G.A.; Coelho, M.G.; Reis Salles, I.; Michel, A.; Oliveira, P.L.; Cunha-E-Silva, N.; Salmon, D.; et al. Heme requirement and intracellular trafficking in *Trypanosoma cruzi* epimastigotes. *Biochem. Biophys. Res. Commun.* **2007**, *355*, 16–22. [CrossRef]
21. Cupello, M.P.; Souza, C.F.; Buchensky, C.; Soares, J.B.; Laranja, G.A.; Coelho, M.G.; Cricco, J.A.; Paes, M.C. The heme uptake process in *Trypanosoma cruzi* epimastigotes is inhibited by heme analogues and by inhibitors of ABC transporters. *Acta Trop.* **2011**, *120*, 211–218. [CrossRef]
22. Flannery, A.R.; Huynh, C.; Mittra, B.; Mortara, R.A.; Andrews, N.W. LFR1 ferric iron reductase of *Leishmania amazonensis* is essential for the generation of infective parasite forms. *J. Biol. Chem.* **2011**, *286*, 23266–23279. [CrossRef]

23. Huynh, C.; Andrews, N.W. Iron acquisition within host cells and the pathogenicity of *Leishmania*. *Cell Microbiol.* **2008**, *10*, 293–300. [CrossRef] [PubMed]
24. Wilson, M.E.; Lewis, T.S.; Miller, M.A.; McCormick, M.L.; Britigan, B.E. *Leishmania chagasi*: Uptake of iron bound to lactoferrin or transferrin requires an iron reductase. *Exp. Parasitol.* **2002**, *100*, 196–207. [CrossRef]
25. Dick, C.F.; De Moura Guimarães, L.; Carvalho-Kelly, L.F.; Cortes, A.L.; Morcillo, L.S.L.; Sampaio, L.S.; Meyer-Fernandes, J.R.; Vieyra, A. A ferric reductase of *Trypanosoma cruzi* (TcFR) is involved in iron metabolism in the parasite. *Exp. Parasitol.* **2020**, *217*, 107962. [CrossRef] [PubMed]
26. Mach, J.; Tachezy, J.; Sutak, R. Efficient iron uptake via a reductive mechanism in procyclic *Trypanosoma brucei*. *J. Parasitol.* **2013**, *99*, 363–364. [CrossRef]
27. Huynh, C.; Sacks, D.L.; Andrews, N.W. A *Leishmania amazonensis* ZIP family iron transporter is essential for parasite replication within macrophage phagolysosomes. *J. Exp. Med.* **2006**, *203*, 2363–2375. [CrossRef] [PubMed]
28. Li, S.; Zhou, X.; Huang, Y.; Zhu, L.; Zhang, S.; Zhao, Y.; Guo, J.; Chen, J.; Chen, R. Identification and characterization of the zinc-regulated transporters, iron-regulated transporter-like protein (ZIP) gene family in maize. *BMC Plant Biol.* **2013**, *13*, 114. [CrossRef] [PubMed]
29. Dick, C.F.; Rocco-Machado, N.; Dos-Santos, A.L.A.; Carvalho-Kelly, L.F.; Alcantara, C.L.; Cunha-E-Silva, N.L.; Meyer-Fernandes, J.R.; Vieyra, A. An Iron Transporter Is Involved in Iron Homeostasis, Energy Metabolism, Oxidative Stress, and Metacyclogenesis in *Trypanosoma cruzi*. *Front. Cell Infect. Microbiol.* **2022**, *11*, 789401. [CrossRef]
30. Atkins, P.; de Paula, J. *Physical Chemistry for the Life Sciences*; Oxford University Press: Oxford, UK, 2006; pp. 200–236.
31. Krumova, K.; Cosa, G. Overview of reactive oxygen species. In *Singlet Oxygen: Applications in Biosciences and Nanosciences*; Nonell, S., Flors, C., Eds.; Royal Society of Chemistry: London, UK, 2016; Volume 1, pp. 1–21.
32. Corrêa, A.F.; Andrade, L.R.; Soares, M.J. Elemental composition of acidocalcisomes of *Trypanosoma cruzi* bloodstream trypomastigote forms. *Parasitol. Res.* **2002**, *88*, 875–880. [CrossRef]
33. Ferella, M.; Nilsson, D.; Darban, H.; Rodrigues, C.; Bontempi, E.J.; Docampo, R.; Andersson, B. Proteomics in *Trypanosoma cruzi*—Localization of novel proteins to various organelles. *Proteomics* **2008**, *8*, 2735–2749. [CrossRef] [PubMed]
34. Paiva, C.N.; Feijó, D.F.; Dutra, F.F.; Carneiro, V.C.; Freitas, G.B.; Alves, L.S.; Mesquita, J.; Fortes, G.B.; Figueiredo, R.T.; Souza, H.S.; et al. Oxidative Stress Fuels *Trypanosoma cruzi* Infection in Mice. *J. Clin. Investig.* **2012**, *122*, 2531–2542. [CrossRef]
35. Koeller, C.M.; van der Wel, H.; Feasley, C.L.; Abreu, F.; da Rocha, J.D.; Montalvão, F.; Fampa, P.; Dos Reis, F.C.; Atella, G.C.; Souto-Padrón, T.; et al. Golgi UDP-GlcNAc:Polypeptide O-α-N-Acetyl-d-Glucosaminyltransferase 2 (TcOGNT2) Regulates Trypomastigote Production and Function in *Trypanosoma cruzi*. *Eukaryot. Cell* **2014**, *13*, 1312–1327. [CrossRef]
36. Mittra, B.; Cortez, M.; Haydock, A.; Ramasamy, G.; Myler, P.J.; Andrews, N.W. Iron Uptake Controls the Generation of *Leishmania* Infective Forms through Regulation of ROS Levels. *J. Exp. Med.* **2013**, *210*, 401–416. [CrossRef]
37. Rozen, S.; Skaletsky, H.J. Primer3 on the WWW for general users and for biologist programmers. *Methods Mol. Biol.* **2000**, *132*, 365–386. [CrossRef] [PubMed]
38. Livak, K.J.; Schmittgen, T.D. Analysis of relative gene expression data using real-time quantitative PCR and the $2^{-\Delta\Delta C(T)}$ method. *Methods* **2001**, *25*, 402–408. [CrossRef] [PubMed]
39. Alcantara, C.L.; Vidal, J.C.; de Souza, W.; Cunha-e-Silva, N.L. The three-dimensional structure of the cytostome-cytopharynx complex of Trypanosoma cruzi epimastigotes. *J. Cell Sci.* **2014**, *127*, 2227–2237. [CrossRef]
40. Nogueira, N.P.; Saraiva, F.M.S.; Oliveira, M.P.; Mendonça, A.P.M.; Inacio, J.D.F.; Almeida-Amaral, E.E.; Menna-Barreto, R.F.; Laranja, G.A.T.; Torres, E.J.L.; Oliveira, M.F.; et al. Heme Modulates *Trypanosoma cruzi* Bioenergetics Inducing Mitochondrial ROS Production. *Free Radic. Biol. Med.* **2017**, *108*, 183–191. [CrossRef]
41. Ferguson, M.; Mockett, R.J.; Shen, Y.; Orr, W.C.; Sohal, R.S. Age-associated decline in mitochondrial respiration and electron transport in *Drosophila melanogaster*. *Biochem. J.* **2005**, *390*, 501–511. [CrossRef]
42. Bradford, M.M. A rapid and sensitive method for the quantitation of microgram quantities of protein utilizing the principle of protein-dye binding. *Anal. Biochem.* **1976**, *72*, 248–254. [CrossRef]
43. Lacerda-Abreu, M.A.; Russo-Abrahão, T.; Rocco-Machado, N.; Cosentino-Gomes, D.; Dick, C.F.; Carvalho-Kelly, L.F.; Cunha Nascimento, M.T.; Rocha-Vieira, T.C.; Meyer-Fernandes, J.R. Hydrogen Peroxide Generation as an Underlying Response to High Extracellular Inorganic Phosphate (Pi) in Breast Cancer Cells. *Int. J. Mol. Sci.* **2021**, *22*, 10096. [CrossRef]
44. Carvalho-Kelly, L.F.; Dick, C.F.; Rocco-Machado, N.; Gomes-Vieira, A.L.; Paes-Vieira, L.; Meyer-Fernandes, J.R. Anaerobic ATP synthesis pathways and inorganic phosphate transport and their possible roles in encystment in *Acanthamoeba castellanii*. *Mol. Cell Biol.* **2022**, *46*, 1288–1298. [CrossRef] [PubMed]
45. Bressi, J.C.; Verlinde, C.L.; Aronov, A.M.; Shaw, M.L.; Shin, S.S.; Nguyen, L.N.; Suresh, S.; Buckner, F.S.; Van Voorhis, W.C.; Kuntz, I.D.; et al. Adenosine analogues as selective inhibitors of glyceraldehyde-3-phosphate dehydrogenase of Trypanosomatidae via structure-based drug design. *J. Med. Chem.* **2001**, *44*, 2080–2093. [CrossRef] [PubMed]
46. Wiggers, H.J.; Cheleski, J.; Zottis, A.; Oliva, G.; Andricopulo, A.D.; Montanari, C.A. Effects of organic solvents on the enzyme activity of *Trypanosoma cruzi* glyceraldehyde-3-phosphate dehydrogenase in calorimetric assays. *Anal. Biochem.* **2007**, *370*, 107–114. [CrossRef]

47. Gomes, M.T.; Paes-Vieira, L.; Gomes-Vieira, A.L.; Cosentino-Gomes, D.; da Silva, A.P.P.; Giarola, N.L.L.; Da Silva, D.; Sola-Penna, M.; Galina, A.; Meyer-Fernandes, J.R. 3-Bromopyruvate: A new strategy for inhibition of glycolytic enzymes in *Leishmania amazonensis*. *Exp. Parasitol.* **2021**, *229*, 108154. [CrossRef] [PubMed]
48. Dos-Santos, A.L.A.; Dick, C.F.; Lopes, L.R.; Rocco-Machado, N.; Muzi-Filho, H.; Freitas-Mesquita, A.L.; Paes-Vieira, L.; Vieyra, A.; Meyer-Fernandes, J.R. Tartrate-resistant phosphatase type 5 in *Trypanosoma cruzi* is important for resistance to oxidative stress promoted by hydrogen peroxide. *Exp. Parasitol.* **2019**, *205*, 107748. [CrossRef]
49. Cunha-e-Silva, N.; Sant'Anna, C.; Pereira, M.G.; Porto-Carreiro, I.; Jeovanio, A.L.; de Souza, W. Reservosomes: Multipurpose organelles? *Parasitol. Res.* **2006**, *99*, 325–327. [CrossRef]
50. Sant'Anna, C.; Nakayasu, E.S.; Pereira, M.G.; Lourenço, D.; de Souza, W.; Almeida, I.C.; Cunha-E-Silva, N.L. Subcellular proteomics of *Trypanosoma cruzi* reservosomes. *Proteomics* **2009**, *9*, 1782–1794. [CrossRef]
51. Pereira, M.G.; Nakayasu, E.S.; Sant'Anna, C.; De Cicco, N.N.; Atella, G.C.; de Souza, W.; Almeida, I.C.; Cunha-e-Silva, N. *Trypanosoma cruzi* epimastigotes are able to store and mobilize high amounts of cholesterol in reservosome lipid inclusions. *PLoS ONE* **2011**, *6*, e22359. [CrossRef]
52. Pereira, M.G.; Visbal, G.; Salgado, L.T.; Vidal, J.C.; Godinho, J.L.; De Cicco, N.N.; Atella, G.C.; de Souza, W.; Cunha-e-Silva, N. *Trypanosoma cruzi* Epimastigotes Are Able to Manage Internal Cholesterol Levels under Nutritional Lipid Stress Conditions. *PLoS ONE* **2015**, *10*, e0128949. [CrossRef] [PubMed]
53. Zhang, S.; Macias-Garcia, A.; Ulirsch, J.C.; Velazquez, J.; Butty, V.L.; Levine, S.S.; Sankaran, V.G.; Chen, J.J. HRI Coordinates Translation Necessary for Protein Homeostasis and Mitochondrial Function in Erythropoiesis. *eLife* **2019**, *8*, e46976. [CrossRef] [PubMed]
54. Pakos-Zebrucka, K.; Koryga, I.; Mnich, K.; Ljujic, M.; Samali, A.; Gorman, A.M. The Integrated Stress Response. *EMBO Rep.* **2016**, *17*, 1374–1395. [CrossRef] [PubMed]
55. Caza, M.; Kronstad, J.W. The cAMP/Protein Kinase A Pathway Regulates Virulence and Adaptation to Host Conditions in *Cryptococcus neoformans*. *Front. Cell Infect. Microbiol.* **2019**, *9*, 212. [CrossRef]
56. Gahura, O.; Hierro-Yap, C.; Zíková, A. Redesigned and Reversed: Architectural and Functional Oddities of the Trypanosomal ATP Synthase. *Parasitology* **2021**, *148*, 1151–1160. [CrossRef] [PubMed]
57. Maugeri, D.A.; Cannata, J.J.; Cazzulo, J.J. Glucose metabolism in *Trypanosoma cruzi*. *Essays Biochem.* **2011**, *51*, 15–30. [CrossRef]
58. Heneberg, P. Redox Regulation of Hexokinases. *Antioxid. Redox Signal* **2019**, *30*, 415–442. [CrossRef]
59. Mateo-Carrasco, H.; Serrano-Castro, P.J.; Molina-Cuadrado, E.; Goodwin, M.; Nguyen, T.V.; Kotecha, P.N. Role of high-dose levetiracetam as add-on therapy for intractable epilepsy: Case report and brief review of the literature. *Int. J. Clin. Pharm.* **2015**, *37*, 559–562. [CrossRef]
60. Wilkinson, S.R.; Obado, S.O.; Mauricio, I.L.; Kelly, J.M. *Trypanosoma cruzi* expresses a plant-like ascorbate-dependent hemoperoxidase localized to the endoplasmic reticulum. *Proc. Natl. Acad. Sci. USA* **2002**, *99*, 13453–13458. [CrossRef]
61. Nogueira, F.B.; Rodrigues, J.F.; Correa, M.M.; Ruiz, J.C.; Romanha, A.J.; Murta, S.M. The level of ascorbate peroxidase is enhanced in benznidazole-resistant populations of *Trypanosoma cruzi* and its expression is modulated by stress generated by hydrogen peroxide. *Mem. Inst. Oswaldo Cruz* **2012**, *107*, 494–502. [CrossRef] [PubMed]
62. Nogueira, N.P.; Saraiva, F.M.; Sultano, P.E.; Cunha, P.R.; Laranja, G.A.; Justo, G.A.; Sabino, K.C.; Coelho, M.G.; Rossini, A.; Atella, G.C.; et al. Proliferation and Differentiation of *Trypanosoma cruzi* Inside Its Vector Have a New Trigger: Redox Status. *PLoS ONE* **2015**, *10*, e0116712. [CrossRef]
63. Warburg, O. On the Origin of Cancer Cells. *Science* **1956**, *123*, 309–314. [CrossRef] [PubMed]
64. Morales, J.; Mogi, T.; Mineki, S.; Takashima, E.; Mineki, R.; Hirawake, H.; Sakamoto, K.; Omura, S.; Kita, K. Novel Mitochondrial Complex II Isolated from *Trypanosoma cruzi* Is Composed of 12 Peptides Including a Heterodimeric Ip Subunit. *J. Biol. Chem.* **2009**, *284*, 7255–7263. [CrossRef] [PubMed]
65. Racker, E. Reconstitution of Cytochrome Oxidase Vesicles and Conferral of Sensitivity to Energy Transfer Inhibitors. *J. Membr. Biol.* **1972**, *10*, 221–235. [CrossRef]
66. Paes, M.C.; Saraiva, F.M.S.; Nogueira, N.P.; Vieira, C.S.D.; Dias, F.A.; Rossini, A.; Coelho, V.L.; Pane, A.; Sang, F.; Alcocer, M. Gene Expression Profiling of *Trypanosoma cruzi* in the Presence of Heme Points to Glycosomal Metabolic Adaptation of Epimastigotes Inside the Vector. *PLoS Negl. Trop. Dis.* **2020**, *14*, e0007945. [CrossRef]
67. Bao, Y.; Weiss, L.M.; Braunstein, V.L.; Huang, H. Role of Protein Kinase A in *Trypanosoma cruzi*. *Infect. Immun.* **2008**, *76*, 4757–4763. [CrossRef]
68. Quiñones, W.; Acosta, H.; Gonçalves, C.S.; Motta, M.C.M.; Gualdrón-López, M.; Michels, P.A.M. Structure, Properties, and Function of Glycosomes in *Trypanosoma cruzi*. *Front. Cell Infect. Microbiol.* **2020**, *10*, 25. [CrossRef] [PubMed]
69. Acosta, H.; Dubourdieu, M.; Quiñones, W.; Cáceres, A.; Bringaud, F.; Concepción, J.L. Pyruvate phosphate dikinase and pyrophosphate metabolism in the glycosome of Trypanosoma cruzi epimastigotes. *Comp. Biochem. Physiol. B Biochem. Mol. Biol.* **2004**, *138*, 347–356. [CrossRef] [PubMed]
70. Rivière, L.; van Weelden, S.W.; Glass, P.; Vegh, P.; Coustou, V.; Biran, M.; van Hellemond, J.J.; Bringaud, F.; Tielens, A.G.; Boshart, M. Acetyl:succinate CoA-transferase in procyclic *Trypanosoma brucei*. Gene identification and role in carbohydrate metabolism. *J. Biol. Chem.* **2004**, *279*, 45337–45346. [CrossRef]

71. Martínez, A.; Prolo, C.; Estrada, D.; Rios, N.; Alvarez, M.N.; Piñeyro, M.D.; Robello, C.; Radi, R.; Piacenza, L. Cytosolic Fe-superoxide dismutase safeguards *Trypanosoma cruzi* from macrophage-derived superoxide radical. *Proc. Natl. Acad. Sci. USA* **2019**, *116*, 8879–8888. [CrossRef]
72. Paula, J.I.O.; Pinto, J.D.S.; Rossini, A.; Nogueira, N.P.; Paes, M.C. New perspectives for hydrogen peroxide in the amastigogenesis of *Trypanosoma cruzi* in vitro. *Biochim. Biophys. Acta Mol. Basis Dis.* **2020**, *1866*, 165951. [CrossRef]

Antioxidants 2022, 11, 277

Article

Nitric Oxide Resistance in *Leishmania (Viannia) braziliensis* Involves Regulation of Glucose Consumption, Glutathione Metabolism and Abundance of Pentose Phosphate Pathway Enzymes

Nathalia Pinho [1,†], Ana Cristina Bombaça [2,†], Jacek R. Wiśniewski [3], Geovane Dias-Lopes [4], Leonardo Saboia-Vahia [1,‡], Elisa Cupolillo [1], José Batista de Jesus [5], Roque P. de Almeida [6], Gabriel Padrón [1,§], Rubem Menna-Barreto [2,*] and Patricia Cuervo [1,*]

1 Laboratório de Pesquisa em Leishmanioses, Instituto Oswaldo Cruz, Fiocruz, Rio de Janeiro 21040-360, RJ, Brazil; nathps11@gmail.com (N.P.); leovahia@gmail.com (L.S.-V.); elisa.cupolillo@ioc.fiocruz.br (E.C.); gpadronpalomares@gmail.com (G.P.)
2 Laboratório de Biologia Celular, Instituto Oswaldo Cruz, Fiocruz, Rio de Janeiro 21040-360, RJ, Brazil; anabombaca@gmail.com
3 Biochemical Proteomics Group, Department of Proteomics and Signal Transduction, Max-Planck-Institute of Biochemistry, 82152 Planegg, Germany; jwisniew@biochem.mpg.de
4 Laboratório de Biologia Molecular e Doenças Endêmicas, Instituto Oswaldo Cruz, Fiocruz, Rio de Janeiro 21040-360, RJ, Brazil; geovane.dl@gmail.com
5 Departamento de Medicina, Universidade Federal de São João Del Rei, São João del Rei 35501-296, MG, Brazil; jbj@ufsj.edu.br
6 Department of Medicine, Hospital Universitário, EBSERH, Universidade Federal de Sergipe, Aracaju 49100-000, SE, Brazil; roquepachecoalmeida@gmail.com
* Correspondence: rubemb@ioc.fiocruz.br (R.M.-B.); patricia.cuervo@fiocruz.br (P.C.); Tel.: +55-21-25621393 (R.M.-B.); +55-21-38658224 (P.C.); Fax: +55-21-25621432 (R.M.-B.); +55-21-38658195 (P.C.)
† These authors contributed equally to this work.
‡ Current address: Laboratório de Virus Respiratórios e Sarampo, Instituto Oswaldo Cruz, Fiocruz, Rio de Janeiro 21040-360, RJ, Brazil.
§ Current address: Center for Genetic Engineering & Biotechnology, La Habana 10600, Cuba.

Citation: Pinho, N.; Bombaça, A.C.; Wiśniewski, J.R.; Dias-Lopes, G.; Saboia-Vahia, L.; Cupolillo, E.; de Jesus, J.B.; de Almeida, R.P.; Padrón, G.; Menna-Barreto, R.; et al. Nitric Oxide Resistance in *Leishmania (Viannia) braziliensis* Involves Regulation of Glucose Consumption, Glutathione Metabolism and Abundance of Pentose Phosphate Pathway Enzymes. *Antioxidants* 2022, 11, 277. https://doi.org/10.3390/antiox11020277

Academic Editor: Serge Ankri

Received: 30 December 2021
Accepted: 23 January 2022
Published: 29 January 2022

Abstract: In American Tegumentary Leishmaniasis production of cytokines, reactive oxygen species and nitric oxide (NO) by host macrophages normally lead to parasite death. However, some *Leishmania braziliensis* strains exhibit natural NO resistance. NO-resistant strains cause more lesions and are frequently more resistant to antimonial treatment than NO-susceptible ones, suggesting that NO-resistant parasites are endowed with specific mechanisms of survival and persistence. To tests this, we analyzed the effect of pro- and antioxidant molecules on the infectivity in vitro of *L. braziliensis* strains exhibiting polar phenotypes of resistance or susceptibility to NO. In addition, we conducted a comprehensive quantitative mass spectrometry-based proteomics analysis of those parasites. NO-resistant parasites were more infective to peritoneal macrophages, even in the presence of high levels of reactive species. Principal component analysis of protein concentration values clearly differentiated NO-resistant from NO-susceptible parasites, suggesting that there are natural intrinsic differences at molecular level among those strains. Upon NO exposure, NO-resistant parasites rapidly modulated their proteome, increasing their total protein content and glutathione (GSH) metabolism. Furthermore, NO-resistant parasites showed increased glucose analogue uptake, and increased abundance of phosphotransferase and G6PDH after nitrosative challenge, which can contribute to NADPH pool maintenance and fuel the reducing conditions for the recovery of GSH upon NO exposure. Thus, increased glucose consumption and GSH-mediated redox capability may explain the natural resistance of *L. braziliensis* against NO.

Keywords: *Leishmania braziliensis*; nitric oxide resistance; American Tegumentary Leishmaniasis (ATL); nitrosative stress; reactive oxygen species; reactive nitrogen species; quantitative proteomics; FASP; mass spectrometry; glycolysis; glutathione metabolism

1. Introduction

Leishmania (*Viannia*) *braziliensis*, a New World Leishmania species, is an important etiological agent of American Tegumentary Leishmaniasis (ATL) in the Americas [1]. Infection with this species may have clinical outcomes, ranging from self-healing localized cutaneous lesions (LCL) to severe disseminated cutaneous and mucocutaneous forms that may result in facial mutilation by destruction of the palate and nose cartilage. Furthermore, disseminated cutaneous leishmaniasis (DL) and mucocutaneous leishmaniasis (MCL) caused by *L. braziliensis* are frequently refractory to treatment and result from the metastatic dissemination of the parasite [2]. Clinical manifestations result from both the host immune responses and the infecting parasite [3–5]. Indeed, intrinsic characteristics of parasites enable the subversion and/or immunomodulation of host immune responses, resulting in the inactivation of cell pathways crucial for parasite elimination [6–9].

Successful establishment of infection and further parasite persistence depend on a complex interaction between the *Leishmania*'s immune subversion arsenal and the microbicidal mechanisms of mononuclear phagocytes [10]. Such mechanisms include lysosomal enzymes, reactive oxygen species (ROS), and reactive nitrogen species (RNS). Oxidative burst and release of nitric oxide (NO) are the most effective mechanisms against *Leishmania* spp. [11,12]. ROS and RNS can be released when the macrophages are stimulated by TNF-α and IFN-γ, and their production is induced by the activation of routes that involve NADPH oxidases and inducible nitric oxide synthase (iNOS). When it happens, parasites are exposed to superoxide anion ($O_2^{\bullet -}$), hydrogen peroxide (H_2O_2), peroxynitrite ($ONOO^-$) and NO [13], which usually results in parasites elimination [11,14,15]. However, some *Leishmania* parasites present natural resistance to NO, helping them to escape the macrophage microbicidal responses and promoting their survival and persistence [16–19]. Indeed, *L. braziliensis* strains isolated from patients with different clinical forms of ATL exhibit different levels of natural resistance to NO. In vitro assays showed that amastigotes from NO-resistant strains survived and multiplied more in human macrophages. In addition, patients infected with NO-resistant parasites presented significantly more severe cutaneous lesions than those infected with NO-susceptible parasites [17].

In vitro infections of human macrophages with *L. braziliensis* NO-resistant strains showed a higher percentage of infected cells and reduced levels of TNF-α production than infections with NO-sensitive parasites. Furthermore, these NO-resistant strains were also correlated with higher refractoriness to pentavalent antimony, the first-line treatment for ATL, suggesting that NO resistance may be related to antimony resistance [20]. In addition, BALB/c mice infected with NO-resistant strains produce more IL-4, which stimulates increased expression of arginase-1, favoring parasites survival and severe forms of disease [21]. Additionally, *L. infantum* field isolates from relapse cases of visceral leishmaniasis are more resistant to antimonial and NO, as well as are more infective to macrophages in vitro, than parasites isolated from responsive patients [22]. However, the mechanisms by which *L. braziliensis* strains can resist/evade the nitrosative stress imposed by host cells have not been clearly defined. Potential mechanisms might include increased abundance of glucose-6-phosphate dehydrogenase (G6PDH), as shown in the study by 2DE-MS of the proteome of *L. infantum* strains resistant or susceptible to NO [23]. Nevertheless, an in-depth study of NO resistance in *Leishmania* is missing.

Based on the previous evidence, we hypothesize that the proteome of *L. braziliensis* NO-resistant strains is tailored to deal with nitrosative stress and, that upon NO challenge, it can be rapidly regulated to minimize the damages caused by nitric oxide. Such adaptations grant the survival and persistence of parasites, leading to chronic infections that are refractory to treatment. To test this, we performed an unbiased and comprehensive quantitative proteomic analysis of NO-resistant and NO-susceptible *L. braziliensis* strains. By using previously reported procedures for parasite sample preparation and absolute label-free protein quantification [24,25], we were able to compare those parasites before and after stimulus with an NO donor, identifying ~6300 proteins and estimating absolute concentrations of ~6000 of these proteins. We also evaluated the effect of nitrosative and oxidative stresses, as well as the effect of antioxidant molecules, on the infection index of each strain. Together, our data provide new evidence on the potential mechanisms underlying the NO resistance in *L. braziliensis*, including increased antioxidant capability involving the glutathione (GSH) system and rapid regulation of glycolysis and pentose phosphate pathway (PPP). Data are available via ProteomeXchange with identifier PXD029462.

2. Materials and Methods

2.1. Ethics Statement

All the protocols were carried out in accordance with the recommendations of the Guide for the Care and Use of Laboratory Animals, according to resolution 196/96 of the National Council for Animal Experimentation—COBEA (https://sbcal.org.br/ (accessed on 2 March 2020)). All the procedures used in this study were approved by the Animal Use Ethics Committee of IOC/Fiocruz (L-005/2017). According to the Brazilian Law of Biodiversity, this study was registered at SisGen (AA2236F).

2.2. Parasite Culture and Growth Curve

The *L. braziliensis* strains IOC/L2853 (MHOM/BR/2004/LTCP 393) and IOC/L2856 (MHOM/BR/2003/LTCP 15171) used in this study were provided by the Collection of *Leishmania* of the Instituto Oswaldo Cruz (CLIOC, http://clioc.fiocruz.br/ (accessed on 2 March 2020)). The IOC/L2853 strain is resistant to NO and was isolated from a MCL patient, who was refractory to treatment. On the other hand, the IOC/L2856 strain is susceptible to NO and was isolated from LCL patient responsive to the treatment [17,20]. Both strains were isolated from the same geographical region and belong to the same zymodeme. Through the text, tables, and figures, those strains will be mentioned as 2853 and 2856.

Promastigotes were cultivated at 25 °C in Schneider's medium (Vitrocell, Campinas, Brazil) supplemented with 20% fetal bovine serum (FBS; Vitrocell, heat-inactivated at 56 °C for 50 min) and 2% urine. To analyze the growth curve of these strains, promastigotes (1×10^5 parasites) were incubated in the culture medium described above, and parasite density was determined every 24 h during 16 days by counting in hemocytometer under light microscopy. Parasites of three-days-old culture (log phase promastigotes) were used for all experiments. The in vitro passages of parasites were controlled, and parasites' infectious capacity was maintained through inoculation in golden hamsters.

2.3. Inhibitory Activity of NO Donor on Promastigotes of *L. braziliensis*

To evaluate the cytotoxicity of $NaNO_2$ (NO donor) on NO-resistant (2853) or NO-susceptible (2856) *L. braziliensis* strains, promastigotes were resuspended in Hanks balanced solution (HBBs, pH 5.0; Sigma-Aldrich, St. Louis, MO, USA), and 100 μL (2×10^7 protozoa/mL) was added to the same volume of freshly $NaNO_2$ previously prepared at twice the desired final concentrations (32 to 0.06 mM). Microplates were incubated at 25 °C for 4 h. Together with the NO donor, 20 μL of PrestoBlue (Invitrogen, Carlsbad, CA, USA) was added to the final concentration of 10%. The measurements were performed at 560 and 590 nm, as recommended by the manufacturer at SpectraMax M3 fluorimeter (Molecular Devices, San Jose, CA, USA), and the result was expressed as IC_{50}/4 h, which corresponds to concentration that led to 50% lysis of the parasites within 4 h.

2.4. Effect of Pro-Oxidant and Antioxidant Molecules on Intracellular Amastigotes

To analyze the effect of oxidative burst on the infection, macrophages were collected from the peritoneal cavity of uninfected male BALB/c mice (5–6 weeks) after the injection of 10 mL of RPMI medium (LGC Biotecnologia, Cotia, Brazil). Peritoneal macrophages were resuspended in RPMI supplemented with 10% FBS and plated in 24-well plates (3×10^5 cells/well) for 24 h at 37 °C and 5% CO_2. After that, culture medium was replaced, and cells were infected or not with promastigotes (10:1 parasites/host cell) of each strain. After 4 h of incubation, cultures were washed to remove non-internalized parasites and maintained under the same conditions until complete 24 h of infection. Infected macrophages were treated or not with 150 μM hydrogen peroxide (H_2O_2), 2 mM $NaNO_2$, 30 U/mL superoxide dismutase from bovine erythrocytes (SOD; Sigma-Aldrich, St. Louis, MO, USA), 40 U/mL catalase from bovine liver (Sigma-Aldrich, St. Louis, MO, USA), 1 μM mitoTEMPO (Santa Cruz Biotechnology, Dallas, TX, USA), or 2 mM Nω-Nitro-L-arginine methyl ester hydrochloride (L-NAME; Sigma-Aldrich, St. Louis, MO, USA), for 48 h. The culture was then stained with fast panoptic (Laborclin, Pinhais, Brazil), and percentage of infected cells, number of protozoa per cells, and infection index (percentage of infected host cells multiplied by the number of parasites per cell) were calculated.

As host toxicity control, non-infected macrophages (5×10^4 cells/well) were also treated with the compounds for 48 h at 37 °C and 5% CO_2. After the treatment, 10 μL PrestoBlue was added to the final concentration of 10% and cells were incubated for 2 h at 37 °C and 5% CO_2. The measurements were performed at 560 and 590 nm in a SpectraMax M3 fluorimeter.

2.5. ROS and RNS Release by Infected Macrophages

To evaluate the levels of reactive species released from infected cells, peritoneal macrophages were obtained as described above and incubated with LPS (100 ng/mL) plus IFN-γ (10 ng/mL) for 30 min. Cells were then infected for 72 h, and culture supernatants were collected for RNS detection by Griess reagent (Sigma-Aldrich, St. Louis, MO, USA) according to manufacturer's instructions. Briefly, 100 μL of culture supernatant was added to the same volume of colorimetric reagent and incubated at room temperature for 30 min, and absorbance was read at 540 nm. Values were compared against an $NaNO_2$ standard curve, ranging between 50 and 3 μM. In parallel, macrophages were resuspended in 100 μL respiration buffer consisting of 65 mM KCl, 10 mM Tris-HCl (pH 7.2), 1 mM $MgCl_2$, and 2.5 mM potassium phosphate monobasic [26]. After that, 100 μM Amplex Red reagent (Invitrogen, Carlsbad, CA, USA) and 50 U/mL horseradish peroxidase (HRP; Sigma-Aldrich, St. Louis, MO, USA) were added and the cells were incubated for 30 min at 37 °C. The supernatants were then collected and analyzed to ROS presence by a SpectraMax M3 fluorimeter.

2.6. Protein Extraction and Sample Preparation

For proteomic analysis, promastigotes were treated or not with $^1/_5$ IC_{50}/4 h $NaNO_2$ (for each strain) in HBSS for 4 h at 25 °C. Then, parasites were washed three times with PBS and resuspended in lysis buffer for proteomics sample preparation. Because we are interested in the early effects of NO on proteome remodeling, we choose a sublethal dose to challenge the parasites. Additionally, with this way we avoid results related to parasites' death. These assays were performed in quadruplicate (four independent biological replicates), and samples were prepared accordingly to previous reports [25]. Briefly, parasites were lysed in a buffer containing 0.05 M Tris-HCl (pH 7.6), 0.05 M DTT and 2% SDS (*w*/*v*) and boiled in a water bath for 5 min. After chilling to room temperature, the SDS lysates were clarified by centrifugation at 10,000×g for 5 min and processed in 30k filtration units (Millipore, Burlington, MA, USA) using the MED-FASP (Multi-Enzyme Digestion—Filter Aided Sample Preparation) protocol. Proteins were digested sequentially with endoproteinase LysC and trypsin in a 1/100 enzyme to protein ratio [27,28]. Peptides were collected, concentrated, and desalted on a C18 reversed phase column.

2.7. LC-MS/MS Analysis

Analysis of the peptide mixtures was performed as described previously [25]. Briefly, the peptides were fractionated on a reversed phase column (50 cm × 75 μm inner diameter) packed with 1.8 μm diameter C18 particles (100 Å pore size; Dr. Maisch, Ammerbuch-Entringen, Germany) using a 105 min acetonitrile gradient in 0.1% formic acid at a flow rate of 250 nL/min. Peptide masses were analyzed using a Q-Exactive HF mass spectrometer (Thermo-Fisher Scientific, Palo Alto, CA, USA) operated in data-dependent mode with survey scans acquired at a resolution of 50,000 at m/z 400 (transient time 256 ms). Fifteen of the most abundant isotope patterns with charge $\geq$ +2 from the survey scan (300–1650 m/z) were selected with an isolation window of 1.6 m/z and fragmented by HCD with normalized collision energies of 25. The maximum ion injection times for the survey scan and the MS/MS scans were 20 and 60 ms, respectively. The ion target values for MS1 and MS2 scan modes were set to 3×10^6 and 1×10^5, respectively. The dynamic exclusion was 25 s and 10 ppm.

2.8. Data Analysis

The mass spectra were searched against a database containing *L. braziliensis* sequences available at UniProtKB/Swiss-Prot, using the Andromeda search engine included in the MaxQuant Software (Ver. 1.2.6.20). Reversed proteins were used as decoys. The option "matching between runs" was used for searching, and parameters such as fragment ion mass tolerance of 0.5 Da and parent ion tolerance of 20 ppm were included. Cysteine carbamidomethylation was set as fixed modification, methionine oxidation as variable modification, and up to two missed cleavages were allowed. The maximum false peptide and protein discovery rate was set as 1%. Protein absolute abundances were calculated based on the spectral protein intensity (raw intensities) using the 'Total Protein Approach' (TPA), and absolute protein copy numbers per cell were estimated using the 'Proteomic Ruler' approach [29]. Calculations of total protein, protein concentration, and copy number were performed in Microsoft Excel. Perseus software (Ver. 1.6.5.0) [30] was used for statistical analysis. Minimal number of valid values was set to 4 per protein in at least one group, and missing values were imputed from a normal distribution. Significances were calculated using the t test for pairwise comparisons, with a threshold of false discovery rate (FDR) of 3%. The mass spectrometry proteomics data were deposited to the ProteomeXchange Consortium via the PRIDE [31] partner repository with the dataset identifier PXD029462.

2.9. Analysis of 2-NBDG Uptake

Promastigotes (5×10^6 protozoa/mL) were treated with $^1/_5$ IC_{50}/4 h $NaNO_2$ in HBSS at 25 °C for 4 h. After that, protozoa were washed with PBS and incubated with 300 μM 2-(N-(7-Nitrobenz-2-oxa-1,3-diazol-4-yl)Amino)-2-Deoxyglucose) (2-NBDG, Molecular Probes, Eugene, OR, USA), a fluorescent glucose analogue, for 30 min. The 2-NBDG specificity was monitored by the parasite incubation at 4 °C for the same amount of time, to decrease the compound uptake. 2-NBDG+ parasites were analyzed using a CytekDxP multi-color upgrade flow cytometer (Cytek, Fremont, CA, USA). A total of 10,000 events were acquired in the region previously established by protozoa morphology, and analyses were performed in Summit 6.1 software (Beckman Coulter, Brea, CA, USA).

2.10. Analysis of Oxygen Uptake

Promastigotes (5×10^6 protozoa/mL) were treated with $^1/_5$ IC_{50}/4 h $NaNO_2$ in HBSS at 25 °C for 4 h. After that, protozoa were washed with PBS and resuspended (2.5×10^7 protozoa/mL) in a respiration buffer at 25 °C with continuous stirring in a high-resolution Oxygraph-2K (Oroboros Instruments, Innsbruck, Austria) [26]. Mitochondrial respiration was confirmed by the addition of 2 μM antimycin A (AA, Sigma-Aldrich, St. Louis, MO, USA) to obtain the residual O_2 consumption (ROX). O_2 concentration and flux data were acquired using DatLab software (Oroboros Instruments, Innsbruck, Austria).

2.11. Analysis of Enzymatic Activities

Promastigotes (5×10^6 protozoa/mL) were treated with $^1/_5$ $IC_{50}/4$ h $NaNO_2$ in HBSS at 25 °C for 4 h, washed with PBS, and kept dry at −20 °C until use to ensure that all biological replicas were extracted and analyzed at the same time and under the same experimental conditions. Protein homogenates were prepared by sonication, as previously described [32]. Briefly, pellet was resuspended in cold PBS containing protease inhibitor cocktail (Sigma-Aldrich, St. Louis, MO, USA) and disrupted on ice by sonication for 10 cycles of 7 s with intervals of 7 s, using a Markson GE50 Ultrasonic Processor. Amplitude was set to 70%, and parasite disruption was monitored by light microscopy. Soluble fraction was obtained, and protein concentration was measured using Pierce™ BCA protein assay kit (Thermo Fisher Scientific, Palo Alto, CA, USA). For glutathione peroxidase (GPx) assay, soluble protein extract (0.5 mg/mL) was incubated in 100 mM KH_2PO_4 (pH 7.8) buffer, supplemented with 1 mM reduced glutathione (Sigma-Aldrich, St. Louis, MO, USA), 5 U/mL glutathione reductase (Sigma-Aldrich, St. Louis, MO, USA), 200 μM NADPH (Sigma-Aldrich, St. Louis, MO, USA), and 300 μM H_2O_2. The rate of the NADP ($\varepsilon = 6.22\ M^{-1}cm^{-1}$) reduction was determined at 340 nm. Lactate dehydrogenase activity was evaluated using a commercial kit (Doles, Goiânia, Brazil) following the manufacturer's protocol, with some modifications. Briefly, 0.5 mg/mL of protein was added to 100 μL manufacturer's substrate solution and 10 μL ferric alum for 2 min at 37 °C. After that, 10 μL NAD+ + phenazine methosulfate (PMS) was added, and absorbance measured for 45 min (every 1 min) at 510 nm. A standard curve was generated with lactate dehydrogenase (Sigma-Aldrich, St. Louis, MO, USA). All enzyme activities were measured at 37 °C in a total reaction volume of 200 μL using a SpectraMax Plus384 spectrophotometer (Molecular Devices, Sunnyvale, CA, USA).

2.12. Statistical Analysis

Analyses were performed with GraphPad Prism version 8.0 for Windows (GraphPad Software, San Diego, CA, USA) or IBM SPSS Statistics 22.0 software (IBM Corporation, New York, NY, USA). Asterisks indicate significant differences with the threshold for significance set at $p \leq 0.05$. Student's t test or two-way ANOVA were used to analyze the statistical significance between the strains. The pairwise comparisons and the number of biological replicates are described in figure legend.

3. Results

3.1. The NO-Resistant 2853 *L. braziliensis* Strain Is More Infective to Macrophages In Vitro

To verify the difference of resistance or susceptibility to NO of the two previously characterized *L. braziliensis* strains [17,20], the inhibitory concentration ($IC_{50}/4$ h) of $NaNO_2$ (NO donor) was evaluated in vitro using replicative promastigotes (logarithmic phase), collected at third day of growth (Figure S1). Confirming the previously reported phenotypes, 2853 strain was significantly more resistant to NO than 2856, exhibiting $IC_{50}/4$ h of 27.6 ± 1.7 and 21.2 ± 0.9 mM, respectively (Figure 1A). The infection index of peritoneal macrophages, which corresponds to percentage of infected host cells × number of parasites per 100 cells, was significantly higher with the NO-resistant parasites (2853) than with the NO-susceptible (2856), even after the treatment of infected macrophages (24 h post-infection) with the 2 mM $NaNO_2$. Additionally, the treatment with 2 mM L-NAME, a specific NOS inhibitor, led to a significant increase of the infection index observed in 2856-infected macrophages, in comparison to 2853-infected ones (Figure 1B).

Furthermore, analysis of peritoneal macrophages supernatants showed that both *L. braziliensis* strains triggered NO-mediated response in infected cells; however, the RNS levels were significantly higher after the infection with NO-resistant parasites, reaching an increase of 3.5-fold in comparison to NO-susceptible strain. Treatment with $NaNO_2$ also potentiated the nitrosative stress started by parasite infection, with such increase being more pronounced in 2853-infected cells. Interestingly, treatment with the NOS inhibitor decreased the production of RNS in both infections, abolishing its detection in the supernatant of 2856-infected cells but not in 2853-infected ones (Figure 1C).

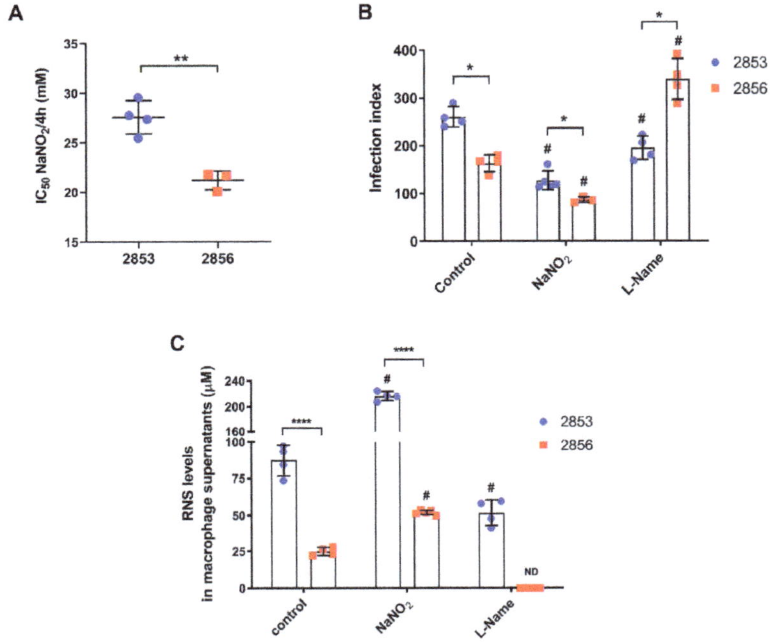

Figure 1. In vitro evaluation of L. braziliensis NO resistance and RNS levels under nitrosative stress. (**A**) Parasites naturally resistant (2853 strain) or susceptible to NO (2856 strain) were exposed to $NaNO_2$ (ranging 0.06–32 mM) during 4 h. Values correspond to the $NaNO_2$ concentration that reduces 50% parasite viability at 4 h of exposure. Dot plots represent mean ± SD of four independent experiments. Statistical differences by t test (** $p = 0.002$). (**B**) Peritoneal macrophages obtained from BALB/c mice were infected for 24 h with promastigotes of each strain. After that, cells were treated with 2 mM $NaNO_2$ or 2 mM L-NAME for 48 h, completing 72 h of infection. Controls correspond to untreated cells infected for 72 h; infection index = percentage of infected host cells × number of parasites per 100 cells. (**C**) Macrophage supernatants were collected, and RNS levels were analyzed by Griess Reagent according to manufacturer's instructions. Graphs represent mean ± SD of at least three independent experiments. Significance of differences between strains were determined by t test using the Holm–Sidak method for multiple comparisons (* $p < 0.05$; **** $p < 0.0001$). Significance of differences between treatments in comparison to control were determined by two-way ANOVA followed by Dunnett's multiple comparisons test (# $p \leq 0.01$); ND = no detected.

3.2. NO-Susceptible Parasites Trigger a More Intense ROS-Dependent Response in Peritoneal Macrophages

Since nitrosative and oxidative stresses work together for host cell's microbicidal mechanisms [13], we evaluate whether resistance to NO in L. braziliensis strains can influence the infective capacity after oxidative burst. To test this, infected peritoneal macrophages were treated with 150 µM H_2O_2 and several antioxidants for 48 h. Treatment with the pro-oxidant molecule significantly decreased the infection index observed in NO-susceptible group, amplifying the differences found between the infections caused by 2853 and 2856. On the other hand, the antioxidants presence increased the infection by NO-susceptible parasites, leading 2856 strain to reach infection index higher than those observed with the NO-resistant parasites (Figure 2A). Treatment with H_2O_2 also decreased by 52.2% the infection caused by the NO-susceptible parasites in comparison to control group, while antioxidants tested increased up to 2.5-fold the infection index of 2856-infected cells. In contrast, neither H_2O_2 nor most of the antioxidant molecules altered the infection index by NO-resistant strain; only the incubation with SOD decreased by 48% the 2853-infection index in relation to control (Figure 2A).

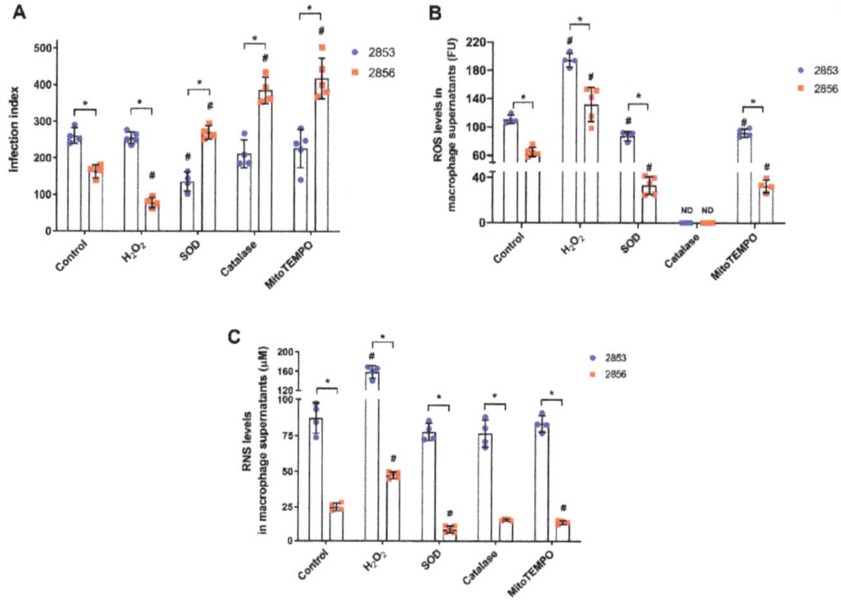

Figure 2. Effect of ROS and RNS levels in peritoneal macrophages infected in vitro with *L. braziliensis* strains resistant and susceptible to NO. Peritoneal macrophages obtained from BALB/c mice were infected with promastigotes for 24 h. After that, cells were treated with 150 μM H_2O_2, 30 U/mL SOD, 40 U/mL catalase, or 1 μM mitoTEMPO for 48 h, completing 72 h of infection. Controls correspond to untreated cells infected for 72 h. (**A**) Infection index, which corresponds to percentage of infected host cells × number of parasites per 100 cells. (**B**) Host cells were incubated with 100 μM Amplex Red and 50 U/mL HRP for 30 min in respiration buffer and analyzed for detection of ROS levels. (**C**) Macrophage supernatants were collected, and the production of RNS was analyzed by Griess Reagent, according to manufacturer's instructions. Graphs represent mean ± SD of at least four independent experiments. Significance of differences between strains were determined by *t* test using the Holm-Sidak method for multiple comparisons (* $p \leq 0.01$). Significance of differences between treatments in comparison to control was determined by two-way ANOVA followed by Dunnett's multiple comparisons test (# $p \leq 0.05$); ND = no detected.

Besides RNS, both *L. braziliensis* strains also trigger ROS-mediated responses in peritoneal macrophages; however, ROS levels were 1.7-fold higher in 2853-infected cells than in 2856-infected ones (Figure 2B). As expected, the treatment with H_2O_2 increased ROS levels in supernatants of peritoneal macrophages (up to 2-fold), and antioxidant molecules significantly decreased the oxidative burst in all analyzed conditions. Nevertheless, even in the presence of antioxidants, ROS levels were significantly higher (~3-fold) in macrophages infected with the NO-resistant strain than in those infected with the NO-susceptible. Interestingly, the treatment with catalase was able to abolish completely the H_2O_2 detection in both infection conditions (Figure 2B).

To test the existence of a relation between the oxidative and nitrosative metabolism during *L. braziliensis* infection, supernatants of those peritoneal macrophages were also evaluated regarding to RNS levels. Remarkably, in all conditions, after treatment with H_2O_2 or antioxidants, macrophages infected with the NO-resistant strain produce among

~3.3-fold to ~9-fold higher levels of RNS than those infected with the NO-susceptible strain (Figure 2C). Treatment with H_2O_2 significantly increased the production of RNS (up to 1.9-fold) in cells infected by both strains, in comparison to infected control group, while antioxidants modulated RNS production only in 2856-infected macrophages, decreasing nitrosative levels. Interestingly, macrophages infected with the 2853 did not suffer such modulation, maintaining high levels of RNS even in the presence of antioxidant molecules (Figure 2C).

3.3. Difference in Protein Abundance Is Observed between NO-Resistant and NO-Susceptible L. braziliensis Strains and it Is Significantly Modulated in Response to NO

As we observed that the NO-resistant strain was significantly more resistant to NO and more infectious for the host cells even after treatment with H_2O_2 and $NaNO_2$, we decided to evaluate the protein abundance profile of this strain with and without $NaNO_2$ challenge and compare it to the profile of the NO-susceptible strain. Mass spectrometry analysis of four biological replicates (independent biological assays) of each strain, challenged or not with $^1/_5$ IC_{50}/4 h $NaNO_2$, allows for the identification of 6296 protein groups, encompassing ~80% of the *Leishmania* predicted proteome (~8000 protein-coding genes predicted, considering one protein per gene) (Table S1). More than 5700 protein groups were identified in each replicate, and 5010 protein groups were identified in all 16 samples (Table S2).

The total protein contents per cell were calculated as previously described [25] using the histone ruler method based on the DNA content reported for *L. braziliensis* reference (strain M2904). Similar to the estimates previously reported [25,33], the 2853 strain contains 3.3 ± 0.11 pg of protein per cell. Remarkably, the protein content of this strain increased significantly when parasites were challenged with the NO donor, reaching 4.3 ± 0.08 pg of protein per parasite (Figure 3A). Interestingly, the 2856 strain exhibited 4.4 ± 0.2 pg of total protein per cell, and this value did not change after challenge with NO (4.4 ± 0.1 pg) (Figure 3A).

Using the total protein approach (TPA) method, we calculated the absolute protein concentrations for each strain. In agreement with previous reports [25], we observed that protein concentration values span 6 orders of magnitude; 90% of the proteome extends over ~3 orders of abundance; and histones, alpha- and beta-tubulin, elongation factor 1-alpha, HSP70, and calmodulin are among the top 20 most abundant proteins (Figure 3B). Remarkably, principal component analysis (PCA) of the protein concentrations revealed a consistent separation between the NO-resistant and NO-susceptible strains, as well as between these strains after the NO challenge (2853+NO and 2856+NO) (Figure 3C), showing a clear clustering for each set of biological replicates. Statistical analysis by PCA also showed that the proteome of the NO-resistant strain is clearly modulated, in terms of protein concentrations, after the $NaNO_2$ challenge.

Using Perseus, a total of 6022 proteins were statistically validated in at least 12 out of the 16 samples (FDR 0.01). The statistical significance of differences in protein abundance among the strains was determined by Student's *t* test at FDR of 3%. In total, the concentrations of 1320 proteins were significantly different between the NO-resistant strain and the NO-susceptible strain; the concentration of 474 proteins was different between these strains treated with $NaNO_2$ (2853+NO and 2856+NO), 850 between 2853 and 2853+NO, and 122 between 2856 and 2856+NO (Table S3 and Figure S2). These results reveal natural intrinsic differences in the proteome between the resistant and susceptible strains and show that the NO-resistant strain more actively modulates its proteome in response to the NO challenge than the NO-susceptible one.

3.4. The Nitrosative Challenge Negatively Modulated Antioxidant Proteins of NO-Susceptible L. braziliensis Strain

As the 2853 and 2856 strains have different responses to pro-oxidant and antioxidant stimuli and exhibit significant differences in their proteomes, we analyzed the abundance levels of proteins involved in the response to oxidative stress and in the maintenance of parasites' redox homeostasis in our proteomics dataset. First, we observed that cumulative concentration of proteins involved in those processes is significantly higher in NO-resistant parasites than in the NO-susceptible. Interestingly, upon nitrosative challenge, the resistant 2853 strain maintains higher concentrations of those proteins, while the susceptible 2856 exhibits a significant decrease of them (Figure 4A).

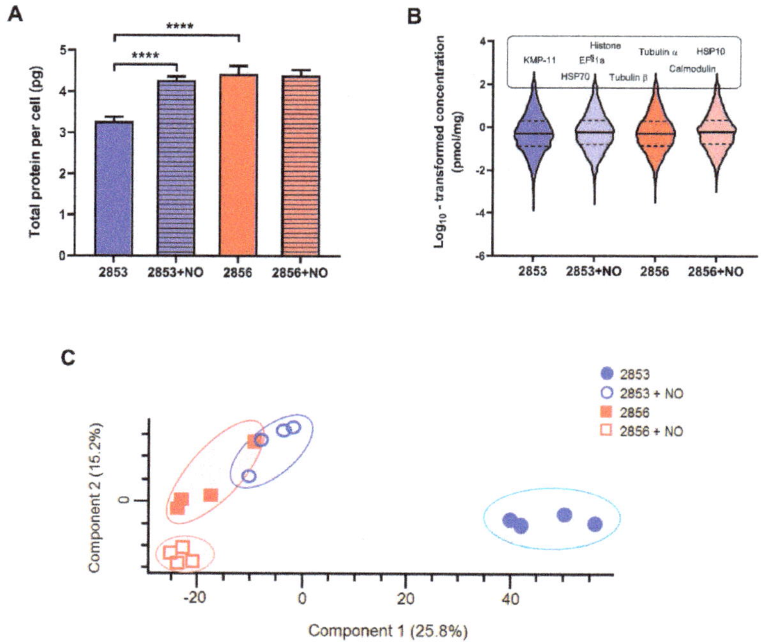

Figure 3. NO-resistance modulates protein abundance of *L. braziliensis* strains. Whole cell lysates from 2853, 2853+NO (2853 challenged with the $^1/_5$ IC_{50}/4 h $NaNO_2$), 2856, and 2856+NO (2856 challenged with the $^1/_5$ IC_{50}/4 h $NaNO_2$) strains were processed by MED-FASP and analyzed by LC-MS/MS. Protein abundances were calculated based on the raw spectral intensities. (**A**) Total protein content per cell of four biological replicates for each strain. Graphs represent mean ± SD of four independent experiments. Statistical differences by *t* test (**** $p < 0.0001$). (**B**) Violin plots depicting distribution of log10 transformed protein concentration values of proteins identified and quantified in each strain. Protein names above violin plots represent most abundant proteins common for all strains. (**C**) Principal component analysis of the absolute protein concentration values of all quantified proteins determined by the Total Protein Approach (TPA) method.

We also analyzed the concentration of specific proteins that are directly involved in the response to oxidative stress. The abundance of ascorbate peroxidase (APx), a heme-containing enzyme that catalyzes the conversion of H_2O_2 into water, was oppositely modulated by NO challenge, increasing in the 2853 strain and decreasing in the 2856 (Figure 4B). The NO-susceptible strain challenged with $NaNO_2$ also exhibited a significant decrease in the absolute concentration of enzymes related to trypanothione-dependent hydroperoxide metabolism, such as tryparedoxin 1 (TXN1) (Figure 4C), tryparedoxin peroxidase (TXNPx) (Figure 4D), and trypanothione reductase (TR) (Figure 4E). Despite the impairment of TXNPx and TR abundance caused by nitrosative stress in NO-resistant parasites, the concentration of these proteins was higher than that observed in the NO-susceptible ones challenged with NO donor (Figure 4D,E). Interestingly, in conditions without NO challenge, NO-resistant strain also exhibited significant higher concentration of superoxide dismutase (SOD) when compared to the NO-susceptible one; however, treatment with $NaNO_2$ significantly decreased cumulative SOD concentration to levels detected in 2856 strain (Figure 4F). These results show that upon NO challenge, the NO-resistant strain is more able to sustain the levels of trypanothione system's enzymes than the NO-susceptible strain.

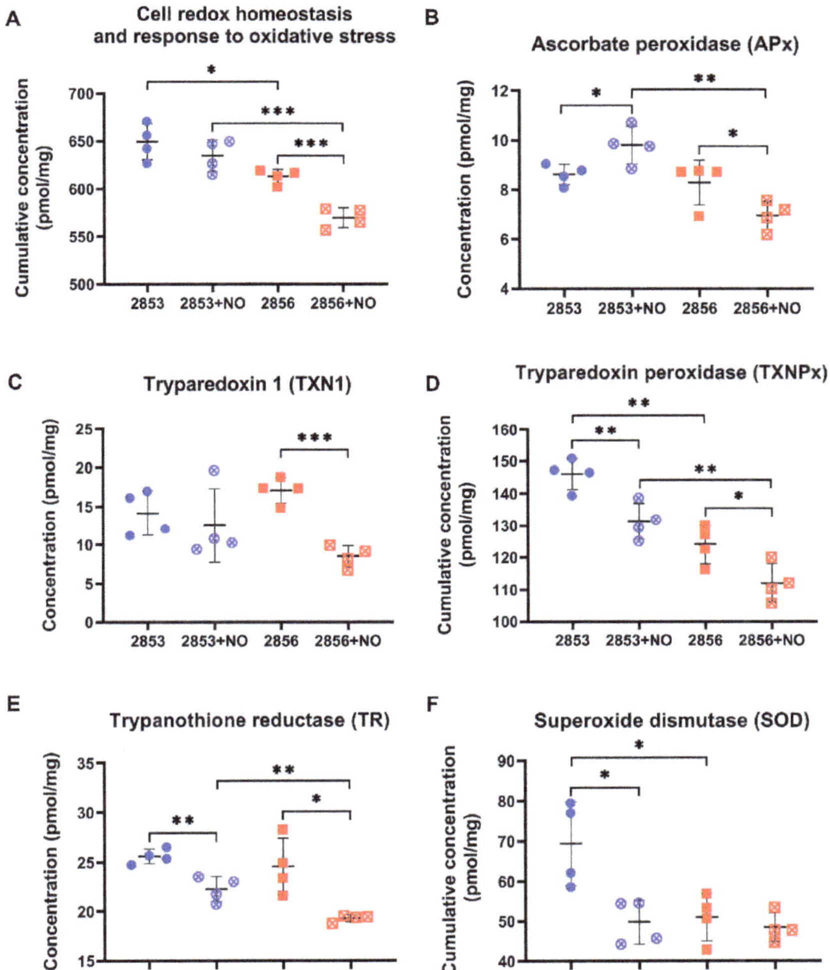

Figure 4. The antioxidant proteins of NO-susceptible L. braziliensis strain are negatively modulated after the nitrosative challenge. (**A**) Cumulative concentration of all identified proteins involved in biological processes of *"cell redox homeostasis and response to oxidative stress"*. Each symbol shows the total sum of the concentration values of proteins involved in those processes in each of the four biological replicates. Concentrations of specific proteins involved in response to oxidative stress: (**B**) ascorbate peroxidase (APx); (**C**) tryparedoxin 1 (TXN1); (**D**) tryparedoxin peroxidase (TXNPx); (**E**) trypanothione reductase (TR); and (**F**) superoxide dismutase (SOD). Graphs represent mean ± SD of four independent experiments. Statistical differences by Student's *t* test (* $p < 0.05$; ** $p < 0.01$; *** $p < 0.001$). 2853: NO-resistant strain; 2853+NO: NO-resistant strain challenged with $^{1}/_{5}$ IC_{50}/4 h $NaNO_2$; 2856: NO-susceptible strain; and 2856+NO: NO-susceptible strain challenged with $^{1}/_{5}$ IC_{50}/4 h $NaNO_2$.

3.5. NO-Resistant L. braziliensis Strain Challenged with NO Increases the Abundance of Enzymes Involved in the Glutathione Pathway

Since glutathione (GSH) is involved in the antioxidant response against oxidative and nitrosative stresses in eukaryotes [34,35], we analyzed the abundance levels of proteins involved in its metabolism in our dataset. Interestingly, although differences in glutathione peroxidase (GPx) cumulative abundance levels were detected only in NO-susceptible parasites challenged with NO donor (Figure 5A), the activity of this enzyme was ~7-fold higher in NO-resistant parasites than in NO-susceptible ones. Additionally, after exposure to theNO donor, GPx activity was significantly upregulated in both strains but was significantly higher in NO-resistant strain (Figure 5B). In addition, upon NO challenge, the concentration of glutamate-cysteine ligase (GSH1), which is an important enzyme responsible for the de novo synthesis of GSH, was significantly increased in NO-resistant parasites (Figure 5C). GSH may also interact directly with NO producing S-nitrosoglutathione (GSNO) [36–38], which can be decomposed by GSNO reductases to oxidized glutathione (GSSG). Alcohol dehydrogenase class III enzymes have GSNO reductase activity and function as a protection against nitrosative stress [39,40]. Notably, we found an alcohol dehydrogenase class III significantly decreased in the 2856 NO-susceptible strain challenged with $NaNO_2$, in relation to 2853 treated with NO donor (Figure 5D). Altogether, these results suggest that GSH metabolism may be triggered by NO in L. braziliensis strains, especially in NO-resistant ones. In addition, alcohol dehydrogenase class III could also have GSNO reductase activity in L. braziliensis, and its downregulation may contribute to the inefficient NO detoxification in NO-susceptible parasites.

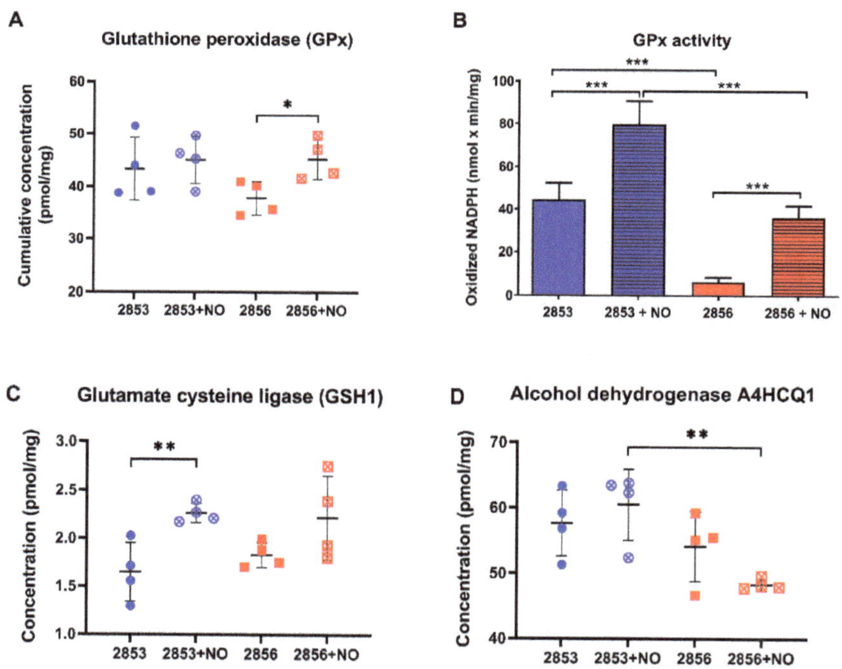

Figure 5. The NO-resistant strain increases glutathione metabolism enzymes in response to •NO challenge. (**A**) Cumulative concentration of all glutathione peroxidases (GPx) identified in our dataset. Each symbol shows the total sum of the GPx concentration values in each of the four biological replicates. (**B**) GPx activity was measured in parasite homogenates. (**C**) Absolute concentration of glutamate cysteine ligase (GSH1) in each strain. (**D**) Absolute concentration of alcohol dehydrogenase A4HCQ1 in each strain. Graphs represent mean ± SD of four independent experiments. Statistical

differences by Student's *t* test (* $p < 0.05$; ** $p < 0.01$; *** $p < 0.001$). 2853: NO-resistant strain; 2853+NO: NO-resistant strain challenged with $^1/_5$ IC_{50}/4 h $NaNO_2$; 2856: NO-susceptible strain; and 2856+NO: NO-susceptible strain challenged with $^1/_5$ IC_{50}/4 h $NaNO_2$.

3.6. Uptake of Glucose Analog and G6PDH Protein Levels Increase in NO-Resistant *L. braziliensis* Strain after the Nitrosative Challenge

Because we observed that NO-resistant strain exhibits more robust antioxidant defenses and responds more efficiently to the nitrosative challenge, mainly through the GSH metabolism, we analyzed if proteins involved in GSH reduction are also modified in those parasites, particularly those committed to NADPH production. First, we examined the abundance of enzymes involved in glycolysis and PPP (Figures S3 and S4). Cumulative concentration of glycolytic enzymes was significantly higher in the 2853 strain than in 2856. Resistant strain significantly diminished protein concentration levels of glycolytic enzymes upon NO challenge, but the glycolytic pathway titers remained similar in NO-susceptible parasites after NO exposure (Figure 6A). Notably, cumulative concentration of PPP enzymes was significantly reduced in NO-susceptible parasites but not in NO-resistant ones in response to NO stimulus (Figure 6B). Analysis of phosphotransferase, the first enzyme of the glycolytic pathway, revealed that 2856 strain has higher protein abundance in comparison to 2853; however, after nitrosative challenge, NO-resistant parasites exhibited a significant increase in this enzyme, while the same did not happen in NO-susceptible ones (Figure 6C). Interestingly, such increase is reflected in significantly elevated glucose analogue uptake in those parasites, which was ~2.5-fold higher in 2853 strain than in 2856 after NO exposure (Figure 6D).

Glucose consumed by parasites is rapidly converted to glucose 6-phosphate, which can follow the glycolytic pathway or be driven to PPP and used for maintaining the NADPH pool. In fact, we observed a significant increase in the concentration of G6PDH in the NO-resistant strain upon NO challenge, whereas the enzyme concentration was reduced in the NO-susceptible one under the same stress condition (Figure 6E). In addition, a significant increase in transaldolase (TAL), an enzyme of non-oxidative branch of PPP, was observed in NO-resistant parasites, while a significant reduction of this protein was detected in the NO-susceptible ones (Figure 6F). Analysis of protein abundance of other glycolytic enzymes supports the idea that glucose 6-phosphate may be entering the PPP pathway in NO-resistant parasites (Figures S3 and S4). Particularly, although the levels of 6-phosphofructokinase-1, an important enzyme to glycolytic pathway, were higher in the NO-resistant parasites than in NO-susceptible ones, they were not modulated by the $NaNO_2$ treatment (Figure 6G).

3.7. Nitrosative Challenge Increases D-Lactate Dehydrogenase Abundance in NO-Resistant *L. braziliensis* Strain

As NO challenge seems to positively regulate the first steps of glycolysis, we were interested in analyzing the status of the enzymes at the pathway exit. The concentration levels of pyruvate kinase, the only known regulated glycolytic enzyme in *Leishmania*, did not suffer significant alterations upon exposure to NO (Figure 7A). In addition, the allosteric regulator of pyruvate kinase, 6-phosphofructo-2-kinase, the enzyme responsible for fructose-2,6-bisphosphate biosynthesis, did not suffer significant changes upon exposure to NO (Figure 7B). Intriguingly, we observed a significant increase in the concentration and activity of D-lactate dehydrogenase (D-LDH) in NO-resistant strain upon NO exposure (Figure 7C,D). Indeed, LDH activity was increased 2-fold in NO-resistant parasites after nitrosative challenge and it was 3.3-fold higher in NO-resistant-challenged parasites than in NO-susceptible-challenged ones (Figure 7D).

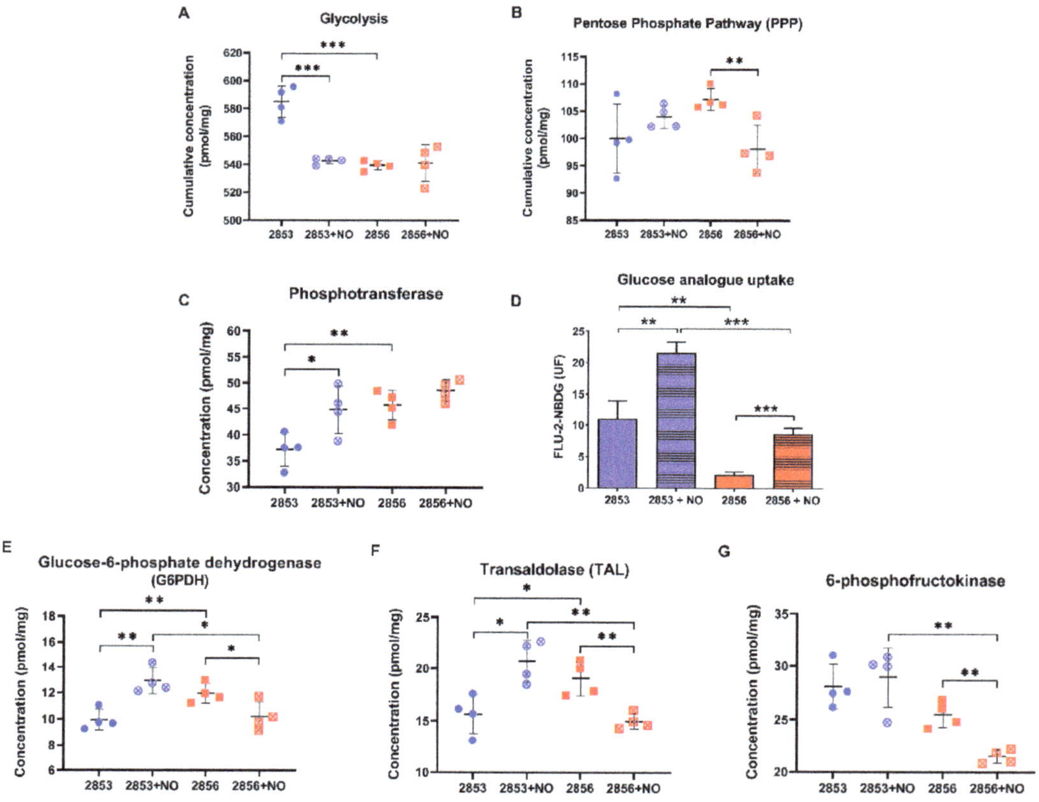

Figure 6. 2-NBDG uptake and G6PDH activity increase in NO-resistant L. braziliensis strain after the nitrosative challenge. Cumulative concentration of all the identified proteins involved in (**A**) glycolysis and (**B**) pentose phosphate pathway (PPP). Each dot represents the total sum of the concentration values of proteins involved in those processes in each of the four biological replicates. Absolute protein concentration of (**C**) phosphotransferase; (**E**) glucose-6-phosphate dehydrogenase (G6PDH); (**F**) transaldolase; and (**G**) 6-phosphofructokinase-1. (**D**) 2-NBDG uptake after parasite incubation for 30 min at 25 °C. Median values were obtained by subtraction of fluorescence at 4 °C. Graphs represent mean ± SD of at least four independent experiments. Statistical differences by Student's *t* test (* $p < 0.05$; ** $p < 0.01$; *** $p < 0.001$). 2853: NO-resistant strain; 2853+NO: NO-resistant strain challenged with $^1/_5$ IC_{50}/4 h $NaNO_2$; 2856: NO-susceptible strain; and 2856+NO: NO-susceptible strain challenged with $^1/_5$ IC_{50}/4 h $NaNO_2$.

3.8. Nitrosative Challenge Impairs Mitochondrial O_2 Consumption by L. braziliensis Strains

Mitochondrial respiration and, specifically, proteins involved in oxidative phosphorylation (OXPHOS) are inhibited by NO due to competition with oxygen [41]. Here, we observed that oxygen consumption in routine condition is 61.5% lower in 2856 strain than in 2853. Additionally, the mitochondrial respiration is more affected by $NaNO_2$ treatment in NO-susceptible parasites than in NO-resistant. Thus, although nitrosative stress induced a significant decrease in both strains, NO-resistant parasites can maintain oxygen consumption 1.7-fold higher than NO-susceptible (Figure 8). Interestingly, ROX state, which indicates mitochondrial-independent oxygen consumption, increased~3.5-fold in 2856 strain when compared to 2853, suggesting higher ROS production by NO-susceptible parasites. Besides that, the NO exposure decreased the oxygen consumption to similar levels during ROX state in both strains (Figure 8).

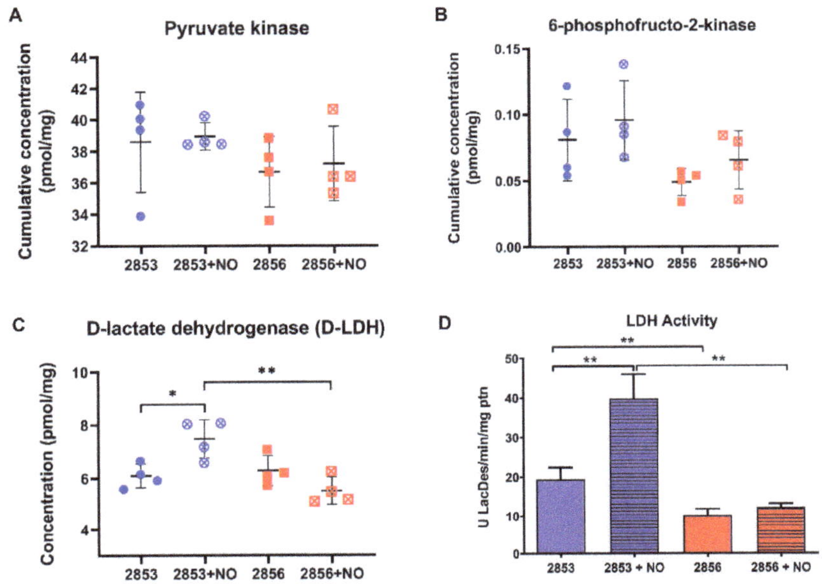

Figure 7. Nitrosative challenge increases D-lactate dehydrogenase (D-LDH) abundance in NO-resistant *L. braziliensis* strain. (**A**) Protein concentration of pyruvate kinase; (**B**) cumulative concentration of 6-phosphofructo-2-kinase; (**C**) protein concentration of D-lactate dehydrogenase (D-LDH); and (**D**) LDH activity in parasite homogenates. Graphs represent mean ± SD of at least four independent experiments. Statistical differences by Student's *t* test (* $p < 0.05$; ** $p < 0.01$). 2853: NO-resistant strain; 2853+NO: NO-resistant strain challenged with $^1/_5$ IC_{50}/4 h $NaNO_2$; 2856: NO-susceptible strain; and 2856+NO: NO-susceptible strain challenged with $^1/_5$ IC_{50}/4 h $NaNO_2$.

Based on these results, the abundance of mitochondrial complexes, as well as the concentration levels of several proteins, were analyzed. First, we observed that there were no differences in the concentration of citrate synthase, an enzyme associated with mitochondrial integrity, among the experimental groups (Figure 9A). Notably, we observed a significant decrease in the accumulated abundance of the proteins involved in OXPHOS in the NO-resistant parasites after the NO challenge (Figure 9B). Intriguingly, the protein concentration of complexes I, II, and IV is significantly higher in NO-susceptible parasites than in NO-resistant, and the exposure to nitrosative stress does not affect the abundance of these proteins (Figure 9C,D,F). In contrast, protein concentration of complex V is significantly lower in NO-susceptible parasites than in NO-resistant ones (Figure 9G). Interestingly, upon nitrosative challenge, parasites of 2853 strain increase the concentration of molecules related to complex I and decrease the abundance of complexes III and V (Figure 9C,E,G). Ubiquinone acts as an electron carrier from complexes I and II to complex III. Interestingly, we found that the concentration of a protein that participates in the biosynthesis of this molecule is significantly decreased in NO-susceptible parasites during nitrosative condition (Figure 9H), suggesting that 2856 strain may be unable to maintain the ubiquinone pool under NO exposure.

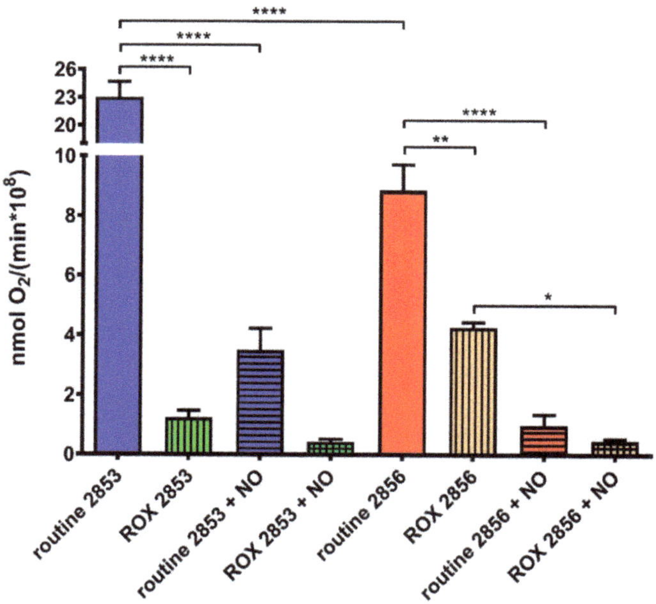

Figure 8. Nitrosative challenge impairs mitochondrial O_2 consumption by *L. braziliensis* strains. Parasites were incubated in the respiration buffer at 25 °C to evaluate the O_2 uptake under routine condition and after the addition of 2 μM AA (ROX state). Graphs represent mean ± SD of four independent experiments. Significance of differences between treatments were determined by two-way ANOVA followed by Tukey's multiple comparisons test (* $p < 0.05$; ** $p < 0.001$; and **** $p < 0.0001$). 2853: NO-resistant strain; 2853+NO: NO-resistant strain challenged with $^1/_5$ IC_{50}/4 h $NaNO_2$; 2856: NO-susceptible strain; and 2856+NO: NO-susceptible strain challenged with $^1/_5$ IC_{50}/4 h $NaNO_2$.

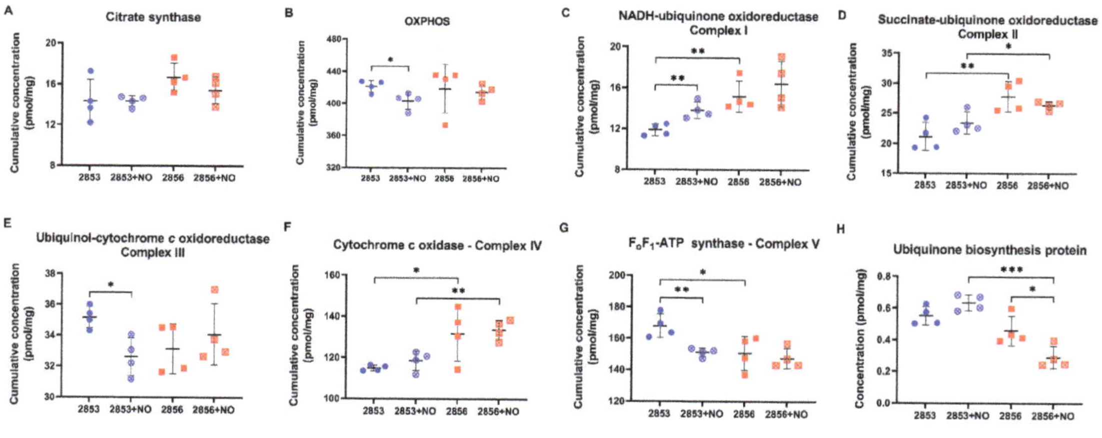

Figure 9. Proteins of mitochondrial oxidative phosphorylation (OXPHOS) are differentially modulated in *L. braziliensis* strains. (**A**) Cumulative concentration of all the identified proteins as citrate synthase. Cumulative concentration of all the identified proteins involved in (**B**) OXPHOS; (**C**) NADH-ubiquinone oxidoreductase (complex I); (**D**) succinate ubiquinone oxidoreductase (complex II); (**E**) ubiquinol:cytochrome c oxidoreductase (complex III); (**F**) cytochrome c oxidase (complex IV); and (**G**) F_oF_1-ATP synthase (complex V). Each dot represents the total sum of the concentration values

of proteins involved in those processes or complexes in each of the four biological replicates. (**H**) Absolute protein concentration of ubiquinone biosynthesis protein. Graphs represent mean ± SD of at least four independent experiments. Statistical differences by Student's *t* test (* $p < 0.05$; ** $p < 0.01$; and *** $p < 0.001$). 2853: NO-resistant strain; 2853+NO: NO-resistant strain challenged with $^1/_5$ IC_{50}/4 h $NaNO_2$; 2856: NO-susceptible strain; and 2856+NO: NO-susceptible strain challenged with $^1/_5$ IC_{50}/4 h $NaNO_2$.

4. Discussion

To complete its life cycle, *Leishmania* must overcome several barriers found in the vertebrate host, such as increased temperature (25 °C in insect vector to 37 °C in mammalian host), pH acidification in phagolysosome, and changes of available carbon sources [42]. Moreover, the parasite must survive the oxidative burst and NO production, two main microbicidal mediators that started after macrophage activation [43]. Although the host immune response is directly related to the prognosis of leishmaniasis [44], intrinsic virulence features of *Leishmania* strains are also decisive for infection and clinical outcome [21]. In the present study, we evaluated two strains of *L. braziliensis* with polarized phenotype of susceptibility or resistance to NO, which were also associated with responsiveness or refractoriness to antimony treatment [17,20]. Aspects related to nitrosative and oxidative stresses suffered by the parasites during host cells infection, in addition to a deep analysis of parasites' proteome-mediated mechanisms for resisting to NO were considered. It is important to point out that during the very initial steps of infection, promastigotes are subjected to the early oxidative burst triggered into the innate immunity cells -at the same time host and effectors- in response to the invasion. To cope with this, promastigotes should display their repertoire of antioxidant responses. Depending on the success during such very early responses, parasites will survive, and differentiate, determining the successful parasite colonization and further persistence. Our proteomics and biochemical data on promastigotes challenged with $NaNO_2$ would reflect these very initial parasite responses.

First, in vitro infection allowed for the identification of differences between both strains, corroborating the higher infection capability of NO-resistant parasites, both in terms of number of infected macrophages, as well as in number of amastigotes per cell. Our findings were similar to those of Souza et al. [20], who used human macrophages for in vitro evaluation of *L. braziliensis* infection. Interestingly, we also observed that both strains trigger host microbicidal responses through increased production of ROS and RNS, with levels exacerbated in the infection produced by NO-resistant strain. In cutaneous leishmaniasis, the development of a non-healing course of infection has been linked to an immunosuppressive profile, with increased arginase 1 activity [45], a cytosolic enzyme that favors parasites' antioxidant metabolism and that, at the same time, competes with iNOS for the common substrate L-arginine, decreasing NO production [46,47]. Although Costa et al. [21] demonstrated that this process also occurs during in vivo infection of BALB/c mice with the NO-resistant *L. braziliensis* strain, it is only elicited in late infection times (seven weeks post-infection). All together, these data suggest that NO-resistant parasites are endowed with specific mechanisms to evade host defenses, raising a question about the phenotypic adaptations that allow this strain to survive in the presence of high concentrations of toxic molecules, especially during early infection course.

According to previous studies, *L. braziliensis* amastigotes survive and replicate much better in the absence of ROS, identifying these molecules as important regulators of the parasite proliferation inside the host cells [48]. In addition, supplementation of cell cultures with pro-oxidants and antioxidants modulates *Leishmania* infection, resulting in a reduced parasite load in stressful conditions [49]. Moreover, sensitivity to distinct reactive species may vary in the parasite stages, with H_2O_2 being more toxic to *Leishmania* amastigotes than $O_2^{\bullet-}$ [43,50]. In line with those observations, infection with the NO-susceptible strain reproduces all phenotypes previously described, including the high sensitivity to H_2O_2 (comparing the infection index of catalase- and SOD-treated cells). In contrast, the infection with NO-resistant parasites has unique features; only increased nitrosative stress derived from $NaNO_2$-treated macrophages was able to downregulate the infection by this strain. Resistant parasites also appear to be unaffected by H_2O_2 since neither this reactive species (even with the increase in RNS production) nor the presence of catalase were able to change the infection index. In contrast, $O_2^{\bullet-}$ metabolism might be essential for NO-resistant strain because only SOD-treated cells showed reduction in the parasite load. The use of tempol, an antioxidant able to promote $O_2^{\bullet-}$ dismutation at rates similar to SOD, pointed to the dual role of this free radical to control the infection, highlighting differences in the pathogenesis of cutaneous leishmaniasis caused by different species of *Leishmania* [49,51,52].

To shed light on the molecular mechanisms underlying the NO-resistant phenotype in *L. braziliensis*, we performed an unbiased and comprehensive quantitative analysis of parasites' proteome. Notably, differences in total protein per cell and absolute protein concentrations clearly distinguished NO-resistant from NO-susceptible parasites, corroborating the notion that there are natural intrinsic differences at proteome level among *L. braziliensis* strains circulating in a same geographical region [9]. In addition, we demonstrated that the rapid modulation of NO-resistant parasites' proteome upon NO challenge involves an increase in total protein content, which is suggestive of ploidy alterations in response to nitrosative stress. In line with this proposal, recently it was reported that aneuploidy in *L. donovani* is followed by proteome modulation and could explain metabolic differences between strains [53]. Aneuploidy and karyotypic mosaicism are common in *Leishmania* spp., and such genome plasticity allow parasites to explore fitness possibilities for survival [54–56]. In *Leishmania*, aneuploidy is a species- and strain-specific trait that varies according to external stimuli, including drug pressure, enabling rapid adaptation to hostile conditions [57–60]. However, aneuploidy may also result in metabolic alterations that lead to oxidative and proteotoxic stresses [61]. Interestingly, we observed that in contrast to NO-resistant strain, the NO-susceptible parasites naturally (without NO stimulus) present a protein content higher than expected and are not able to modulate it upon NO challenge, suggesting that NO-susceptible parasites are "naturally stressed" and that proteome (and genome) plasticity in these parasites probably reached a limit. Remarkably, the deep proteomics approach conducted in this study allows one to clearly differentiate resistant parasites from susceptible ones and reveals the subset of proteins that explain those phenotypes.

Proteomic profiling of *L. donovani* parasites adapted to sub-lethal doses of NO donor showed the upregulation of several parasite's proteins involved in the ROS detoxification pathway, suggesting a cross-resistance to both nitrosative and oxidative stresses [62]. Accordingly, we observed a positive modulation in the concentration of proteins associated with *"cell redox homeostasis and response to oxidative stress"* in NO-resistant parasites, even in the condition without the NO donor. In addition, the challenge with $NaNO_2$ specifically affected the concentration of APx, increasing protein abundance in NO-resistant parasites and decreasing in NO-susceptible ones. Sardar et al. [61] also demonstrated that exposure of *L. donovani* promastigotes to ROS or RNS elevates APx protein abundance up to 2.5-fold, whereas combination of both stresses produced an additive effect, increasing this protein in 3.2-fold. Additionally, ROS-inducible APx of *L. amazonensis* is essential for parasite infectivity, replication, and virulence in vitro and in vivo models [63]. Collectively, these data support the idea that NO-resistance in *L. braziliensis* is associated with APx increased abundance as a response to stressful conditions, which also explains the successful intracellular replication of NO-resistant parasites in macrophages overproducing H_2O_2. Interestingly, *L. braziliensis* APx-overexpressing parasites have an 8-fold increase in the antimony-resistance index [64]. As *L. braziliensis* NO-resistant parasites used here are also refractory to antimony treatment, our data reinforce the idea that APx could participate in both NO- and Sb-resistance.

Regarding the $T(SH)_2$/TR system, we observed a decrease in TR abundance after NO challenge. Although this phenomenon occurred to a lesser extent in NO-resistant parasites, such reduction indicates that other mechanisms are concurring to maintain the pool of reducing intermediates. Despite the fact that $T(SH)_2$ is a much more efficient scavenger of hydroperoxides than other thiols present in trypanosomatids [65], all these protozoa possess glutathione-dependent enzymes [66]. Susceptibility of different *Leishmania* species to the NO donor S-nitroso-N-acetyl-D,L-penicillamine (SNAP) was inversely correlated with the levels of GSH but not with their total thiol content (including $T(SH)_2$ levels) [67]. In addition, *L. donovani* showed ~2.9-fold upregulation of glutathione peroxidase-like protein in detrimental to TR in response to nitrosative stress [62]. In agreement with those observations, our results show that *L. braziliensis* NO-resistant parasites have higher concentrations of proteins involved in GSH biosynthesis pathway, particularly GSH1, and GPx activity than NO-susceptible parasites upon NO exposure. Such increased levels can avoid the toxic effects of nitrosative stress and oxidative burst found after treatment with NO donor, or during host infection. Additionally, we cannot rule out the role of GSH pool in regeneration of ascorbate, a cofactor of APx, since in several organisms this is the main ascorbate recycling system [68]. Although in trypanosomatids this process seems to be achieved via $T(SH)_2$ [69], further analyses should be performed to evaluate the activity of GSH in *L. braziliensis* parasites, especially in NO-resistant ones.

In *L. donovani*, the response to oxidative stress requires a rapid metabolic reconfiguration of glucose metabolism, involving a shift from glycolysis toward PPP for replenishment of NADPH pool when parasites are exposed to oxidants. mRNA and protein levels of PPP enzymes of the oxidative and nonoxidative branch, such as G6PDH and transaldolase, were up-regulated in promastigotes exposed to sublethal doses of pro-oxidants, while the viability of promastigotes treated with G6PD inhibitor and sublethal doses of ROS was restored by coincubation with N-acetyl cysteine or GSH [70]. In addition, cell lines overexpressing G6PDH and transaldolase are also more resistant to antimonial, amphotericin B, and miltefosine [70]. In agreement, our dataset indicates that NO resistance in *L. braziliensis* also involves a rapid shift in glucose metabolism from glycolysis to PPP in response to nitrosative stress, without the detriment of glycolysis per se. Protein concentration of G6PDH and transaldolase were increased in NO-resistant parasites and decreased in NO-susceptible ones in response to the NO challenge. Such increase in protein concentration enables NO-resistant parasites to replenish intracellular NADPH levels and, consequently, the GSH pool, to maintain its cellular redox balance. The proteome of promastigotes of *L. infantum* strains resistant to NO also showed increased abundance of G6PDH, while amastigotes of *L. infantum* cell lines selected for NO-resistant exhibited overexpression of 6-phosphogluconate dehydrogenase mRNA levels [23,71], reinforcing the notion that the PPP is an indispensable pathway for resistance to nitrosative stress.

Intriguingly, we observed a significant increase in the concentration of D-LDH in NO-resistant parasites after exposure to NO, and such increase was accompanied by increased LDH enzymatic activity. In *L. major*, D-lactate produced via methylglyoxal metabolism would be converted to pyruvate by a D-LDH, with the resulting molecule feeding the tricarboxylic acid cycle (TCA) and other pathways [72–74]. As the methylglyoxal is a toxic glycolytic metabolite, we hypothesize that the increase in glucose uptake and, consequently, in the production of methylglyoxal may occur in response to nitrosative stress in NO-resistant parasites challenged with NO, leading to upregulation of D-LDH abundance and activity in these parasites. However, further assays need to be done to demonstrate the accumulation of that metabolite in parasites under nitrosative stress. Interestingly, and supporting once again the idea that NO- and drug-resistance are related, recently it was observed that transcript levels coding for D-LDH increased 56-fold in a *L. donovani* cell line selected for paromomycin resistance and that wild-type parasites transfected with *D-LDH* acquired a significant resistance against the drug [75].

It has been suggested that a metabolic shift from glycolysis to mitochondrial respiration is detrimental for *L. mexicana* virulence in vivo, leading to high ROS production and increased sensitivity of parasites to NO. Therefore, the maintenance of glucose uptake would be an advantage in the oxidative environment of the phagolysosome [76]. Accordingly, in NO-resistant parasites, apart the complex I, the concentration of all components of mitochondrial electron transport system was not modified or was even diminished (complexes

III and V) after NO challenge. Such results may explain the drastic decrease in oxygen consumption by NO-resistant strain upon NO exposure. This scenario is extremely interesting and raises several possibilities about the metabolic adaptations suffered by *L. braziliensis*. For an efficient microbicidal response, macrophages need to activate both NADPH oxidases and iNOS, which can cause hypoxia conditions and lead to expression of hypoxia-inducible factor-1α (HIF-1α) by host cells [45]. In patients with ATL caused by *L. braziliensis*, the expression of this transcription factor has already been described, suggesting hypoxia participation during cutaneous and mucocutaneous clinical outcomes [77]. As described by Degrossoli et al. [78], low oxygen tension, derived from enhanced ROS generation, also leads to the reduction of intracellular parasites in *L. amazonensis*-infected macrophages. Together, these data may suggest that the phagolysosomal environment, rich in NO and ROS, is not an attractive place to perform OXPHOS and derived metabolisms since they are dependent on oxygen availability. A model reconstruction of energy metabolism in *L. infantum* suggested a reduction in oxygen intake in the amastigote scenario, in comparison to promastigotes, probably indicating the adaptation of amastigote metabolism to hypoxic environment of the macrophage [79]. Hence, the observed increase in glucose uptake and the decrease in mitochondrial oxygen consumption after NO challenge could reflect the adaptations that amastigote forms of NO-resistant *L. braziliensis* should undergo seek to survive in host cells. In addition, complex I would be increased to maintenance of NADH/NAD^+ ratio since complex I seems to conserve all subunits containing the redox centers required to ubiquinone reduction [80,81].

5. Conclusions

Together, our data show that NO resistance in *L. braziliensis* involves rapid remodeling of the parasites' proteome, resulting in increased protein content as well as in increased GSH metabolism, higher levels of glucose consumption, elevated abundance of PPP enzymes, and lower mitochondrial respiration, all of which can contribute to thiol and NADPH pool maintenance in these parasites, enabling them to successfully colonize and persist in host cells.

Supplementary Materials: The following supporting information can be downloaded at: https://www.mdpi.com/article/10.3390/antiox11020277/s1, Figure S1: Growth curves of *L. braziliensis* strains; Figure S2: Volcano plot representation of differences in protein concentration between NO-resistant and NO-susceptible *L. braziliensis* strains; Figure S3: Overview of the glycolysis pathway in *L. braziliensis* and identified enzymes; Figure S4: Overview of the pentose phosphate pathway (PPP) in *L. braziliensis* and identified enzymes; Table S1: Protein Groups identified; Table S2: Summary of general information about the proteomes of *Leishmania braziliensis* strains resistant or susceptible to NO; Table S3: Statistical differences among groups.

Author Contributions: Conceptualization, P.C. and R.M.-B.; methodology, J.R.W.; validation, P.C., N.P., R.M.-B. and A.C.B.; formal analysis, N.P., A.C.B., J.R.W., G.D.-L., L.S.-V., J.B.d.J., E.C., G.P., R.M.-B. and P.C.; investigation, N.P., A.C.B., J.R.W., G.D.-L., L.S.-V., R.P.d.A., G.P., J.B.d.J. and P.C.; resources, R.P.d.A., J.R.W., E.C., J.B.d.J., R.M.-B. and P.C.; writing—original draft preparation, P.C., N.P. and A.C.B.; writing—review and editing, P.C., R.M.-B.; N.P., A.C.B., J.R.W., E.C., J.B.d.J., G.P.; supervision, P.C. and R.M.-B.; project administration, P.C.; funding acquisition, P.C. All authors have read and agreed to the published version of the manuscript.

Funding: This research was funded by Conselho Nacional De Desenvolvimento Científico E Tecnológico—CNPq (P.C.—Universal grant No. 423300/2018-0); FIOCRUZ (P.C., N.P., L.S.-V, E.C., G.P.—PAEF grant No. IOC-023-FIO-18-2-63); Fundação De Amparo À Pesquisa Do Estado De Rio De Janeiro—FAPERJ (P.C.—JCNE E-26/203.253/2017, N.P.—TCT No. E-26/202.464/2017); Max-Planck Society For The Advancement Of Science; and Coordenação De Aperfeiçoamento De Pessoal De Nível Superior—CAPES, Brasil—Finance Code 001 (P.C.—Process No. 88887.374332/2019-00). G.P. was a CAPES fellow of the Visitant Professor Program (Process No. 0344141). P.C. and R.P.d.A. are CNPq PQ-fellows (PQ-P.C. Process No. 305796/2017-8 and 312573/2020-0, R.P.d.A.—PQ Process No. 309776/2018-0).

Institutional Review Board Statement: All the protocols were carried out in accordance with the recommendations of the Guide for the Care and Use of Laboratory Animals, according to resolution 196/96 of the National Council for Animal Experimentation—COBEA (https://sbcal.org.br/ (accessed on 2 March 2020)). The animal study protocol was approved by the Animal Use Ethics Committee of INSTITUTO OSWALDO CRUZ-IOC/Fiocruz (L-005/2017; 07/02/2017). According to the Brazilian Law of Biodiversity, this study was registered at SisGen (AA2236F).

Data Availability Statement: The mass spectrometry proteomics data have been deposited to the ProteomeXchange Consortium via the PRIDE [31] partner repository with the dataset identifier PXD029462.

Acknowledgments: The authors are grateful to Matthias Mann for continuous support. We thank Katharina Zettl for technical help with mass spectrometric measurements, and Rosane Temporal, quality manager of LPL-FIOCRUZ-RJ and all the staff of CLIOC, for technical assistance and quality advice.

Conflicts of Interest: The authors declare no conflict of interest. The funders had no role in the design of the study; in the collection, analyses, or interpretation of data; in the writing of the manuscript; or in the decision to publish the results.

References

1. Alvar, J.; Vélez, I.D.; Bern, C.; Herrero, M.; Desjeux, P.; Cano, J.; Jannin, J.; de Boer, M. Leishmaniasis worldwide and global estimates of its incidence. *PLoS ONE* **2012**, *7*, e35671. [CrossRef] [PubMed]
2. Jirmanus, L.; Glesby, M.J.; Guimaraes, L.H.; Lago, E.; Rosa, M.E.; Machado, P.R.; Carvalho, E.M. Epidemiological and clinical changes in American tegumentary leishmaniasis in an area of *Leishmania (viannia) braziliensis* transmission over a 20-year period. *Am. J. Trop. Med. Hyg.* **2012**, *86*, 426–433. [CrossRef] [PubMed]
3. Pearson, R.D.; de Queiroz Sousa, A. Clinical spectrum of leishmaniasis. *Clin. Infect. Dis.* **1996**, *22*, 1–13. [CrossRef] [PubMed]
4. Novais, F.O.; Scott, P. CD8$^+$ T cells in cutaneous leishmaniasis: The good, the bad, and the ugly. *Semin. Immunopathol.* **2015**, *37*, 251–259. [CrossRef]
5. Christensen, S.M.; Dillon, L.A.L.; Carvalho, L.P.; Passos, S.; Novais, F.O.; Hughitt, V.K.; Beiting, D.P.; Carvalho, E.M.; Scott, P.; El-Sayed, N.M.; et al. Meta-Transcriptome profiling of the human-*Leishmania braziliensis* cutaneous lesion. *PLoS Negl. Trop. Dis.* **2016**, *10*, e0004992. [CrossRef] [PubMed]
6. Gomez, M.A.; Contreras, I.; Hallé, M.; Tremblay, M.L.; McMaster, R.W.; Olivier, M. *Leishmania* GP63 alters host signaling through cleavage-activated protein tyrosine phosphatases. *Sci. Signal.* **2009**, *2*, ra58. [CrossRef]
7. Silverman, J.M.; Reiner, N.E. *Leishmania* exosomes deliver preemptive strikes to create an environment permissive for early infection. *Front. Cell. Infect. Microbiol.* **2011**, *1*, 26. [CrossRef]
8. De Carvalho, R.V.H.; Lima-Junior, D.S.; da Silva, M.V.G.; Dilucca, M.; Rodrigues, T.S.; Horta, C.V.; Silva, A.L.N.; da Silva, P.F.; Frantz, F.G.; Lorenzon, L.B.; et al. *Leishmania* RNA virus exacerbates leishmaniasis by subverting innate immunity via TLR3-mediated NLRP3 inflammasome inhibition. *Nat. Commun.* **2019**, *10*, 5273. [CrossRef]
9. Rodríguez-Vega, A.; Losada-Barragán, M.; Berbert, L.R.; Mesquita-Rodrigues, C.; Bombaça, A.C.S.; Menna-Barreto, R.; Aquino, P.; Carvalho, P.C.; Padrón, G.; de Jesus, J.B.; et al. Quantitative analysis of proteins secreted by *Leishmania (Viannia) braziliensis* strains associated to distinct clinical manifestations of American Tegumentary Leishmaniasis. *J. Proteomics* **2021**, *232*, 104077. [CrossRef]
10. Kaye, P.; Scott, P. Leishmaniasis: Complexity at the host-pathogen interface. *Nat. Rev. Microbiol.* **2011**, *9*, 604–615. [CrossRef]
11. Bogdan, C. Nitric oxide and the immune response. *Nat. Immunol.* **2001**, *2*, 907–916. [CrossRef] [PubMed]
12. Sacks, D.; Sher, A. Evasion of innate immunity by parasitic protozoa. *Nat. Immunol.* **2002**, *3*, 1041–1047. [CrossRef] [PubMed]
13. Liew, F.Y.; Wei, X.Q.; Proudfoot, L. Cytokines and nitric oxide as effector molecules against parasitic infections. *Philos. Trans. R. Soc. B Biol. Sci.* **1997**, *352*, 1311–1315. [CrossRef] [PubMed]
14. Genestra, M.; de Souza, W.J.S.; Cysne-Finkelstein, L.; Leon, L.L. Comparative analysis of the nitric oxide production by *Leishmania* sp. *Med. Microbiol. Immunol.* **2003**, *192*, 217–223. [CrossRef]
15. Novais, F.O.; Carvalho, L.P.; Graff, J.W.; Beiting, D.P.; Ruthel, G.; Roos, D.S.; Betts, M.R.; Goldschmidt, M.H.; Wilson, M.E.; de Oliveira, C.I.; et al. Cytotoxic T cells mediate pathology and metastasis in cutaneous leishmaniasis. *PLoS Pathog.* **2013**, *9*, e1003504. [CrossRef]
16. Holzmuller, P.; Sereno, D.; Lemesre, J.L. Lower nitric oxide susceptibility of trivalent antimony-resistant amastigotes of *Leishmania infantum*. *Antimicrob. Agents Chemother.* **2005**, *49*, 4406–4409. [CrossRef]
17. Giudice, A.; Camada, I.; Leopoldo, P.T.G.; Pereira, J.M.B.; Riley, L.W.; Wilson, M.E.; Ho, J.L.; de Jesus, A.R.; Carvalho, E.M.; Almeida, R.P. Resistance of *Leishmania (Leishmania) amazonensis* and *Leishmania (Viannia) braziliensis* to nitric oxide correlates with disease severity in Tegumentary Leishmaniasis. *BMC Infect. Dis.* **2007**, *7*, 7. [CrossRef] [PubMed]
18. Santos, P.L.; Costa, R.V.; Braz, J.M.; Santos, L.F.V.C.; Batista, A.C.; Vasconcelos, C.R.O.; Rangel, M.R.; de Jesus, A.R.; de Moura, T.R.; Leopoldo, P.T.G.; et al. *Leishmania chagasi* naturally resistant to nitric oxide isolated from humans and dogs with visceral leishmaniasis in Brazil. *Nitric Oxide Biol. Chem.* **2012**, *27*, 67–71. [CrossRef]

19. Ávila, L.R.; Gomes, C.M.; Oliveira, P.G.; Gomes, R.S.; Vinaud, M.C.; Dorta, M.L.; Uliana, S.R.B.; Ribeiro-Dias, F.; Oliveira, M.A.P. Promastigote parasites cultured from the lesions of patients with mucosal leishmaniasis are more resistant to oxidative stress than promastigotes from a cutaneous lesion. *Free Radic. Biol. Med.* **2018**, *129*, 35–45. [CrossRef]
20. Souza, A.S.; Giudice, A.; Pereira, J.M.B.; Guimarães, L.H.; de Jesus, A.R.; de Moura, T.R.; Wilson, M.E.; Carvalho, E.M.; Almeida, R.P. Resistance of *Leishmania (Viannia) braziliensis* to nitric oxide: Correlation with antimony therapy and TNF-α production. *BMC Infect. Dis.* **2010**, *10*, 209. [CrossRef]
21. Costa, D.L.; Carregaro, V.; Lima-Júnior, D.S.; Silva, N.M.; Milanezi, C.M.; Cardoso, C.R.; Giudice, Â.; de Jesus, A.R.; Carvalho, E.M.; Almeida, R.P.; et al. BALB/c mice infected with antimony treatment refractory isolate of *Leishmania braziliensis* present severe lesions due to IL-4 production. *PLoS Negl. Trop. Dis.* **2011**, *5*, e965. [CrossRef] [PubMed]
22. De Moura, T.R.; Santos, M.L.B.; Braz, J.M.; Santos, L.F.V.C.; Aragão, M.T.; de Oliveira, F.A.; Santos, P.L.; da Silva, Â.M.; de Jesus, A.R.; de Almeida, R.P. Cross-Resistance of *Leishmania infantum* isolates to nitric oxide from patients refractory to antimony treatment, and greater tolerance to antileishmanial responses by macrophages. *Parasitol. Res.* **2016**, *115*, 713–721. [CrossRef] [PubMed]
23. Alcolea, P.J.; Tuñón, G.I.L.; Alonso, A.; García-Tabares, F.; Ciordia, S.; Mena, M.C.; Campos, R.N.S.; Almeida, R.P.; Larraga, V. Differential protein abundance in promastigotes of nitric oxide-sensitive and resistant *Leishmania chagasi* strains. *Proteome Clin. Appl.* **2016**, *10*, 1132–1146. [CrossRef] [PubMed]
24. Dias-Lopes, G.; Wiśniewski, J.R.; de Souza, N.P.; Vidal, V.E.; Padrón, G.; Britto, C.; Cuervo, P.; de Jesus, J.B. In-Depth quantitative proteomic analysis of trophozoites and pseudocysts of Trichomonas vaginalis. *J. Proteome Res.* **2018**, *17*, 3704–3718. [CrossRef]
25. Pinho, N.; Wiśniewski, J.R.; Dias-Lopes, G.; Saboia-Vahia, L.; Bombaça, A.C.S.; Mesquita-Rodrigues, C.; Menna-Barreto, R.; Cupolillo, E.; de Jesus, J.B.; Padrón, G.; et al. In-Depth quantitative proteomics uncovers specie-specific metabolic programs in *Leishmania (Viannia)* species. *PLoS Negl. Trop. Dis.* **2020**, *14*, e0008509. [CrossRef]
26. Gonçalves, R.L.S.; Barreto, R.F.S.M.; Polycarpo, C.R.; Gadelha, F.R.; Castro, S.L.; Oliveira, M.F. A comparative assessment of mitochondrial function in epimastigotes and bloodstream trypomastigotes of Trypanosoma cruzi. *J. Bioenerg. Biomembr.* **2011**, *43*, 651–661. [CrossRef]
27. Wiśniewski, J.R.; Zougman, A.; Nagaraj, N.; Mann, M. Universal sample preparation method for proteome analysis. *Nat. Methods* **2009**, *6*, 359–362. [CrossRef]
28. Wiśniewski, J.R. Filter-Aided Sample Preparation: The versatile and efficient method for proteomic analysis. *Methods Enzymol.* **2017**, *585*, 15–27. [CrossRef]
29. Wiśniewski, J.R. Label-Free and standard-free absolute quantitative proteomics using the "total protein" and "proteomic ruler" approaches. *Methods Enzymol.* **2017**, *585*, 49–60. [CrossRef]
30. Tyanova, S.; Temu, T.; Sinitcyn, P.; Carlson, A.; Hein, M.Y.; Geiger, T.; Mann, M.; Cox, J. The Perseus computational platform for comprehensive analysis of (prote)omics data. *Nat. Methods* **2016**, *13*, 731–740. [CrossRef]
31. Perez-Riverol, Y.; Csordas, A.; Bai, J.; Bernal-Llinares, M.; Hewapathirana, S.; Kundu, D.J.; Inuganti, A.; Griss, J.; Mayer, G.; Eisenacher, M.; et al. The PRIDE database and related tools and resources in 2019: Improving support for quantification data. *Nucleic Acids Res.* **2019**, *47*, D442–D450. [CrossRef] [PubMed]
32. Menna-Barreto, R.F.S.; Goncalves, R.L.S.; Costa, E.M.; Silva, R.S.F.; Pinto, A.V.; Oliveira, M.F.; de Castro, S.L. The effects on *Trypanosoma cruzi* of novel synthetic naphthoquinones are mediated by mitochondrial dysfunction. *Free Radic. Biol. Med.* **2009**, *47*, 644–653. [CrossRef] [PubMed]
33. Aebischer, T. *Leishmania* spp. proteome data sets: A comprehensive resource for vaccine development to target visceral leishmaniasis. *Front. Immunol.* **2014**, *5*, 260. [CrossRef] [PubMed]
34. Aquilano, K.; Baldelli, S.; Ciriolo, M.R. Glutathione: New roles in redox signalling for an old antioxidant. *Front. Pharmacol.* **2014**, *5*, 196. [CrossRef] [PubMed]
35. Aquilano, K.; Baldelli, S.; Ciriolo, M.R. Glutathione is a crucial guardian of protein integrity in the brain upon nitric oxide imbalance. *Commun. Integr. Biol.* **2011**, *4*, 477–479. [CrossRef]
36. Basu, S.; Keszler, A.; Azarova, N.A.; Nwanze, N.; Perlegas, A.; Shiva, S.; Broniowska, K.A.; Hogg, N.; Kim-Shapiro, D.B. A novel role for cytochrome c: Efficient catalysis of S-nitrosothiol formation. *Free Radic. Biol. Med.* **2010**, *48*, 255–263. [CrossRef]
37. Broniowska, K.A.; Diers, A.R.; Hogg, N. S-Nitrosoglutathione. *Biochim. Biophys. Acta Gen. Subj.* **2013**, *1830*, 3173–3181. [CrossRef]
38. Baldelli, S.; Ciccarone, F.; Limongi, D.; Checconi, P.; Palamara, A.T.; Ciriolo, M.R. Glutathione and nitric oxide: Key team players in use and disuse of skeletal muscle. *Nutrients* **2019**, *11*, 2318. [CrossRef]
39. Godoy, L.; Gonzàlez-Duarte, R.; Albalat, R. S-Nitrosogluthathione reductase activity of amphioxus ADH3: Insights into the nitric oxide metabolism. *Int. J. Biol. Sci.* **2006**, *2*, 117–124. [CrossRef]
40. Barnett, S.D.; Buxton, I.L.O. The role of S-nitrosoglutathione reductase (GSNOR) in human disease and therapy. *Crit. Rev. Biochem. Mol. Biol.* **2017**, *52*, 340–354. [CrossRef]
41. Brown, G.C. Nitric oxide and mitochondrial respiration. *Biochim. Biophys. Acta Bioenerg.* **1999**, *1411*, 351–369. [CrossRef]
42. McConville, M.J.; Saunders, E.C.; Kloehn, J.; Dagley, M.J. *Leishmania carbon* metabolism in the macrophage phagolysosome—Feast or Famine? *F1000Research* **2015**, *4*, 938. [CrossRef]
43. Van Assche, T.; Deschacht, M.; da Luz, R.A.I.; Maes, L.; Cos, P. Leishmania-macrophage interactions: Insights into the redox biology. *Free Radic. Biol. Med.* **2011**, *51*, 337–351. [CrossRef]

44. Volpedo, G.; Pacheco-Fernandez, T.; Holcomb, E.A.; Cipriano, N.; Cox, B.; Satoskar, A.R. Mechanisms of immunopathogenesis in cutaneous leishmaniasis and post kala-azar dermal leishmaniasis (PKDL). *Front. Cell. Infect. Microbiol.* **2021**, *11*, 685296. [CrossRef]
45. Bogdan, C. Macrophages as host, effector and immunoregulatory cells in leishmaniasis: Impact of tissue micro-environment and metabolism. *Cytokine X* **2020**, *2*, 100041. [CrossRef] [PubMed]
46. Colotti, G.; Ilari, A. Polyamine metabolism in *Leishmania*: From arginine to trypanothione. *Amino Acids* **2011**, *40*, 269–285. [CrossRef] [PubMed]
47. Wu, G.; Morris, S.M. Arginine metabolism: Nitric oxide and beyond. *Biochem. J.* **1998**, *336*, 1–17. [CrossRef] [PubMed]
48. Novais, F.O.; Nguyen, B.T.; Beiting, D.P.; Carvalho, L.P.; Glennie, N.D.; Passos, S.; Carvalho, E.M.; Scott, P. Human classical monocytes control the intracellular stage of *Leishmania braziliensis* by reactive oxygen species. *J. Infect. Dis.* **2014**, *209*, 1288–1296. [CrossRef] [PubMed]
49. Oliveira, L.B.; Celes, F.S.; Paiva, C.N.; de Oliveira, C.I. The paradoxical leishmanicidal effects of superoxide dismutase (SOD) mimetic tempol in *Leishmania braziliensis* infection in vitro. *Front. Cell. Infect. Microbiol.* **2019**, *9*, 237. [CrossRef]
50. Haidaris, C.G.; Bonventre, P.F. A role for oxygen-dependent mechanisms in killing of *Leishmania donovani* tissue forms by activated macrophages. *J. Immunol.* **1982**, *129*, 850–855.
51. Gantt, K.R.; Goldman, T.L.; McCormick, M.L.; Miller, M.A.; Jeronimo, S.M.B.; Nascimento, E.T.; Britigan, B.E.; Wilson, M.E. Oxidative responses of human and murine macrophages during phagocytosis of *Leishmania chagasi*. *J. Immunol.* **2001**, *167*, 893–901. [CrossRef] [PubMed]
52. Linares, E.; Giorgio, S.; Augusto, O. Inhibition of in vivo leishmanicidal mechanisms by tempol: Nitric oxide down-regulation and oxidant scavenging. *Free Radic. Biol. Med.* **2008**, *44*, 1668–1676. [CrossRef] [PubMed]
53. Cuypers, B.; Meysman, P.; Erb, I.; Bittremieux, W.; Valkenborg, D.; Baggerman, G.; Mertens, I. Four layer multi-omics reveals molecular responses to aneuploidy in *Leishmania*. *BioRxiv* **2021**. [CrossRef]
54. Rogers, M.B.; Hilley, J.D.; Dickens, N.J.; Wilkes, J.; Bates, P.A.; Depledge, D.P.; Harris, D.; Her, Y.; Herzyk, P.; Imamura, H.; et al. Chromosome and gene copy number variation allow major structural change between species and strains of *Leishmania*. *Genome Res.* **2011**, *21*, 2129–2142. [CrossRef] [PubMed]
55. Sterkers, Y.; Lachaud, L.; Crobu, L.; Bastien, P.; Pagès, M. FISH analysis reveals aneuploidy and continual generation of chromosomal mosaicism in *Leishmania major*. *Cell. Microbiol.* **2011**, *13*, 274–283. [CrossRef]
56. Dumetz, F.; Imamura, H.; Sanders, M.; Seblova, V.; Myskova, J.; Pescher, P.; Vanaerschot, M.; Meehan, C.J.; Cuypers, B.; de Muylder, G.; et al. Modulation of aneuploidy in *Leishmania donovani* during adaptation to different in vitro and in vivo environments and its impact on gene expression. *MBio* **2017**, *8*, e00599-17. [CrossRef] [PubMed]
57. Downing, T.; Imamura, H.; Decuypere, S.; Clark, T.G.; Coombs, G.H.; Cotton, J.A.; Hilley, J.D.; de Doncker, S.; Maes, I.; Mottram, J.C.; et al. Whole genome sequencing of multiple *Leishmania donovani* clinical isolates provides insights into population structure and mechanisms of drug resistance. *Genome Res.* **2011**, *21*, 2143–2156. [CrossRef]
58. Imamura, H.; Downing, T.; van den Broeck, F.; Sanders, M.J.; Rijal, S.; Sundar, S.; Mannaert, A.; Vanaerschot, M.; Berg, M.; de Muylder, G.; et al. Evolutionary genomics of epidemic visceral leishmaniasis in the Indian subcontinent. *Elife* **2016**, *5*, e12613. [CrossRef]
59. Bussotti, G.; Gouzelou, E.; Boité, M.C.; Kherachi, I.; Harrat, Z.; Eddaikra, N.; Mottram, J.C.; Antoniou, M.; Christodoulou, V.; Bali, A.; et al. *Leishmania* genome dynamics during environmental adaptation reveal strain-specific differences in gene copy number variation, karyotype instability, and telomeric amplification. *MBio* **2018**, *9*, e01399-18. [CrossRef]
60. Patino, L.H.; Imamura, H.; Cruz-Saavedra, L.; Pavia, P.; Muskus, C.; Méndez, C.; Dujardin, J.C.; Ramírez, J.D. Major changes in chromosomal somy, gene expression and gene dosage driven by Sb^{III} in *Leishmania braziliensis* and *Leishmania panamensis*. *Sci. Rep.* **2019**, *9*, 9485. [CrossRef]
61. Zhu, J.; Tsai, H.J.; Gordon, M.R.; Li, R. Cellular stress associated with aneuploidy. *Dev. Cell.* **2018**, *44*, 420–431. [CrossRef] [PubMed]
62. Sardar, A.H.; Kumar, S.; Kumar, A.; Purkait, B.; Das, S.; Sen, A.; Kumar, M.; Sinha, K.K.; Singh, D.; Equbal, A.; et al. Proteome changes associated with *Leishmania donovani* promastigote adaptation to oxidative and nitrosative stresses. *J. Proteomics* **2013**, *81*, 185–199. [CrossRef] [PubMed]
63. Xiang, L.; Laranjeira-Silva, M.F.; Maeda, F.Y.; Hauzel, J.; Andrews, N.W.; Mittra, B. Ascorbate-Dependent peroxidase (APX) from *Leishmania amazonensis* is a reactive oxygen species-induced essential enzyme that regulates virulence. *Infect. Immun.* **2019**, *87*, e00193-19. [CrossRef] [PubMed]
64. De Moreira, D.S.; Xavier, M.V.; Murta, S.M.F. Ascorbate peroxidase overexpression protects *Leishmania braziliensis* against trivalent antimony effects. *Mem. Inst. Oswaldo Cruz.* **2018**, *113*, 1–5. [CrossRef] [PubMed]
65. Ariyanayagam, M.R.; Fairlamb, A.H. Ovothiol and trypanothione as antioxidants in trypanosomatids. *Mol. Biochem. Parasitol.* **2001**, *115*, 189–198. [CrossRef]
66. Krauth-Siegel, L.R.; Comini, M.A.; Schlecker, T. The trypanothione system. *Subcell. Biochem.* **2007**, *44*, 231–251. [CrossRef] [PubMed]
67. Romão, P.R.T.; Tovar, J.; Fonseca, S.G.; Moraes, R.H.; Cruz, A.K.; Hothersall, J.S.; Noronha-Dutra, A.A.; Ferreira, S.H.; Cunha, F.Q. Glutathione and the redox control system trypanothione/trypanothione reductase are involved in the protection of *Leishmania* spp. against nitrosothiol-induced cytotoxicity. *Braz. J. Med. Biol. Res.* **2006**, *39*, 355–363. [CrossRef]

68. Halliwell, B. Vitamin C: Poison, prophylactic or panacea? *Trends Biochem. Sci.* **1999**, *24*, 255–259. [CrossRef]
69. Krauth-Siegel, R.L.; Lüdemann, H. Reduction of dehydroascorbate by trypanothione. *Mol. Biochem. Parasitol.* **1996**, *80*, 203–208. [CrossRef]
70. Ghosh, A.K.; Sardar, A.H.; Mandal, A.; Saini, S.; Abhishek, K.; Kumar, A.; Purkait, B.; Singh, R.; Das, S.; Mukhopadhyay, R.; et al. Metabolic reconfiguration of the central glucose metabolism: A crucial strategy of *Leishmania donovani* for its survival during oxidative stress. *FASEB J.* **2015**, *29*, 2081–2098. [CrossRef]
71. Holzmuller, P.; Hide, M.; Sereno, D.; Lemesre, J.L. *Leishmania infantum* amastigotes resistant to nitric oxide cytotoxicity: Impact on in vitro parasite developmental cycle and metabolic enzyme activities. *Infect. Genet. Evol.* **2006**, *6*, 187–197. [CrossRef] [PubMed]
72. Darling, T.N.; Blum, J.J. D-Lactate production by *Leishmania braziliensis* through the glyoxalase pathway. *Mol. Biochem. Parasitol.* **1988**, *28*, 121–127. [CrossRef]
73. Wyllie, S.; Fairlamb, A.H. Methylglyoxal metabolism in trypanosomes and *Leishmania*, Semin. *Cell Dev. Biol.* **2011**, *22*, 271–277. [CrossRef]
74. Silva, M.S.; Ferreira, A.E.N.; Gomes, R.; Tomás, A.M.; Freire, A.P.; Cordeiro, C. Glyoxalase Enzymes in Trypanosomatids. In *Trypanosomatid Diseases*; Selzer, P.M., Jäger, T., Koch, O., Flohé, L., Eds.; Wiley-VCH Verlag GmbH & Co. KGaA: Weinheim, Germany, 2013. [CrossRef]
75. Rastrojo, A.; García-Hernández, R.; Vargas, P.; Camacho, E.; Corvo, L.; Imamura, H.; Dujardin, J.C.; Castanys, S.; Aguado, B.; Gamarro, F.; et al. Genomic and transcriptomic alterations in *Leishmania donovani* lines experimentally resistant to antileishmanial drugs. *Int. J. Parasitol. Drugs Drug Resist.* **2018**, *8*, 246–264. [CrossRef] [PubMed]
76. Saunders, E.C.; Naderer, T.; Chambers, J.; Landfear, S.M.; McConville, M.J. *Leishmania mexicana* can utilize amino acids as major carbon sources in macrophages but not in animal models. *Mol. Microbiol.* **2018**, *108*, 143–158. [CrossRef] [PubMed]
77. Charpentier, T.; Hammami, A.; Stäger, S. Hypoxia inducible factor 1α: A critical factor for the immune response to pathogens and *Leishmania*. *Cell. Immunol.* **2016**, *309*, 42–49. [CrossRef] [PubMed]
78. Degrossoli, A.; Arrais-Silva, W.W.; Colhone, M.C.; Gadelha, F.R.; Joazeiro, P.P.; Giorgio, S. The influence of low oxygen on macrophage response to *Leishmania* infection. *Scand. J. Immunol.* **2011**, *74*, 165–175. [CrossRef]
79. Subramanian, A.; Jhawar, J.; Sarkar, R.R. Dissecting *Leishmania* infantum energy metabolism—A systems perspective. *PLoS ONE* **2015**, *10*, e0137976. [CrossRef]
80. Tomás, A.M.; Castro, H. Redox metabolism in mitochondria of trypanosomatids. *Antioxid. Redox Signal.* **2013**, *19*, 696–707. [CrossRef]
81. Chen, M.; Bennedsen, M.; Zhai, L.; Kharazmi, A. Purification and enzymatic activity of an NADH-fumarate reductase and other mitochondrial activities of *Leishmania parasites*. *Apmis* **2001**, *109*, 801–808. [CrossRef]

MDPI
St. Alban-Anlage 66
4052 Basel
Switzerland
Tel. +41 61 683 77 34
Fax +41 61 302 89 18
www.mdpi.com

Antioxidants Editorial Office
E-mail: antioxidants@mdpi.com
www.mdpi.com/journal/antioxidants

www.ingramcontent.com/pod-product-compliance
Lightning Source LLC
Chambersburg PA
CBHW061457210926
51850CB00008B/1747
9783036576176